An Introduction to Stata Programming

An Introduction to Stata Programming

Christopher F. Baum
Boston College

A Stata Press Publication
StataCorp LP
College Station, Texas

Published by Stata Press, 4905 Lakeway Drive, College Station, Texas 77845
Typeset in LaTeX 2_ε
Printed in the United States of America

10 9 8 7 6 5 4 3 2 1

ISBN-10: 1-59718-045-9
ISBN-13: 978-1-59718-045-0

To Paula

Contents

List of tables xv

List of figures xvii

Preface xix

Acknowledgments xxi

Notation and typography xxiii

1 Why should you become a Stata programmer? 1

   Do-file programming 1

   Ado-file programming 2

   Mata programming for ado-files 2

 1.1 Plan of the book 3

 1.2 Installing the necessary software 3

2 Some elementary concepts and tools 5

 2.1 Introduction 5

  2.1.1 What you should learn from this chapter 5

 2.2 Navigational and organizational issues 5

  2.2.1 The current working directory and profile.do 6

  2.2.2 Locating important directories: sysdir and adopath 6

  2.2.3 Organization of do-files, ado-files, and data files 7

 2.3 Editing Stata do- and ado-files 8

 2.4 Data types 9

  2.4.1 Storing data efficiently: The compress command 11

  2.4.2 Date and time handling 11

  2.4.3 Time-series operators 12

 2.5 Handling errors: The capture command 14

2.6   Protecting the data in memory: The preserve and restore commands   14

2.7   Getting your data into Stata   15

    2.7.1   Inputting data from ASCII text files and spreadsheets . . .   15

        Handling text files   16

        Free format versus fixed format   17

        The insheet command   18

        Accessing data stored in spreadsheets   20

        Fixed-format data files   20

    2.7.2   Importing data from other package formats   25

2.8   Guidelines for Stata do-file programming style   26

    2.8.1   Basic guidelines for do-file writers   27

    2.8.2   Enhancing speed and efficiency   29

2.9   How to seek help for Stata programming   29

3   Do-file programming: Functions, macros, scalars, and matrices   33

3.1   Introduction   33

    3.1.1   What you should learn from this chapter   33

3.2   Some general programming details   34

    3.2.1   The varlist   35

    3.2.2   The numlist   35

    3.2.3   The if exp and in range qualifiers   35

    3.2.4   Missing data handling   36

        Recoding missing values: The mvdecode and mvencode
            commands   37

    3.2.5   String-to-numeric conversion and vice versa   37

        Numeric-to-string conversion   38

        Working with quoted strings   39

3.3   Functions for the generate command   40

    3.3.1   Using if exp with indicator variables   42

    3.3.2   The cond() function   44

    3.3.3   Recoding discrete and continuous variables   45

3.4	Functions for the egen command	47
	Official egen functions	47
	egen functions from the user community	49
3.5	Computation for by-groups	50
3.5.1	Observation numbering: _n and _N	50
3.6	Local macros	53
3.7	Global macros	56
3.8	Extended macro functions and macro list functions	56
3.8.1	System parameters, settings, and constants: creturn	57
3.9	Scalars	58
3.10	Matrices	60
4	**Cookbook: Do-file programming I**	**63**
4.1	Tabulating a logical condition across a set of variables	63
4.2	Computing summary statistics over groups	65
4.3	Computing the extreme values of a sequence	66
4.4	Computing the length of spells	67
4.5	Summarizing group characteristics over observations	71
4.6	Using global macros to set up your environment	73
4.7	List manipulation with extended macro functions	74
4.8	Using creturn values to document your work	76
5	**Do-file programming: Validation, results, and data management**	**79**
5.1	Introduction	79
5.1.1	What you should learn from this chapter	79
5.2	Data validation: The assert, count, and duplicates commands	79
5.3	Reusing computed results: The return and ereturn commands	86
5.3.1	The ereturn list command	90
5.4	Storing, saving, and using estimated results	93
5.4.1	Generating publication-quality tables from stored estimates	98
5.5	Reorganizing datasets with the reshape command	99
5.6	Combining datasets	105

5.7   Combining datasets with the append command   107

5.8   Combining datasets with the merge command   108

       5.8.1   The dangers of many-to-many merges   110

5.9   Other data-management commands   111

       5.9.1   The fillin command   112

       5.9.2   The cross command   112

       5.9.3   The stack command   112

       5.9.4   The separate command   114

       5.9.5   The joinby command   115

       5.9.6   The xpose command   115

6   Cookbook: Do-file programming II   **117**

6.1   Efficiently defining group characteristics and subsets   117

       6.1.1   Using a complicated criterion to select a subset of observations   118

6.2   Applying reshape repeatedly   119

6.3   Handling time-series data effectively   123

6.4   reshape to perform rowwise computation   126

6.5   Adding computed statistics to presentation-quality tables   128

       6.5.1   Presenting marginal effects rather than coefficients   130

6.6   Generating time-series data at a lower frequency   132

7   Do-file programming: Prefixes, loops, and lists   **139**

7.1   Introduction   139

       7.1.1   What you should learn from this chapter   139

7.2   Prefix commands   139

       7.2.1   The by prefix   140

       7.2.2   The xi prefix   142

       7.2.3   The statsby prefix   145

       7.2.4   The rolling prefix   146

       7.2.5   The simulate and permute prefix   148

       7.2.6   The bootstrap and jackknife prefixes   151

       7.2.7   Other prefix commands   153

7.3 The forvalues and foreach commands 154

8 Cookbook: Do-file programming III **161**

8.1 Handling parallel lists 161

8.2 Calculating moving-window summary statistics 162

8.2.1 Producing summary statistics with rolling and merge 164

8.2.2 Calculating moving-window correlations 165

8.3 Computing monthly statistics from daily data 166

8.4 Requiring at least n observations per panel unit 167

8.5 Counting the number of distinct values per individual 169

9 Do-file programming: Other topics **171**

9.1 Introduction 171

9.1.1 What you should learn from this chapter 171

9.2 Storing results in Stata matrices 171

9.3 The post and postfile commands 175

9.4 Output: The outsheet, outfile, and file commands 177

9.5 Automating estimation output 181

9.6 Automating graphics 184

9.7 Characteristics 188

10 Cookbook: Do-file programming IV **191**

10.1 Computing firm-level correlations with multiple indices 191

10.2 Computing marginal effects for graphical presentation 194

10.3 Automating the production of LaTeX tables 197

10.4 Tabulating downloads from the Statistical Software Components archive 202

10.5 Extracting data from graph files' sersets 204

10.6 Constructing continuous price and returns series 209

11 Ado-file programming **215**

11.1 Introduction 215

11.1.1 What you should learn from this chapter 216

11.2 The structure of a Stata program 216

11.3   The program statement   217

11.4   The syntax and return statements   218

11.5   Implementing program options   221

11.6   Including a subset of observations   222

11.7   Generalizing the command to handle multiple variables   224

11.8   Making commands byable   226

           Program properties   228

11.9   Documenting your program   228

11.10   egen function programs   231

11.11   Writing an e-class program   232

           11.11.1   Defining subprograms   234

11.12   Certifying your program   234

11.13   Programs for ml, nl, nlsur, simulate, bootstrap, and jackknife   236

           Writing an ml-based command   237

           11.13.1   Programs for the nl and nlsur commands   240

           11.13.2   Programs for the simulate, bootstrap, and jackknife prefixes   242

11.14   Guidelines for Stata ado-file programming style   244

           11.14.1   Presentation   244

           11.14.2   Helpful Stata features   245

           11.14.3   Respect for datasets   246

           11.14.4   Speed and efficiency   246

           11.14.5   Reminders   247

           11.14.6   Style in the large   247

           11.14.7   Use the best tools   248

12   Cookbook: Ado-file programming   249

12.1   Retrieving results from rolling: 249

12.2   Generalization of egen function pct9010() to support all pairs of
           quantiles   252

12.3   Constructing a certification script   254

12.4 Using the ml command to estimate means and variances 259

   12.4.1 Applying equality constraints in ml estimation 261

12.5 Applying inequality constraints in ml estimation 262

12.6 Generating a dataset containing the single longest spell 267

13 Mata functions for ado-file programming **271**

13.1 Mata: First principles 271

   13.1.1 What you should learn from this chapter 272

13.2 Mata fundamentals 272

   13.2.1 Operators 272

   13.2.2 Relational and logical operators 274

   13.2.3 Subscripts 274

   13.2.4 Populating matrix elements 275

   13.2.5 Mata loop commands 276

   13.2.6 Conditional statements 278

13.3 Function components 279

   13.3.1 Arguments 279

   13.3.2 Variables 280

   13.3.3 Saved results 280

13.4 Calling Mata functions 281

13.5 Mata's st_ interface functions 283

   13.5.1 Data access 283

   13.5.2 Access to locals, globals, scalars, and matrices 285

   13.5.3 Access to Stata variables' attributes 286

13.6 Example: st_ interface function usage 286

13.7 Example: Matrix operations 288

   13.7.1 Extending the command 293

13.8 Creating arrays of temporary objects with pointers 295

13.9 Structures 299

13.10 Additional Mata features 302

   13.10.1 Macros in Mata functions 302

13.10.2 Compiling Mata functions 303

13.10.3 Building and maintaining an object library 304

13.10.4 A useful collection of Mata routines 305

14 Cookbook: Mata function programming   307

14.1 Reversing the rows or columns of a Stata matrix 307

14.2 Shuffling the elements of a string variable 311

14.3 Firm-level correlations with multiple indices with Mata 312

14.4 Passing a function to a Mata function 316

14.5 Using subviews in Mata 319

14.6 Storing and retrieving country-level data with Mata structures . . . 321

14.7 Locating nearest neighbors with Mata 327

14.8 Computing the seemingly unrelated regression estimator 331

14.9 A GMM-CUE estimator using Mata's optimize() functions 337

References   349

Author index   353

Subject index   355

Tables

2.1   Numeric data types 9

5.1   Models of sulphur dioxide concentration 99

9.1   Grunfeld company statistics 174
9.2   Grunfeld company estimates 175
9.3   Wage equations for 1984 184

10.1 Director-level variables 202

11.1 MCAS percentile ranges 226

Figures

5.1   Superimposed scatterplots 113

7.1   Rolling robust regression coefficients 148

7.2   Distribution of the sample median via Monte Carlo simulation . . . 150

7.3   Q–Q plot of the distribution of the sample median 151

8.1   Growth rate histories of several firms 164

8.2   Estimated monthly volatility from daily data 167

9.1   Automated graphics 188

10.1   Point and interval elasticities computed with `mfx` 197

10.2   Air quality in U.S. cities 206

12.1   Rolling `lincom` estimates 252

Preface

This book is a concise introduction to the art of Stata programming. It covers three types of programming that can be used in working with Stata: do-file programming, ado-file programming, and Mata functions that work in conjunction with do- and ado-files. Its emphasis is on the automation of your work with Stata and how programming on one or more of these levels can help you use Stata more effectively.

In the development of these concepts, I do not assume that you have prior experience with Stata programming, although familiarity with the command-line interface is helpful. Examples are drawn from several disciplines, although my background as an applied econometrician is evident in the selection of some sample problems. The introductory chapter motivates the *why*: why should you invest time and effort into learning Stata programming? In chapter 2, I discuss elementary concepts of the command-line interface and describe some commonly used tools for working with programs and datasets.

The format of the book may be unfamiliar to readers who have some familiarity with other books that help you learn how to use Stata. Beginning with chapter 3, each odd-numbered chapter is followed by a "cookbook" chapter containing several "recipes", 40 in total. Each recipe poses a problem: how can I perform a certain task with Stata programming? The recipe then provides a complete solution to the problem and describes how the features presented in the previous chapter can be put to good use. As in the kitchen, you may not want to follow a recipe exactly from the cookbook; just as in cuisine, a minor variation on the recipe may meet your needs, or the techniques presented in that recipe can help you see how Stata programming applies to your specific problem.

Most Stata users who delve into programming make use of do-files to automate and document their work. Consequently, the major focus of the book is do-file programming, covered in chapters 3, 5, 7, and 9. Some users will find that writing formal Stata programs, or ado-files, meets their needs. Chapter 11 is a concise summary of ado-file programming, with the following cookbook chapter presenting several recipes that contain developed ado-files. Stata's matrix programming language, Mata, can also be helpful in automating certain tasks. Chapter 13 presents a summary of Mata concepts and the key features that allow interchange of variables, scalars, macros, and matrices. The last chapter presents several examples of Mata functions developed to work with ado-files. All the do-files, ado-files, Mata functions, and datasets used in the book's examples and recipes are available from the Stata Press web site, as discussed in *Notation and typography*.

Acknowledgments

I must acknowledge many intellectual debts that have been incurred during the creation of this book. I am most indebted to Nicholas J. Cox, who served as a technical reviewer of the manuscript, both for his specific contributions to this project and his willingness to share his extensive understanding of Stata with all of us in the Stata user community. His *Speaking Stata* columns alone are worth the cost of a subscription to the *Stata Journal*. Studying Nick's many routines and working with him on developing several Stata commands has taught me a great deal about how to program Stata effectively.

My collaboration with Mark E. Schaffer on the development of `ivreg2` has led to fruitful insights into programming techniques and to the coauthored section 14.9. At StataCorp, Bill Gould, David Drukker, Alan Riley, and Vince Wiggins have been enthusiastic about the potential for a primer on Stata programming, and Bill's *Mata Matters* columns have been invaluable in helping me understand the potential of this language. Other members of the Stata user community, including my coauthors Mark E. Schaffer, Steven Stillman, Bill Rising, and Vince Wiggins, as well as Austin Nichols and David Roodman, have contributed a great deal to my development of the topics covered in this manuscript. Ben Jann has been particularly helpful in sharing his understanding of Mata.

Many members of Statalist and participants in Stata Users Group meetings in Boston, London, Essen, and Berlin have raised issues that have led to several of the "cookbook recipes" in this book. I also thank several generations of PhD students at Boston College and workshop participants at Smith College, University of Nevada–Las Vegas, Rensselaer Polytechnic Institute, University of Sheffield, University of Nottingham, and DIW Berlin for useful feedback on my presentations of Stata usage.

I am deeply grateful to my wife, Paula Arnold, for graciously coping with my seemingly 24/7 efforts in completing this book.

Oak Square School
Brighton, Massachusetts
February 2009

Christopher F. Baum

Notation and typography

In this book, I assume that you are somewhat familiar with Stata, that you know how to input data, and that you know how to use previously created datasets, create new variables, run regressions, and the like.

I designed this book for you to learn by doing, so I expect you to read this book while sitting at a computer and using the sequences of commands contained in the book to replicate my results. In this way, you will be able to generalize these sequences to suit your own needs.

Generally, I use the `typewriter font` to refer to Stata commands, syntax, and variables. A "dot" prompt followed by a command indicates that you can type verbatim what is displayed after the dot (in context) to replicate the results in the book.

I use the *italic font* for words that are not supposed to be typed; instead, you are to substitute another word or words for them. For example, if I said to type `by(`*groupvar*`)`, you should replace "*groupvar*" with the actual name of the group variable.

All the datasets and do-files for this book are freely available for you to download. You can also download all the user-written commands described in this book. At the Stata dot prompt, type

```
. net from http://www.stata-press.com/data/itsp/
. net install itsp
. net get itsp
```

After installing the files, type `spinst_itsp` to obtain all the user-written commands used in the book's examples. You should check the messages produced by the `spinst_itsp` command. If there are any error messages, follow the instructions at the bottom of the output to complete the download.

In a net-aware Stata, you can also load the dataset by specifying the complete URL of the dataset. For example,

```
. use http://www.stata-press.com/data/itsp/air2
```

This text complements the material in the Stata manuals but does not replace it, so I often refer to the Stata manuals by using [R], [P], etc. For example, [R] **xi** refers to the *Stata Base Reference Manual* entry for **xi**, and [P] **syntax** refers to the entry for **syntax** in the *Stata Programming Reference Manual*.

1 Why should you become a Stata programmer?

This book provides an introduction to several contexts of Stata programming. I must first define what I mean by *programming*. You can consider yourself a Stata programmer if you write *do-files*: sequences of Stata commands that you can execute with the do command, by double-clicking on the file, or by running them in the Do-file Editor. You might also write what Stata formally defines as a *program*: a set of Stata commands that includes the program ([P] **program**) command. A Stata program, stored in an *ado-file*, defines a new Stata command. You can also use Stata's matrix programming language, Mata, to write routines in that language that are called by ado-files. Any of these tasks involves Stata programming.[1]

With that set of definitions in mind, we must deal with the *why*: why should you become a Stata programmer? After answering that essential question, this text takes up the *how*: how you can become a more efficient user of Stata by using programming techniques, be they simple or complex.

Using any computer program or language is all about *efficiency*: getting the computer to do the work that can be routinely automated, allowing you to make more efficient use of your time and reducing human errors. Computers are excellent at performing repetitive tasks; humans are not. One of the strongest rationales for learning how to use programming techniques in Stata is the potential to shift more of the repetitive burden of data management, statistical analysis, and the production of graphics to the computer. Let's consider several specific advantages of using Stata programming techniques in the three contexts listed above.

Do-file programming

Using a do-file to automate a specific data-management or statistical task leads to reproducible research and the ability to document the empirical research process. This

---

1. There are also specialized forms of Stata programming, such as dialog programming, scheme programming, and class programming. A user-written program can present a dialog, like any official Stata command, if its author writes a dialog file. The command can also be added to the User menu of Stata's graphical interface. For more information, see [P] **dialog programming** and [P] **window programming**. Graphics users can write their own schemes to set graphic defaults. See [G] **schemes intro** for details. Class programming allows you to write object-oriented programs in Stata. As [P] **class** indicates, this has primarily been used in Stata's graphics subsystem and graphical user interface. I do not consider these specialized forms of programming in this book.

reduces the effort needed to perform a similar task at a later point or to document for your coworkers or supervisor the specific steps you followed. Ideally, your entire research project should be defined by a set of do-files that execute every step from the input of the raw data to the production of the final tables and graphs. Because a do-file can call another do-file (and so on), a hierarchy of do-files can be used to handle a complex project.

The beauty of this approach is its flexibility. If you find an error in an earlier stage of the project, you need only to modify the code and then rerun that do-file and those following to bring the project up to date. For instance, a researcher may need to respond to a review of her paper—submitted months ago to an academic journal—by revising the specification of variables in a set of estimated models and estimating new statistical results. If all the steps that produce the final results are documented by a set of do-files, her task is straightforward. I argue that all serious users of Stata should gain some facility with do-files and the Stata commands that support repetitive use of commands.

That advice does not imply that Stata's interactive capabilities should be shunned. Stata is a powerful and effective tool for exploratory data analysis and ad hoc queries about your data. But data-management tasks and the statistical analyses leading to tabulated results should not be performed with "point-and-click" tools that leave you without an audit trail of the steps you have taken.

Ado-file programming

On a second level, you may find that despite the breadth of Stata's official and user-written commands, there are tasks you must repeatedly perform that involve variations on the same do-file. You would like Stata to have a *command* to perform those tasks. At that point, you should consider Stata's ado-file programming capabilities. Stata has great flexibility: a Stata command need be no more than a few lines of Stata code. Once defined, that command becomes a "first-class citizen". You can easily write a Stata program, stored in an ado-file, that handles all the features of official Stata commands such as if *exp*, in *range*, and command options. You can (and should) write a help file that documents the program's operation for your benefit and for those with whom you share the code. Although ado-file programming requires that you learn how to use some additional commands used in that context, it can help you become more efficient in performing the data-management, statistical, or graphical tasks that you face.

Mata programming for ado-files

On a third level, your ado-files can perform some complicated tasks that involve many invocations of the same commands. Stata's ado-file language is easy to read and write, but it is *interpreted*. Stata must evaluate each statement and translate it into machine code. The Mata programming language (`help mata`) creates *compiled* code, which can run much faster than ado-file code. Your ado-file can call a Mata routine to carry out a computationally intensive task and return the results in the form of Stata variables, scalars, or matrices. Although you may think of Mata solely as a matrix language, it is actually a general-purpose programming language, suitable for many nonmatrix-oriented tasks, such as text processing and list management.

The level of Stata programming that you choose to attain and master depends on your needs and skills. As I have argued, the vast majority of interactive Stata users can and should take the next step of learning how to use do-files efficiently to take full advantage of Stata's capabilities and to save time. A few hours of investment in understanding the rudiments of do-file programming—as covered in the chapters to follow—will save you days or weeks over the course of a sizable research project.

A smaller fraction of users may choose to develop ado-files. Many users find that those features lacking in official Stata are adequately provided by the work of members of the Stata user community who have developed and documented ado-files, sharing them via the *Stata Journal*, the Statistical Software Components (SSC) archive,[2] or their own user site. However, developing a reading knowledge of ado-file code is highly useful for many Stata users. It permits you to scrutinize ado-file code—either that of official Stata or user-written code—and more fully understand how it performs its often minor modifications to existing code to meet your needs.

Mata was new to Stata as of version 9. It has already been embraced by a number of programmers wishing to take advantage of its many features and its speed. As an introduction to Stata programming, this book places the least emphasis on Mata subroutines for ado-file programming, but I do discuss the basic elements of that environment. I do not discuss interactive use of Mata in this book.

1.1 Plan of the book

The chapters of this book present the details of the three types of Stata programming discussed above, placing the greatest emphasis on the first topic: effective use of do-file programming. Each fairly brief chapter on the structure of programming techniques is followed by a "cookbook" chapter. These chapters contain several "recipes" for the solution of a particular, commonly encountered problem, illustrating the necessary programming techniques. Like in a literal cookbook, the recipes here are illustrative examples; you are free to modify the ingredients to produce a somewhat different dish. The recipes may not address your precise problem, but they should prove helpful in devising a solution as a variation on the same theme.

1.2 Installing the necessary software

This book uses Stata to illustrate many aspects of programming. Stata's capabilities are not limited to the commands of official Stata documented in the manuals and in online help. The capabilities include a wealth of commands documented in the *Stata*

---

2. For details on the SSC ("Boston College") archive of user-contributed routines, type `help ssc`.

Journal, the *Stata Technical Bulletin*, and the SSC archive. Those commands will not be available in your copy of Stata unless you have already located and installed them. To locate a user-written command (such as *thatcmd*), use `findit` *thatcmd* (see [R] **search**).

Newer versions of the user-written commands that you install today may become available. The official Stata command `adoupdate` ([R] **adoupdate**), which you can use at any time, will check to see whether newer versions of any user-written commands are available. Just as the command `update query` will determine whether your Stata executable and official ado-files are up to date, `adoupdate` will determine whether any user-written commands installed in your copy of Stata are up to date.

2 Some elementary concepts and tools

2.1 Introduction

This chapter lays out some of the basics that you will need to be an effective Stata programmer. The first section discusses navigational and organizational issues: How should you organize your files? How will Stata find a do-file or an ado-file? The following sections describe how to edit files, appropriate data types, several useful commands for programmers, and some guidelines for Stata programming style. The penultimate section suggests how you can seek help for your programming problems.

2.1.1 What you should learn from this chapter

- Know where your files are: master the current working directory and the ado-path
- Learn how to edit do- and ado-files effectively
- Use appropriate data types for your variables: use `compress` when useful
- Use time-series operators effectively
- Use `capture`, `preserve`, and `restore` to work efficiently
- Use Stata's data input commands effectively
- Adopt a good style for do-file programming and internal documentation
- Know where (and when) to seek help with your Stata programming
- Know how to trace your do-file's execution to diagnose errors

2.2 Navigational and organizational issues

We are all familiar with the colleague whose computer screen resembles a physical desk in a cluttered office: a computer with icons covering the screen in seemingly random order. If you use Stata only interactively, storing data files on the desktop might seem like a reasonable idea. But when you start using do-files to manage a project or writing your own ado-files, those files' locations become crucial; they will not work if they are in the wrong place on your computer. This section describes those navigational and organizational issues to help ensure that your files are in the best places.

2.2.1   The current working directory and profile.do

Like most programs, Stata has a concept of the *current working directory* (CWD). At
any point in time, Stata is referencing a specific directory or folder accessible from your
computer. It may be a directory on your own hard disk, or one located on a network
drive or removable disk. In interactive mode, Stata displays the CWD in its toolbar.
Why is the CWD important? If you save a file—a .dta file or a log file—it will be placed
in the CWD unless you provide a full file specification directing it to another directory.
That is, `save myfile, replace` will save that file in the CWD. Likewise, if you attempt
to use a file with the syntax `use myfile, clear`, it will search for that file in the CWD,
returning an error if the file is not located.

One of the most common problems beginning users of Stata face is saving a data
file and not knowing where it was saved. Of course, if you never change the CWD, all
your materials will be in the same place, but do you really want everything related
to your research to be located in one directory? On my computer, the directory is
`/Users/baum/Documents`, and I would rather not commingle documents with Stata
data files, log files, and graph files. Therefore, you probably should change the CWD to
a directory, or folder, dedicated to your research project and set up multiple directories
for separate projects. You can change your CWD with the `cd` command; for example, `cd`
`/data/city`, `cd d:/project`, or `cd "My Documents/project1"`.[1] You can use the `pwd`
command to display the CWD at any time. Both `cd` and `pwd` are described in [D] **cd**.

You may want Stata to automatically change the CWD to your preferred location
when you start Stata. You can accomplish this with `profile.do`. This file, placed
in your home directory,[2] will execute a set of commands when you invoke Stata. You
might include the command `cd c:/data/NIHproject` in `profile.do` to direct Stata to
automatically change the CWD to that location.

2.2.2   Locating important directories: sysdir and adopath

The `sysdir` command provides a list of seven directories or folders on your computer
that are important to Stata. The `STATA`, `UPDATES`, and `BASE` directories contain the
Stata program itself and the official ado-files that make up most of Stata. You should
not tamper with the files in these directories. Stata's `update` ([R] **update**) command
will automatically modify the contents of these three directories. The `SITE` directory
may reference a network drive in a university or corporate setting where a system
administrator places ado-files to be shared by several users.

---

1. Even in a Windows environment, you can use the forward slash (/) as a directory separator. The
   backslash (\), usually used in Windows file specifications, can be problematic in Stata do-files,
   because Stata interprets the backslash as an escape character modifying Stata's handling of the
   following character. You can avoid problems (and enhance cross-platform compatibility of your
   programs) by using the forward slash (/) as a directory separator in all Stata programs.
2. See `help profilew` (Windows), `help profilem` (Mac OS X), or `help profileu` (Unix/Linux) for
   specific details about your operating system.

The PLUS directory is automatically created when you download any user-written materials. If you use findit to locate and install user-written programs from the *Stata Journal* or *Stata Technical Bulletin*, their ado-files and help files will be located in a subdirectory of the PLUS directory.[3] If you use the ssc ([R] **ssc**) command to download user-written materials from the Statistical Software Components (SSC) ("Boston College") archive or net install to download materials from a user's site, these materials will also be placed in a subdirectory of the PLUS directory.

The PERSONAL directory is, as its name suggests, personal. You can place your own ado-files in that directory. If you want to modify an official Stata ado-file, you should make a copy of it, change its name (for instance, rename sureg.ado to sureg2.ado) and place it it your PERSONAL directory.

Why are there all these different places for Stata's ado-files? The answer lies in the information provided by the adopath command:

```
. adopath
  [1]  (UPDATES)    "/Applications/Stata/ado/updates/"
  [2]  (BASE)       "/Applications/Stata/ado/base/"
  [3]  (SITE)       "/Applications/Stata/ado/site/"
  [4]               "."
  [5]  (PERSONAL)   "~/Library/Application Support/Stata/ado/personal/"
  [6]  (PLUS)       "~/Library/Application Support/Stata/ado/plus/"
  [7]  (OLDPLACE)   "~/ado/"
```

Like sysdir, this command lists seven directories. The order of these directories is important because it defines how Stata will search for a command. It will attempt to find foo.ado in UPDATES, then in BASE, the possible locations for Stata's official ado-files.[4] The fourth directory is ".", that is, the CWD. The fifth is PERSONAL, and the sixth is PLUS.[5] This pecking order implies that if foo.ado is not to be found among Stata's official ado-files or the SITE directory, Stata will examine the CWD. If that fails, it will look for foo.ado in PERSONAL (and its subdirectories). If that fails, it will look in PLUS (and its subdirectories) and, as a last resort, in OLDPLACE. If foo.ado is nowhere to be found, Stata will generate an unrecognized command error.

This search hierarchy indicates that you can locate an ado-file in one of several places. In the next section, I discuss how you might choose to organize ado-files, as well as do-files and data files related to your research project.

2.2.3 Organization of do-files, ado-files, and data files

It is crucially important that you place ado-files on the ado-path. You can place them in your CWD ([4] above in the ado-path listing), but that is generally a bad idea because if you work in any other directory, those ado-files will not be found. If the ado-files are your own or have been written by a coworker, place them in PERSONAL. If you

---

3. For example, if sysdir shows that PLUS is ~/Library/Application Support/Stata/ado/plus/, foo.ado and foo.hlp will be placed in ~/Library/Application Support/Stata/ado/plus/f/.

4. As mentioned above, you can ignore SITE unless you are accessing Stata in a networked environment.

5. We also ignore OLDPLACE, which is only of historical interest.

download ado-files from the SSC archive, please heed the advice that you should always use Stata—not a web browser—to perform the download and locate the files into the correct directory (in PLUS).

What about your do-files, data files, and log files? It makes sense to create a directory, or folder, in your home directory for each separate project and to store all project-related files in that directory. You can always fully qualify a data file when you use it in a do-file, but if you move that data file to another computer, the do-file will fail to find it. Referencing files in the same directory simplifies making a copy of that directory for a coworker or collaborator and makes it possible to run the do-files from an external drive such as a flash disk or "memory key".

It is also a good idea to include a cd command at the top of each do-file, referencing the CWD. Although this command would have to be altered if you moved the directory to a different computer, it will prevent a common mistake: saving data files or log files to a directory other than the project directory.

You might also have several projects that depend on one or two key data files. Rather than duplicating possibly large data files in each project directory, you can refer to them with a relative file specification. Say that your research directory is d:/data/research with subdirectories d:/data/research/project1 and d:/data/research/project2. Place the key data file master.dta in the research directory, and refer to it in the project directories with use ../master, clear. The double dot indicates that the file is to be found in the parent (enclosing) directory, while allowing you to move your research directory to a different drive (or even to a Mac OS X or a Linux computer) without having to alter the use statement.

2.3   Editing Stata do- and ado-files

If a do-file or ado-file is relatively small[6] and you are working in Stata's graphical user interface, you should use Stata's Do-file Editor to create or modify the file. The Do-file Editor has an advantage over most external editors: it allows you to execute only a section of the file by selecting those lines and clicking on the **Do** icon.

You should recognize that do- and ado-files are merely text files with file types of .do or .ado rather than .txt. Even still, it is a very poor idea to edit them in a word processor such as Microsoft Word. A word processor must read the text file and convert it into its own binary format, and when the file is saved, it must reverse the process.[7] Furthermore, a word processor will usually present the file in a variable-width character format, which is harder to read. But the biggest objection to word processing a do-file or ado-file is the waste of your time. It is considerably faster to edit a file in the Do-file Editor and execute it immediately without the need to translate it back into text.

---

6. You can edit files smaller than 128 kilobytes if you are on a Windows system. There are no limits in Stata for Macintosh or Stata for Unix/Linux.
7. This is the same translation process that takes place when you insheet a text file and outsheet it back to text within Stata.

What if the file is too large to be edited in the Do-file Editor? Every operating system supports a variety of text editors, many of them free. A useful compendium of information on text editors has been collected and organized by Nicholas J. Cox and is available from the SSC archive as `texteditors`, an HTML document. You can also view the latest version of that regularly updated document in your web browser.[8] Many of the text editors described in the document have "Stata editing modes" that colorize Stata commands, indicate different elements of syntax, and even flag apparent errors in the code. In any text editor, it is not necessary to close the file after modifying it. You merely save the file, and its revised version can then be executed by Stata.

2.4   Data types

Stata, as a programming language, supports more data types than do many statistical packages. The major distinction to consider is between numeric and string data types. Data-management tasks often involve conversions between numeric and string variables. For instance, data read from a text file (such as a `.csv` or tab-delimited file created by a spreadsheet) are often considered to be a string variable by Stata even though most of the contents are numeric. The commands `destring` ([D] **destring**) and `tostring` are helpful in this regard, as are `encode` ([D] **encode**) and `decode`.

String variables can hold values up to 244 characters in length, one byte for each character. You usually do not need to declare their length, because Stata's string functions (`help string functions`) will generate a string variable long enough to hold the contents of any `generate` ([D] **generate**) operation. String variables require as many bytes of storage per observation as their declaration; for example, a `str20` variable requires 20 bytes per observation.

Stata's numeric data types include `byte`, `int`, `long`, `float`, and `double`. The `byte`, `int`, and `long` data types can hold only integer contents. See table 2.1 for more information on numeric data types.

Table 2.1. Numeric data types

Storage type	Minimum	Maximum	Bytes
byte	-127	100	1
int	$-32,767$	$32,740$	2
long	$-2,147,483,647$	$2,147,483,620$	4
float	-1.701×10^{38}	1.701×10^{38}	4
double	-8.988×10^{307}	8.988×10^{307}	8

---

8. To view the document, visit http://ideas.repec.org/c/boc/bocode/s423801.html.

The `long` integer data type can hold all signed nine-digit integers but only some ten-digit integers. Integers are held in their exact representation by Stata so that you can store a nine-digit integer (such as a U.S. Social Security number) as a `long` data type. However, lengthy identification numbers also can be stored as a `double` data type or as a string variable. Often that will be a wise course of action, because then you need not worry about possible truncation of values. You also will find it useful to use string variables when a particular identification code could contain characters. For instance, the CUSIP (Committee on Uniform Security Identification Procedures) code used to identify U.S. security issues used to be wholly numeric but now may contain one or more nonnumeric characters. Storing these values as strings avoids later problems with numeric missing values.

As displayed above, the two floating-point data types, `float` and `double`, can hold very large numbers. But many users encounter problems with much smaller floating-point values if they mistakenly assume that floating-point arithmetic operations are exact. Floating-point numbers (those held as mantissa and exponent, such as 1.056×10^3), expressed in base 10, must be stored as base 2 (binary) numbers. Although 1/10 is a rational fraction in base 10, it is not so in the binary number system used in a computer:

```
. display %21x 1/10
+1.999999999999aX-004
```

Further details of this issue can be found in [U] **12.2.2 Numeric storage types**, Gould (2006b), and Cox (2006b). The implications should be clear: an `if` condition that tests some floating-point value for equality, such as `if diff == 0.01`, is likely to fail when you expect that it would succeed.[9] A `float` contains approximately seven significant digits in its mantissa. This implies that if you read a set of nine-digit U.S. Social Security numbers into a `float`, they will not be held exactly. A `double` contains approximately 15 significant digits. We know that residuals computed from a linear regression using `regress` and `predict eps, residual` should sum to exactly zero. In Stata's finite-precision computer arithmetic using the default `float` data type, residuals from such a regression will sum to a value in the range of 10^{-7} rather than 0.0. Thus discussions of the `predict` ([R] **predict**) command often advise using `predict double eps, residual` to compute more accurate residuals.

What are the implications of finite-precision arithmetic for Stata programming?

- You should store ID numbers with many digits as string variables, not as integers, floats, or doubles.

- You should not rely on exact tests of a floating-point value against a constant, not even zero. The `reldif()` function (`help math functions`) can be used to test for approximate equality.

---

9. The alternative syntax `if diff == float(0.01)` will solve this problem.

- As suggested above, use `double` floating-point values for any generated series where a loss of precision might be problematic, such as residuals, predicted values, scores, and the like.

- You should be wary of variables' values having very different scales, particularly when a nonlinear estimation method is used. Any regression of `price` from the venerable `auto.dta` reference dataset on a set of regressors will display extremely large sums of squares in the analysis of variance table. Scaling `price` from dollars to thousands of dollars obviates this problem. The scale of this variable does not affect the precision of linear regression, but it could be problematic for nonlinear estimation techniques.

- Use integer data types where it is appropriate to do so. Storing values as `byte` or `int` data types when feasible saves disk space and memory.

2.4.1   Storing data efficiently: The compress command

`compress` ([D] **compress**) is a useful command, particularly when working with datasets acquired from other statistical packages. This command will examine each variable in memory and determine whether it can be stored more efficiently. It is guaranteed never to lose information or reduce the precision of measurements. The advantage of storing indicator (0/1) variables as a `byte` data type rather than as a four-byte `long` data type is substantial for survey datasets with many indicator variables. It is an excellent idea to apply `compress` when performing the initial analysis of a new dataset. Alternatively, if you are using the third-party Stat/Transfer application to convert data from an SAS or SPSS format, use the Stat/Transfer optimize option.

2.4.2   Date and time handling

Stata does not have a separate data type for calendar dates. Dates are represented, as they are in a spreadsheet program, by numbers known as `%t` values measuring the time interval from a reference date or *epoch*. For example, the epoch for Stata (and SAS) is midnight on 1 January 1960. Days following that date have positive integer values, while days prior to it have negative integer values. Dates represented in days are known as `%td` values. Other calendar frequencies are represented by the number of weeks, months, quarters, or half-years since that reference date: `%tw`, `%tm`, `%tq`, and `%th` values, respectively. The year is represented as a `%ty` value, ranging from AD 100 to AD 9999. For time-series data represented by other calendar schemes (e.g., business-daily data), you can use consecutive integers and the *generic* form, `%tg`.

As of version 10, Stata provides support for accurate intradaily measurements of time, down to the millisecond. A date-and-time variable is known as a %tc (clock) value, and it can be defined to any intraday granularity: hours, minutes, seconds, or milliseconds.[10] For more information, see [U] **12.3 Dates and times**.

It is important when working with variables containing dates and times to ensure that the proper Stata data type is used for their storage. Weekly and lower-frequency values (including generic values) can be stored as data type int or as data type float. Daily (%td) values should be stored as data type long or as data type float. If the int data type is used, dates more than 32,740 days from 1 January 1960 (i.e., beyond 21 August 2049) cannot be stored.

More stringent requirements apply to clock (date-and-time) values. These values *must* be stored as data type double to avoid overflow conditions. Clock values, like other time values, are integers, and there are 86,400,000 milliseconds in a day. The double data type is capable of precisely storing date-and-time measurements within the range of years (AD 100–9999) defined in Stata.

Although it is important to use the appropriate data type for date-and-time values, you should avoid using a larger data type than needed. The int data type requires only two bytes per observation, the long and float data types require four bytes, and the double data type requires eight bytes. Although every date value could be stored as a double, that would be wasteful of memory and disk storage, particularly in a dataset with many observations.

A package of functions (see help dates and times) is available to handle the definition of date variables and date/time arithmetic. Display of date variables in calendar formats (such as 08 Nov 2006) and date-and-time variables with the desired intraday precision is handled by the definition of proper formats. Because dates and times are numeric variables, you should distinguish between the content or value of a date/time variable and the format in which it will be displayed.

One helpful hint: if you are preparing to move data from a spreadsheet into Stata with the insheet ([D] **insheet**) command, make sure that any date variables in the spreadsheet display as four-digit years. It is possible to deal with two-digit years, such as in 11/08/06, in Stata, but it is easier to format the dates with four-digit years (for example, 11/08/2006) before reading those data into Stata.

2.4.3 Time-series operators

Stata provides the *time-series operators* L., F., D., and S., which allow the specification of *lags*, *leads* (forward values), *differences*, and *seasonal differences*, respectively.[11] The time-series operators make it unnecessary to create a new variable to use a lag, difference, or lead. When combined with a numlist, they allow the specification of a set of these

---

10. There are also %tC values that take *leap seconds* into account for very precise measurement of time intervals.

11. This section is adapted from section 3.2.1 of Baum (2006a).

constructs in a single expression. Consider the lag operator, L., which when added to the beginning of a variable name refers to the (first-)lagged value of that variable: L.x. A number can follow the operator, so that L4.x refers to the fourth lag of x. More generally, a numlist can be used, so that L(1/4).x refers to the first through fourth lags of x, and L(1/4).(x y z) defines a list of four lagged values of each of the variables x, y, and z. These expressions can be used almost anywhere that a varlist is required.

Similarly to the lag operator, the lead operator, F., allows the specification of future values of one or more variables. Strictly speaking, the lead operator is unnecessary because a lead is a negative lag, and an expression such as L(-4/4).x will work, labeling the negative lags as leads. The difference operator, D., can be used to generate differences of any order. The first difference, D.x, is Δx or $x_t - x_{t-1}$. The second difference, D2.x, is not $x_t - x_{t-2}$ but rather $\Delta(\Delta x_t)$; that is, $\Delta(x_t - x_{t-1})$, or $x_t - 2x_{t-1} + x_{t-2}$. You can also combine the time-series operators so that LD.x is the lag of the first difference of x (that is, $x_{t-1} - x_{t-2}$) and refers to the same expression as DL.x. The seasonal difference operator, S., is used to compute the difference between the value in the current period and the period one year ago. For quarterly data, you might type S4.x to generate $x_t - x_{t-4}$ (the seasonal change), and S8.x generates $x_t - x_{t-8}$.

In addition to being easy to use, time-series operators will never misclassify an observation. You could refer to a lagged value as x[_n-1] or a first difference as x[_n] - x[_n-1], but that construction is not only cumbersome but also dangerous. Consider an annual time-series dataset in which the 1981 and 1982 data are followed by the data for 1984, 1985, ..., with the 1983 data not appearing in the dataset (i.e., they are physically absent, not simply recorded as missing values). The observation-number constructs above will misinterpret the lagged value of 1984 to be 1982, and the first difference for 1984 will incorrectly span the two-year gap. The time-series operators will not make this mistake. Because tsset ([TS] **tsset**) has been used to define year as the time-series calendar variable, the lagged value or first difference for 1984 will be properly coded as missing whether or not the 1983 data are stored as missing in the dataset. Thus you should always use the time-series operators when referring to past or future values or when computing differences in a time-series dataset.

The time-series operators also provide an important benefit in the context of *longitudinal* or *panel* datasets ([XT] **xt**), where each observation, $x_{i,t}$, is identified with both an i and a t subscript. If those data are xtset ([XT] **xtset**) or tsset, using the time-series operators will ensure that references will not span panels. For instance, z[_n-1] in a panel context will allow you to reference $z_{1,T}$ (the last observation of panel 1) as the prior value of $z_{2,1}$ (the first observation of panel 2). In contrast, L.z (or D.z) will never span panel boundaries. Panel data should always be xtset or tsset, and any time-series references should use the time-series operators.

2.5   Handling errors: The capture command

When an error is encountered in an interactive Stata session, it is displayed on the screen. When a do-file is being executed, however, Stata's default behavior causes the do-file to abort when an error occurs.[12] There are circumstances when a Stata error should be ignored, for example, when calculating a measure from each by-group that can be computed only if there are more than 10 observations in the by-group. Rather than programming conditional logic that prevents that calculation from taking place with insufficient observations, you could use capture ([P] **capture**) as a prefix on that command. For instance, `capture regress y x1-x12` will prevent the failure of one regression from aborting the do-file. If you still would like to see the regression results for those regressions which are feasible, use `noisily capture ....` The `capture` command can also be used to surround a block of statements, as in

```
capture {
regress y x1-x12
regress y x13-x24
regress y x25-x36
}
```

rather than having to repeat `capture` on each `regress` command.

2.6   Protecting the data in memory: The preserve and restore commands

Several Stata commands replace the data in memory with a new dataset. For instance, the `collapse` ([D] **collapse**) command makes a dataset of summary statistics, whereas `contract` ([D] **contract**) makes a dataset of frequencies or percentages. In a program, you may want to invoke one of these commands, but you may want to retain the existing contents of memory for further use in the do-file. You need the `preserve` ([P] **preserve**) and `restore` commands, which will allow you to set aside the current contents of memory in a temporary file and bring them back when needed. For example,

```
. sysuse auto
(1978 Automobile Data)

. generate lprice = log(price)

. preserve

. collapse (max) max_lprice=lprice max_mpg=mpg (iqr) iqr_lprice=lprice
> iqr_mpg=mpg if !missing(rep78), by(rep78)

. sort rep78

. save repstats, replace
file repstats.dta saved
```

We use and modify the `auto.dta` dataset, then `preserve` the modified file. The `collapse` command creates a dataset with one observation for each value of `rep78`,

---

12. The `nostop` option of the do ([R] **do**) command can be used to prevent termination. However, there is no way to invoke this option if you launch the do-file by double-clicking on it.

the by() variable. We **sort** that dataset of summary statistics and save it. We are now
ready to return to our main dataset:

```
. restore

. sort rep78

. merge rep78 using repstats, uniqusing
variable rep78 does not uniquely identify observations in the master data

. summarize lprice max_lprice iqr_lprice
    Variable |      Obs        Mean    Std. Dev.       Min        Max

      lprice |       74    8.640633    .3921059   8.098947   9.674452
  max_lprice |       69    9.456724    .2643769   8.503905   9.674452
  iqr_lprice |       69    .4019202    .0425909   .1622562   .4187889
```

The **restore** command brings the **preserved** dataset back into memory. We **sort** by
rep78 and use **merge** to combine the individual automobile data in memory with the
summary statistics from **repstats.dta**. Although these computations could have been
performed without **collapse**,[13] the convenience of that command should be clear. The
ability to set the current dataset aside (without having to explicitly **save** it) and bring
it back into memory when needed is a useful feature.

2.7   Getting your data into Stata

This section discusses data input and manipulation issues.[14] Source data can be downloaded from a web site, acquired in spreadsheet format, or made available in the format
of some other statistical package. The following two subsections deal with those variations.

2.7.1   Inputting data from ASCII text files and spreadsheets

Before carrying out statistical analysis with Stata, many researchers must face several
thorny issues in converting their foreign data into a Stata-usable form. These issues
range from the mundane (e.g., a text-file dataset may have coded missing values as 99)
to the challenging (e.g., a text-file dataset may be in a *hierarchical* format, with master
records and detail records). Although a brief guide to these issues cannot possibly cover
all the ways in which external data can be organized and transformed for use in Stata,
several rules apply:

- You should familiarize yourself with the various Stata commands for data input.
  Each has its use, and in the spirit of "do not pound nails with a screwdriver",
  data handling is much simpler if you use the correct tool. Reading the [U] **21
  Inputting data** section is well worth the investment of your time.

---

13. You can generate these new variables by using the **egen max_lprice = max(lprice), by(rep78)**
    and **egen iqr_lprice = iqr(lprice), by(rep78)** commands. See section 3.4.
14. This section is adapted from appendix A of Baum (2006a).

- When you need to manipulate a text file, use a text editor, not a word processor or a spreadsheet.

- Get the data into Stata as early in the process as you can, and perform all manipulations via well-documented do-files that can be edited and re-executed if need be (or if a similar dataset is encountered). Given this exhortation, we will not discuss `input` ([D] **input**) or the Data Editor, which allows interactive entry of data, or various copy-and-paste strategies involving simultaneous use of a spreadsheet and Stata. Such strategies are not reproducible and should be avoided.

- Keeping track of multiple steps of the data input and manipulation process requires good documentation. If you ever need to replicate or audit the data manipulation process, you will regret it if your documentation did not receive the proper attention.

- Working with anything but a simple rectangular data array will almost always require the use of `append` ([D] **append**), `merge` ([D] **merge**), or `reshape` ([D] **reshape**). You should review the discussion of those commands in chapter 5 and understand their capabilities.

Handling text files

Text files—often described as ASCII files—are the most common source of raw data in microeconomic research. Text files can have any file extension: they can be labeled `.raw` (as Stata would prefer), `.txt`, `.csv`, or `.asc`. A text file is just that: text. Word processing programs, like Microsoft Word, are inappropriate tools for working with text files since they have their own native, binary format and generally use features such as proportional spacing, which causes columns to be misaligned. A word processor uses a considerable amount of computing power to translate a text file into its own native format before it can display it on the screen. The inverse of that transformation must be used to create a text file that subsequently can be read by Stata.

Stata does not read binary files other than those in its own `.dta` format.[15] The second rule above counsels the use of a *text editor*, rather than a word processor or spreadsheet, when manipulating text files. Every operating system supports a variety of text editors, many of which are freely available. A useful summary of text editors of interest to Stata users is edited by Nicholas J. Cox and is available as a web page from `ssc` as the package `texteditors`. You will find that a good text editor—one without the memory limitations present in Stata's do-file editor or in the built-in routines in some operating systems—is much faster than a word processor when scrolling through a large data file. Many text editors colorize Stata commands, making them useful for Stata program development. Text editors are also extremely useful when working with large microeconomic survey datasets that are accompanied with machine-readable codebooks, often many megabytes in size. Searching those codebooks for particular keywords with a robust text editor is efficient.

---

15. Exceptions to that rule: the `fdause` command ([D] **fdasave**) can read an SAS transport (`.xpt`) file, and the `haver` ([TS] **haver**) command can read Haver Analytics datasets on the Windows platform.

Free format versus fixed format

Text files can be *free format* or *fixed format*. A free-format file contains several fields per record, separated by *delimiters*, characters that are not to be found within the fields. A purely numeric file (or one with simple string variables, such as U.S. state codes) can be *space-delimited*. That is, successive fields in the record are separated by one or more space characters:

```
AK 12.34  0.09  262000
AL 9.02 0.075 378000
AZ 102.4  0.1  545250
```

The columns in the file need not be aligned. These data can be read from a text file (by default with extension .raw) with Stata's `infile` ([D] **infile**) command, which must assign names (and if necessary, data types) to the variables:

```
. clear
. infile str2 state members prop potential using appA_1
(3 observations read)
. list
```

	state	members	prop	potential
1.	AK	12.34	.09	262000
2.	AL	9.02	.075	378000
3.	AZ	102.4	.1	545250

Here we must indicate that the first variable is a string variable of maximum length 2 characters (`str2`), or every record will generate an error that says `state` cannot be read as a number. We can even have a string variable with contents of varying length in the record:

```
. clear
. infile str2 state members prop potential str20 state_name key using appA_2
(3 observations read)
. list
```

	state	members	prop	potent~l	state_~e	key
1.	AK	12.34	.09	262000	Alaska	1
2.	AL	9.02	.075	378000	Alabama	2
3.	AZ	102.4	.1	545250	Arizona	3

However, this scheme will break down as soon as we hit New Hampshire. The space within the state name will be taken as a delimiter, and Stata will become befuddled. If string variables with embedded spaces are to be used in a space-delimited file, they themselves must be delimited (usually with quotation marks in the text file):

```
. clear

. type appA_3.raw
AK 12.34  0.09   262000 Alaska 1
AL 9.02 0.075 378000 Alabama 2
AZ 102.4  0.1   545250 Arizona 3
NH 14.9  0.02   212000 "New Hampshire" 4

. infile str2 state members prop potential str20 state_name key using appA_3
(4 observations read)

. list
```

	state	members	prop	potential	state_name	key
1.	AK	12.34	.09	262000	Alaska	1
2.	AL	9.02	.075	378000	Alabama	2
3.	AZ	102.4	.1	545250	Arizona	3
4.	NH	14.9	.02	212000	New Hampshire	4

So what should you do if your text file is space-delimited and contains string variables with embedded spaces? That is a difficult question because no mechanical transformation will generally solve this problem. For instance, using a text editor to change multiple spaces to a single space and then to change each single space to a tab character will not help, because it will then place a tab between New and Hampshire.

If the data are downloadable from a web page that offers formatting choices, you should choose a *tab-delimited* rather than a space-delimited format. The other option, comma-delimited text, or *comma-separated values* (.csv), has its own difficulties. Consider field contents (without quotation marks) such as "College Station, TX", "J. Arthur Jones, Jr.", "F. Lee Bailey, Esq.", or "T. Frank Kennedy, S.J.". If every city name is followed by a comma, then no problem, because the city and state can then be read as separate variables. But if some are written without commas ("Brighton MA"), the problem returns. In any case, parsing proper names with embedded commas is problematic. Tab-delimited text avoids most of these problems.

The insheet command

If we are to read tab-delimited text files, the **infile** command is no longer the right tool for the job; we should now use **insheet**. Two caveats: **insheet**, despite its name, does not read binary spreadsheet files (e.g., files of type .xls), and it reads a tab-delimited (or comma-delimited) text file regardless of whether a spreadsheet program was involved in its creation. For instance, most database programs contain an option to generate a tab-delimited or comma-delimited export file, and many datasets available for web download are in one of these formats.

The **insheet** command is handy. This is the command to use as long as one observation in your target Stata dataset is contained on a single record with tab or comma delimiters. Stata will automatically try to determine which delimiter is in use (but options **tab** and **comma** are available), or any ASCII character can be specified as a delimiter with the **delimiter**(*char*) option. For instance, some European database exports use semicolon (;) delimiters because standard European numeric formats use the comma as the decimal separator. If the first line of the `.raw` file contains valid Stata variable names, they will be used. This is useful because data being extracted from a spreadsheet often have that format. To use the sample dataset above, now tab-delimited with a header record of variable names, type

```
. clear
. insheet using appA_4
(6 vars, 4 obs)
. list
```

	state	members	prop	potent~1	state_name	key
1.	AK	12.34	.09	262000	Alaska	1
2.	AL	9.02	.075	378000	Alabama	2
3.	AZ	102.4	.1	545250	Arizona	3
4.	NH	14.9	.02	212000	New Hampshire	4

The issue of embedded spaces or commas no longer arises in tab-delimited data, and you can rely on the first line of the file to define the variable names.

It is particularly important to heed any information or error messages produced by the data input commands. If you know how many observations are present in the text file, check to see that the number Stata reports is correct. Likewise, the **summarize** ([R] **summarize**) command should be used to discern whether the number of observations, minimum and maximum, for each numeric variable is sensible. Data-entry errors often can be detected by noting that a particular variable takes on nonsensical values, usually denoting the omission of one or more fields on that record. Such an omission can also trigger one or more error messages. For instance, leaving out a numeric field on a particular record will move an adjacent string field into that variable. Stata will then complain that it cannot read the string as a number. A distinct advantage of the tab- and comma-delimited formats is that missing values can be coded with two successive delimiters. As will be discussed in chapter 5, **assert** ([D] **assert**) can be used to good advantage to ensure that reasonable values appear in the data.

An additional distinction exists between **infile** and **insheet**: the former command can be used with **if** *exp* and **in** *range* qualifiers to selectively input data. For instance, with a very large text-file dataset, you could use **in 1/1000** to read only the first 1,000 observations and verify that the input process is working properly. By using **if gender=="M"**, we could read only the male observations; or by using **if runiform() <= 0.15**, we could read each observation with probability 0.15. These qualifiers cannot be used with **insheet**. However, if the text-file dataset is huge and the computer is slow, you could always read the entire dataset and apply **keep** or **drop** ([D] **drop**) conditions to mimic the action of **infile**.

Accessing data stored in spreadsheets

In the third rule on page 16, we counseled that copy-and-paste techniques should not be used to transfer data from another application directly to Stata. Such a technique cannot be reliably replicated. How do you know that the first and last rows or columns of a spreadsheet were selected and copied to the clipboard, without any loss of data or extraneous inclusion of unwanted data? If the data are presently in a spreadsheet, the appropriate portion of that spreadsheet should be copied and pasted (in Excel, *Paste Special* to ensure that only values are stored) into a new blank sheet. If Stata variable names are to be added, leave the first row blank so that they can be filled in. Save that sheet, and that sheet alone, as *Text Only – Tab delimited* to a new filename. If you use the file extension .raw, it will simplify reading the file into Stata.

Two caveats regarding dates: First, both Excel and Stata work with the notion that calendar dates are successive integers from an arbitrary starting point. To read the dates into a Stata date variable, they must be formatted with a four-digit year, preferably in a format with delimiters (e.g., 12/6/2004 or 6-Dec-2004). It is much easier to make these changes in the spreadsheet program before reading the data into Stata. Second, Macintosh OS X users of Excel should note that Excel's default is the 1904 date system. If the spreadsheet was produced in Excel for Windows and the steps above are used to create a new sheet with the desired data, the dates will be off by four years (the difference between Excel for Macintosh and Excel for Windows defaults). Uncheck the preference *Use the 1904 date system* before saving the file as text.

Fixed-format data files

Many text-file datasets are composed of *fixed-format* records: those obeying a strict column format in which a variable appears in a specific location in each record of the dataset. Such datasets are accompanied by *codebooks*, which define each variable's name, data type, location in the record, and possibly other information, such as missing values, value labels, or frequencies for integer variables.[16] We present a fragment of the codebook for the study "National Survey of Hispanic Elderly People, 1988",[17] available from the Inter-University Consortium for Political and Social Research.

---

16. Stata itself has the ability to produce a codebook from a Stata dataset via the codebook ([D] **codebook**) command.

17. Available at http://webapp.icpsr.umich.edu/cocoon/ICPSR-STUDY/09289.xml.

```
VAR 0001        ICPSR STUDY NUMBER-9289      NO MISSING DATA CODES
                REF 0001        LOC    1 WIDTH  4          DK   1 COL  3- 6
VAR 0002        ICPSR EDITION NUMBER-2       NO MISSING DATA CODES
                REF 0002        LOC    5 WIDTH  1          DK   1 COL  7
VAR 0003        ICPSR PART NUMBER-001        NO MISSING DATA CODES
                REF 0003        LOC    6 WIDTH  3          DK   1 COL  8-10
VAR 0004        ICPSR ID                     NO MISSING DATA CODES
                REF 0004        LOC    9 WIDTH  4          DK   1 COL 11-14
VAR 0005        ORIGINAL ID                  NO MISSING DATA CODES
                REF 0005        LOC   13 WIDTH  4          DK   1 COL 15-18
VAR 0006        PROXY                        NO MISSING DATA CODES
                REF 0006        LOC   17 WIDTH  1          DK   1 COL 19
VAR 0007        TIME BEGUN-HOUR                       MD=99
                REF 0007        LOC   18 WIDTH  2          DK   1 COL 20-21
VAR 0008        TIME BEGUN-MINUTE                     MD=99
                REF 0008        LOC   20 WIDTH  2          DK   1 COL 22-23
VAR 0009        TIME BEGUN-AM/PM                      MD=9
                REF 0009        LOC   22 WIDTH  1          DK   1 COL 24
VAR 0010        AGE                          NO MISSING DATA CODES
                REF 0010        LOC   23 WIDTH  3          DK   1 COL 25-27
VAR 0011        HISPANIC GROUP               NO MISSING DATA CODES
                REF 0011        LOC   26 WIDTH  1          DK   1 COL 28
VAR 0012        HISPANIC GROUP-OTHER                  MD=99
                REF 0012        LOC   27 WIDTH  2          DK   1 COL 29-30
VAR 0013        MARITAL STATUS               NO MISSING DATA CODES
                REF 0013        LOC   29 WIDTH  1          DK   1 COL 31

    Q.A3.  ARE YOU NOW MARRIED, WIDOWED, DIVORCED, SEPARATED, OR
                HAVE YOU NEVER MARRIED?
           -------------------------------------------------------
            1083  1.   MARRIED
             815  2.   WIDOWED
             160  3.   DIVORCED
              99  4.   SEPARATED
              14  5.   NOT MARRIED, LIVING WITH PARTNER
             128  6.   NEVER MARRIED

VAR 0014        MARITAL STATUS-YEARS              MD=97 OR GE  98
                REF 0014        LOC   30 WIDTH  2          DK   1 COL 32-33
VAR 0015        RESIDENCE TYPE                        MD=7
                REF 0015        LOC   32 WIDTH  1          DK   1 COL 34
VAR 0016        RESIDENCE TYPE-OTHER               MD=GE  99
                REF 0016        LOC   33 WIDTH  2          DK   1 COL 35-36
VAR 0017        OWN/RENT                             MD=7
                REF 0017        LOC   35 WIDTH  1          DK   1 COL 37
VAR 0018        OWN/RENT-OTHER                       MD=99
                REF 0018        LOC   36 WIDTH  2          DK   1 COL 38-39
VAR 0019        LIVE ALONE                   NO MISSING DATA CODES
                REF 0019        LOC   38 WIDTH  1          DK   1 COL 40
VAR 0020        HOW LONG LIVE ALONE             MD=7 OR GE  8
                REF 0020        LOC   39 WIDTH  1          DK   1 COL 41
VAR 0021        PREFER LIVE ALONE               MD=7 OR GE  8
                REF 0021        LOC   40 WIDTH  1          DK   1 COL 42
```

The codebook specifies the column in which each variable starts (LOC) and the number of columns it spans (WIDTH).[18] In this fragment of the codebook, only integer numeric variables appear. The missing data codes (MD) for each variable are also specified. The

_____

18. The COL field should not be considered.

listing above provides the full codebook details for variable 13, marital status, quoting the question posed by the interviewer, coding of the six possible responses, and the frequency counts of each response.

One of the important notions about a fixed-format data file is that fields need not be separated, as we see above, where the single-column fields of variables 0019, 0020, and 0021 are stored as three successive integers. Stata must be instructed to interpret each of those digits as a separate variable. This is done with a *data dictionary*: a separate Stata file, with file extension `.dct`, specifying the necessary information to read a fixed-format data file. The information in the codebook can be translated, line for line, into the Stata data dictionary. The data dictionary need not be comprehensive. You might not want to read certain variables from the raw data file, so you would merely ignore those columns. This might be particularly important when working with Stata/IC and its limit of 2,047 variables. Many survey datasets contain many more than 2,000 variables. By judiciously specifying only the subset of variables that are of interest in your research, you can read such a text file with Stata/IC.

Stata supports two different formats of data dictionaries. The simpler format, used by `infix` ([D] **infix (fixed format)**), requires only that the starting and ending columns of each variable are given along with any needed data-type information. To illustrate, we specify the information needed to read a subset of fields in this codebook into Stata variables, using the description of the data dictionary in `infix`:

```
. clear
. infix using 09289-infix
infix dictionary using 09289-0001-Data.raw {
* dictionary to read extract of ICPSR study 9289
  int v1      1-4
  int v2      5
  int v3      6-8
  int v4      9-12
  int v5      13-16
  int v6      17
  int v7      18-19
  int v8      20-21
  int v9      22
  int v10     23-25
  int v11     26
  int v12     27-28
  int v13     29
  int v14     30-31
  int v15     32
  int v16     33-34
  int v17     35
  int v18     36-37
  int v19     38
  int v20     39
  int v21     40
}
(2299 observations read)
```

Alternately, we could set up a dictionary file for the fixed-format version of `infile`. This is the more powerful option because it allows us to attach variable labels and specify value labels. However, rather than specifying the column range of each field that you want to read, you must indicate where it starts and its field width, given as the `%infmt` for that variable. With a codebook like the one displayed above, we have the field widths available. We could also calculate the field widths from the starting and ending column numbers. We must not only specify which are string variables but also give their data storage type. The storage type could differ from the `%infmt` for that variable. You might read a six-character code into a ten-character field knowing that other data use the latter width for that variable.

```
. clear
. infile using 09289-0001-Data
infile dictionary using 09289-0001-Data.raw {
_lines(1)
_line(1)
_column(1)    int   V1               %4f    "ICPSR STUDY NUMBER-9289"
_column(5)    int   V2     :V2       %1f    "ICPSR EDITION NUMBER-2"
_column(6)    int   V3               %3f    "ICPSR PART NUMBER-001"
_column(9)    int   V4               %4f    "ICPSR ID"
_column(13)   int   V5               %4f    "ORIGINAL ID"
_column(17)   int   V6     :V6       %1f    "PROXY"
_column(18)   int   V7     :V7       %2f    "TIME BEGUN-HOUR"
_column(20)   int   V8     :V8       %2f    "TIME BEGUN-MINUTE"
_column(22)   int   V9     :V9       %1f    "TIME BEGUN-AM-PM"
_column(23)   int   V10    :V10      %3f    "AGE"
_column(26)   int   V11    :V11      %1f    "HISPANIC GROUP"
_column(27)   int   V12    :V12      %2f    "HISPANIC GROUP-OTHER"
_column(29)   int   V13    :V13      %1f    "MARITAL STATUS"
_column(30)   int   V14    :V14      %2f    "MARITAL STATUS-YEARS"
_column(32)   int   V15    :V15      %1f    "RESIDENCE TYPE"
_column(33)   int   V16    :V16      %2f    "RESIDENCE TYPE-OTHER"
_column(35)   int   V17    :V17      %1f    "OWN-RENT"
_column(36)   int   V18    :V18      %2f    "OWN-RENT-OTHER"
_column(38)   int   V19    :V19      %1f    "LIVE ALONE"
_column(39)   int   V20    :V20      %1f    "HOW LONG LIVE ALONE"
_column(40)   int   V21    :V21      %1f    "PREFER LIVE ALONE"
}
(2299 observations read)
```

The `_column()` directives in this dictionary are used where dictionary fields are not adjacent. You could skip back and forth along the input record because the columns read need not be in ascending order. But then we could achieve the same thing with the `order` ([D] **order**) command after data input. We are able to define variable labels by using `infile`. In both examples above, the dictionary file specifies the name of the data file, which need not be the same as that of the dictionary file. For example, `highway.dct` could read `highway.raw`, and if that were the case, the latter filename need not be specified. But we might want to use the same dictionary to read more than one `.raw` file. To do so, leave the filename out of the dictionary file, and use the `using` modifier to specify the name of the `.raw` file. After loading the data, we can `describe` ([D] **describe**) its contents:

```
. describe

Contains data
  obs:          2,299
  vars:            21
  size:       105,754 (99.9% of memory free)
```

variable name	storage type	display format	value label	variable label
V1	int	%8.0g		ICPSR STUDY NUMBER-9289
V2	int	%8.0g	V2	ICPSR EDITION NUMBER-2
V3	int	%8.0g		ICPSR PART NUMBER-001
V4	int	%8.0g		ICPSR ID
V5	int	%8.0g		ORIGINAL ID
V6	int	%8.0g	V6	PROXY
V7	int	%8.0g	V7	TIME BEGUN-HOUR
V8	int	%8.0g	V8	TIME BEGUN-MINUTE
V9	int	%8.0g	V9	TIME BEGUN-AM-PM
V10	int	%8.0g	V10	AGE
V11	int	%8.0g	V11	HISPANIC GROUP
V12	int	%8.0g	V12	HISPANIC GROUP-OTHER
V13	int	%8.0g	V13	MARITAL STATUS
V14	int	%8.0g	V14	MARITAL STATUS-YEARS
V15	int	%8.0g	V15	RESIDENCE TYPE
V16	int	%8.0g	V16	RESIDENCE TYPE-OTHER
V17	int	%8.0g	V17	OWN-RENT
V18	int	%8.0g	V18	OWN-RENT-OTHER
V19	int	%8.0g	V19	LIVE ALONE
V20	int	%8.0g	V20	HOW LONG LIVE ALONE
V21	int	%8.0g	V21	PREFER LIVE ALONE

```
Sorted by:
     Note:  dataset has changed since last saved
```

The dictionary indicates that value labels are associated with the variables, but it does not define those labels. Commands such as

```
label define V13    1 "MARRIED" 2 "WIDOWED" 3 "DIVORCED" 4 "SEPARATED" ///
                    5 "NOT MAR COHABITG" 6 "NEVER MARRIED"
```

must be given to create those labels.

One other advantage of the more elaborate `infile` data-dictionary format should be noted. Many large survey datasets contain several variables that are real or floating-point values, such as a wage rate in dollars and cents or a percentage interest rate, such as 6.125%. To save space, the decimal points are excluded from the text file and the codebook indicates how many decimal digits are included in the field. You could read these data as integer values and perform the appropriate division in Stata, but a simpler solution would be to build this information into the data dictionary. By specifying that a variable has an `infmt` of, for example, `%6.2f`, a value such as `123456` can be read properly as daily sales of $1,234.56.

Stata's data-dictionary syntax can handle many more complicated text datasets, including those with multiple records per observation or those with header records that are to be ignored. See [D] **infile (fixed format)** for full details.

2.7.2   Importing data from other package formats

The previous section discussed how foreign data files could be brought into Stata. Often the foreign data are already in the format of some other statistical package or application. For instance, several economic- and financial-data providers make SAS-formatted datasets readily available, while socioeconomic datasets are often provided in SPSS format. The most straightforward and inexpensive way to deal with these package formats involves a third-party application: Stat/Transfer, a product of Circle Systems, Inc. It is not the only third-party application to provide such functionality: DBMS/Copy is often mentioned in this context. However, Stat/Transfer has the advantage of a comarketing relationship with StataCorp, so you can acquire a copy of Stat/Transfer at an advantageous price with your Stata order.

The alternative to Stat/Transfer usually involves owning (or having access to) a working copy of the other statistical package and having enough familiarity with the syntax of that package to understand how a dataset can be exported from its own proprietary format to ASCII format.[19] Even for those researchers who have that familiarity and copies of another package, this is a rather cumbersome solution because (like Stata) packages such as SAS and SPSS have their own conventions for missing data formats, value labels, data types, etc. Although the raw data can be readily exported to ASCII format, these attributes of the data will have to be re-created in Stata. For a large survey dataset with many hundred (or several thousand) variables, that is unpalatable. A transformation utility like Stat/Transfer performs all those housekeeping chores, ensuring that any attributes attached to the data (extended missing value codes, value labels, etc.) are placed in the Stata-format file. Of course, the mapping between packages is not always one to one. In Stata, a value label stands alone and can be attached to any variable or set of variables, whereas in other packages, it is generally an attribute of a variable and must be duplicated for similar variables.

An important distinction between Stata and SAS and SPSS is Stata's flexible set of data types. Stata, like the C language in which its core code is written, offers five numeric data types ([D] **data types**): the integer types `byte`, `int`, and `long`, and the floating-point types `float` and `double`. Stata also offers the string types `str1–str244`. Most other packages do not have this broad array of data types and resort to storing all numeric data in a single data type: "Raw data come in many different forms, but SAS simplifies this. In SAS there are just two data types: numeric and character" (Delwiche and Slaughter 1998, 4). This simplicity bears a sizable cost because an indicator variable requires only one byte of storage and a double-precision floating-point variable requires eight bytes to hold up to 15 decimal digits of accuracy. Stata allows the user to specify the data type based on the contents of each variable, which can result in considerable savings in terms of both disk space and execution time when reading or writing those variables to disk. Stat/Transfer can be instructed to optimize a target Stata-format file in the transfer process, or you can use Stata's `compress` command to automatically perform that optimization. In any case, you should always take advan-

---

19. If SAS datasets are available in the SAS transport (`.xpt`) format, they can be read by Stata's `fdause` command.

tage of this optimization because it will reduce the size of files and require less of your computer's memory to work with them.

A useful feature of Stat/Transfer is the ability to generate a subset of a large file while transferring it from SAS or SPSS format. I spoke above of the possibility of reading only certain variables from a text file to avoid Stata/IC's limitation of 2,047 variables. You can always Stat/Transfer a sizable survey data file from SAS to Stata format, but if there are more than 2,047 variables in the file, the target file must be specified as a Stata/SE file. If you do not have access to Stata/SE, the transfer will be problematic. The solution is to use Stat/Transfer's ability to read a list of variables that you would like to keep (or a list of variables to drop), which will generate a subset file "on the fly". Because Stat/Transfer can generate a machine-readable list of variable names, that list can be edited to produce the keep list or drop list.

Although we have spoken of SAS and SPSS, Stat/Transfer is capable of exchanging datasets with a wide variety of additional packages, including GAUSS, Excel, MATLAB, etc.; see http://stattransfer.com for details. Versions of Stat/Transfer for Windows, Mac OS X, and Linux/Unix are available.

We must also mention an alternative solution for data transfer between databases supporting some flavor of structured query language (SQL). Stata can perform Open Database Connectivity (ODBC) operations with databases accessible via that protocol (see [D] **odbc** for details). Because most SQL databases and non-SQL data structures such as Excel and Microsoft Access support ODBC, this is often suggested as a workable solution to dealing with foreign data. It does require that the computer system on which you are running Stata is equipped with ODBC drivers. These are installed by default on Windows systems with Microsoft Office but may require the purchase of a third-party product for Mac OS X or Linux systems. If the necessary database connectivity is available, Stata's **odbc** is a full-featured solution. It allows for both the query of external databases and the insertion or update of records in those databases.

2.8   Guidelines for Stata do-file programming style

As you move away from interactive use of Stata and make greater use of do-files and ado-files in your research, the style of the contents of those files becomes more important. One of the reasons for using do-files is the audit trail that they provide. Are your do-files readable and comprehensible—not only today but also in several months? To highlight the importance of good programming style practices, I present an edited excerpt from Nicholas J. Cox's excellent essay "Suggestions on Stata programming style"[20] (Cox 2005f). The rest of this section is quoted from that essay.

Programming in Stata, like programming in any other computer language, is partly a matter of syntax, as Stata has rules that must be obeyed. It is also partly a matter of style. Good style includes, but is not limited to, writing programs that are, above all else, clear. They are clear to the programmer, who may revisit them repeatedly,

---

20. I am grateful to Nick Cox for permission to reproduce his excellent suggestions.

and they are clear to other programmers, who may wish to understand them, to debug them, to extend them, to speed them up, to imitate them, or to borrow from them.

People who program a great deal know this: setting rules for yourself and then obeying them ultimately yields better programs and saves time. They also know that programmers may differ in style and even argue passionately about many matters of style, both large and small. In this morass of varying standards and tastes, I suggest one overriding rule: Set and obey programming style rules for yourself. Moreover, obey each of the rules I suggest unless you can make a case that your own rule is as good or better.

Enough pious generalities: The devil in programming is in the details. Many of these details reflect longstanding and general advice (e.g., Kernighan and Plauger [1978]).

2.8.1 Basic guidelines for do-file writers

- Use natural names for logical constants or variables. Thus `local OK` should be 1 if true and 0 if false, permitting idioms such as `if 'OK'`. (But beware such logicals taking on missing values.)

- Type expressions so they are readable. Some possible rules are as follows:

  - Put spaces around each binary operator except ^ (`gen z = x + y` is clear, but `x ^ 2` looks odder than `x^2`).
  - `*` and `/` allow different choices. `num / den` is arguably clearer than `num/den`, but readers might well prefer 2/3 to 2 / 3. Overall readability is paramount; compare, for example,

    `hours + minutes / 60 + seconds / 3600`

    with

    `hours + minutes/60 + seconds/3600`
  - Put a space after each comma in a function, etc.
  - Use parentheses for readability.

Such a spaced-out style can, however, make it difficult to fit expressions on one line, another desideratum.

- Adopt a consistent style for flow control. Stata has `if`, `while`, `foreach`, and `forvalues` structures that resemble those in many mainstream programming languages. Programmers in those languages often argue passionately about the best layout. Choose one such layout for yourself. Here is one set of rules:

  - Tab lines consistently after `if`, `else`, `while`, `foreach`, or `forvalues` (the StataCorp convention is that a tab is eight spaces and is greatly preferable if Stata is to show your programs properly).
  - Put a space before braces.

– Align the `i` of `if` and the `e` of `else`, and align closing braces, `}`, with the `i`, `e`, or `f`:

```
if ... {
        ...
        ...
}
else {
        ...
        ...
}
while ... {
        ...
        ...
}
foreach ... {
        ...
        ...
}
```

In Stata 8 and later, putting the opening and closing braces on lines above and below the body of each construct is required (with the exceptions that the whole of an `if` construct or the whole of an `else` construct can legally be placed on one line). For earlier releases, it is strongly advised.

- Write within 80 columns (72 are even better). The awkwardness of viewing (and understanding) long lines outweighs the awkwardness of splitting commands into two or more physical lines.

- Use `#delimit ;` sparingly (Stata is not C). To deal with long lines, use `///` and continue on the next line of your do-file. You can use the triple-slash continuation on any number of consecutive lines of text.

- Use blank lines to separate distinct blocks of code.

- Consider putting `quietly` on a block of statements rather than on each or many of them. An alternative is to use `capture`, which eats what output there might have been and any errors that might occur; this is sometimes the ideal combination.

- You can express logical negation with either `!` or `~`. Choose one and stick with it. StataCorp has changed recently from preferring `~` to preferring `!`.

- Define constants to machine precision. Thus use `_pi` or `c(pi)` rather than some approximation such as 3.14159, or use `-digamma(1)` for the Euler–Mascheroni constant γ rather than 0.57721. Cruder approximations can give results adequate for your purposes, but that does not mean you should eschew wired-in features.

- Avoid "magic numbers". Use a macro or a scalar to hold a number whose meaning is obvious only to the programmer, and add a comment indicating what that number represents.

2.8.2   Enhancing speed and efficiency

Here is a list of basic ways to increase the speed and efficiency of your code:

- Test for fatal conditions as early as possible. Do no unnecessary work before checking that a vital condition has been satisfied.

- Use `summarize, meanonly` for speed when its returned results are sufficient (see Cox [2007e]). Also consider whether `quietly count` fits the purpose better.

- `foreach` and `forvalues` are cleaner and faster than most `while` loops.

- Try to avoid looping over observations, which is slow. Fortunately, it can usually be avoided.

- Avoid using a variable to hold a constant. A macro or a scalar is usually all that is needed. One clear exception is that some graphical effects depend on a variable being used to store a constant.

2.9   How to seek help for Stata programming

Programming in any language can be a frustrating experience. In this section, I discuss several resources that you may find helpful in solving a programming problem, whether it arises as you scratch your head wondering "How can I get Stata to do that?" or after you have coded a trial solution. A do-file or ado-file program can be syntactically correct but not produce the desired results, or correct answers, and that is the more insidious problem: a program that works properly most of the time but may fail under special conditions.

As obvious as it may seem, the most useful resources for Stata programmers are the online help files and excellent printed documentation. The online help files contain a great deal of the information in the manuals, but there is no substitute for the manuals. You should ensure that you have ready access to the latest version of Stata's manuals. That said, recall that many of Stata's commands are enhanced, corrected, or even introduced between major versions of Stata, and the only documentation for new features or new commands appears in the online help files. The `findit` command is often most useful, because it looks for the specified keyword(s) in online help files, frequently asked questions (FAQs) on the Stata web site, *Stata Technical Bulletin* and *Stata Journal* articles, and the SSC archive.[21] A judicious choice of keywords for `findit` will often prove helpful.

What other resources are available? If you believe Stata is not producing the results that it should according to the documentation for a command, you can contact Stata technical support, following the guidelines at
http://www.stata.com/support/tech-support. Like any large, complex software system,

---

21. The `search` command places output in the Results window and scans all these sources except the SSC archive and other user sites. `search..., all`, a synonym for `findit`, will access all the sites mentioned.

Stata has bugs. Thankfully, most bugs in official Stata are short-lived. Stata's `update` capability makes it possible for Stata's developers to provide corrected code within a few days of a bug's diagnosis—but only if you keep your official Stata up to date! Before producing a trouble report for tech support, ensure that your Stata executable and ado-files are fully up to date by using `update query`. Even if you are using an older version of Stata, you may not have the last release of that older version installed. Many reported bugs have already been corrected in the latest downloadable routines.

This caveat also applies, even more emphatically, to the user-written Stata routines that you may have downloaded from the *Stata Journal*, SSC archive, or individual users' web sites. Although a good deal of user-written code can rival official Stata code in terms of reliability, it stands to reason that user-programmers (such as myself) are not as uniformly competent and careful as professional software developers. Thankfully, the `adoupdate` command allows you to check whether user-written code on your machine is up to date. Most user-programmers reliably respond to any reports of bugs in their code and post corrected versions on the SSC archive or their own web sites. Before reporting a bug in a user-written routine, ensure that you have the latest version.

Many users' programming problems are solved expeditiously by a posting on Statalist. Participants in this active discussion list can often suggest a solution to a well-posed programming problem or assist in the diagnosis of your do-file or ado-file problem. To post to Statalist, you must be a member of the list, and to get helpful comments, you should read and heed the Statalist FAQs (http://www.stata.com/support/statalist/faq). A posting that states "The `xyz` command doesn't work" will not attract much attention. If you want Statalisters to help, you should pose your problem in well-defined terms that allow list participants to understand what you are doing (or not doing) and what you are trying to do. You should not send attachments or excessively lengthy output to Statalist, and you should post in ASCII text mail format. Attachments, lengthy messages, and non-ASCII messages may be discarded by the managing software. Several other important admonitions appear in the Statalist FAQs. In summary, Statalisters can be very helpful in solving your problems if you explain them clearly. This caveat also applies to another source of support: local "gurus", either colleagues or research support staff in your institution.

Lastly, I must mention Stata's own capabilities to diagnose your problem with a program you have written. Most of official Stata and user-written Stata routines are written in the interpreted *ado-file language*. When you give a Stata command, only the standard results or error messages usually appear. How can you track down an error in a do-file containing a loop construct (such as `forvalues` or `foreach`) or in an ado-file? You can track it by using `set trace on` ([P] **trace**). This command will instruct Stata to echo both the lines of code that it executes as well as their results (for instance, any substitutions of names or values into a command). By examining the results, which should be directed to a log file, you can often determine where and why an error occurs. Coupled with judicious use of the `display` ([P] **display**) command to print intermediate values, this can often speed the solution of a programming problem.[22]

---

22. You cannot `trace` the execution of a *built-in* Stata command. The `which` ([R] **which**) command will indicate whether a command is a built-in or an ado-file code.

3 Do-file programming: Functions, macros, scalars, and matrices

3.1 Introduction

This chapter describes several elements of *do-file programming*: functions used to generate new variables; macros that store individual results; and lists, scalars, and matrices. Although functions will be familiar to all users of Stata, macros and scalars are often overlooked by interactive users. Because nearly all Stata commands return results in the form of macros and scalars, familiarity with these concepts is useful.

The first section of the chapter deals with several general details: varlists, numlists, **if** *exp* and **in** *range* qualifiers, missing data handling, and string-to-numeric conversion (and vice versa). Subsequent sections present functions for **generate**, functions for **egen** ([D] **egen**), computation with a **by** *varlist*:, and an introduction to macros, scalars, and matrices.

3.1.1 What you should learn from this chapter

- Understand varlists, numlists, and **if** and **in** qualifiers
- Know how to handle missing data and conversion of values to missing and vice versa
- Understand string-to-numeric conversion and vice versa
- Be familiar with functions for use with **generate**
- Understand how to recode discrete and continuous variables
- Be familiar with the capabilities of **egen** functions
- Know how to use by-groups effectively
- Understand the use of **local** and **global** macros
- Be familiar with extended macro functions and macro list functions
- Understand how to use numeric and string scalars
- Know how to use matrices to retrieve and store results

3.2   Some general programming details

In this section, we use the `census2c` dataset of U.S. state-level statistics to illustrate details of do-file programming:

```
. use census2c
(1980 Census data for NE and NC states)
. list, sep(0)
```

	state	region	pop	popurb	medage	marr	divr
1.	Connecticut	NE	3107.6	2449.8	32.00	26.0	13.5
2.	Illinois	N Cntrl	11426.5	9518.0	29.90	109.8	51.0
3.	Indiana	N Cntrl	5490.2	3525.3	29.20	57.9	40.0
4.	Iowa	N Cntrl	2913.8	1708.2	30.00	27.5	11.9
5.	Kansas	N Cntrl	2363.7	1575.9	30.10	24.8	13.4
6.	Maine	NE	1124.7	534.1	30.40	12.0	6.2
7.	Massachusetts	NE	5737.0	4808.3	31.20	46.3	17.9
8.	Michigan	N Cntrl	9262.1	6551.6	28.80	86.9	45.0
9.	Minnesota	N Cntrl	4076.0	2725.2	29.20	37.6	15.4
10.	Missouri	N Cntrl	4916.7	3349.6	30.90	54.6	27.6
11.	Nebraska	N Cntrl	1569.8	987.9	29.70	14.2	6.4
12.	New Hampshire	NE	920.6	480.3	30.10	9.3	5.3
13.	New Jersey	NE	7364.8	6557.4	32.20	55.8	27.8
14.	New York	NE	17558.1	14858.1	31.90	144.5	62.0
15.	N. Dakota	N Cntrl	652.7	318.3	28.30	6.1	2.1
16.	Ohio	N Cntrl	10797.6	7918.3	29.90	99.8	58.8
17.	Pennsylvania	NE	11863.9	8220.9	32.10	93.7	34.9
18.	Rhode Island	NE	947.2	824.0	31.80	7.5	3.6
19.	S. Dakota	N Cntrl	690.8	320.8	28.90	8.8	2.8
20.	Vermont	NE	511.5	172.7	29.40	5.2	2.6
21.	Wisconsin	N Cntrl	4705.8	3020.7	29.40	41.1	17.5

This dataset, `census2c`, is arranged in tabular format, similarly to a spreadsheet. The table rows are the *observations*, cases, or records. The columns are the Stata *variables*, or fields. We see that there are 21 rows, each corresponding to one U.S. state in the North East or North Central regions, and seven columns, or variables: `state`, `region`, `pop`, `popurb`, `medage`, `marr`, and `divr`. The variables `pop` and `popurb` represent each state's 1980 population and urbanized population, respectively, in thousands. The variable `medage`, median age, is measured in years, while the variables `marr` and `divr` represent the number of marriages and divorces, respectively, in thousands.

The Stata variable names must be distinct and follow certain rules of syntax. For instance, they cannot contain embedded spaces, hyphens (-), or characters outside the sets `A-Z`, `a-z`, `0-9`, and `_` . In particular, a full stop, or period (.), cannot appear within a variable name. Variable names must start with a letter or an underscore. Most importantly, *case matters*: `STATE`, `State`, and `state` are three different variables to Stata. The Stata convention, which I urge you to adopt, is to use lowercase names for all variables to avoid confusion and to use uppercase only for some special reason. You can always use variable labels to hold additional information.

3.2.1   The varlist

Many Stata commands accept a varlist, a list of one or more variables to be used. A varlist can contain the variable names, or you can use a wild card (*), such as in *id. The * will stand in for an arbitrary set of characters. In the census2c dataset, pop* will refer to both pop and popurb:

```
. summarize pop*
    Variable |       Obs        Mean    Std. Dev.       Min        Max
-------------+--------------------------------------------------------
         pop |        21    5142.903    4675.152    511.456   17558.07
      popurb |        21    3829.776    3851.458    172.735   14858.07
```

A varlist can also contain a hyphenated list, such as dose1-dose4. This hyphenated list refers to all variables in the dataset between dose1 and dose4, including those two, in the order the variables appear in the dataset. The order of variables is provided by describe and is shown in the Variables window. It can be modified by the order command.

3.2.2   The numlist

Many Stata commands require the use of a numlist, a list of numeric arguments. A numlist can be provided in several ways. It can be spelled out explicitly, as in 0.5 1.0 1.5. It may involve a range of values, such as 1/4 or -3/3; these lists would include the integers between those limits. You could also specify 10 15 to 30, which would count from 10 to 30 by 5s, or you could use a colon to say the same thing: 10 15:30. You can count by steps, as in 1(2)9, which is a list of the first five odd integers, or 9(-2)1, which is the same list in reverse order. Square brackets can be used in place of parentheses.

One thing that generally should not appear in a numlist is a comma. A comma in a numlist will usually cause a syntax error. Other programming languages' loop constructs often spell out a range with an expression, such as 1,10. In Stata, such an expression will involve a numlist of 1/10. One of the primary uses of the numlist is for the forvalues ([P] **forvalues**) statement, which is described in section 7.3 (but not all valid numlists are acceptable in forvalues).

3.2.3   The if exp and in range qualifiers

Stata commands operate on all the observations in memory by default. Almost all Stata commands accept *qualifiers*: if *exp* and in *range* clauses that restrict the command to a subset of the observations. If we wanted to apply a transformation to a subset of the dataset or wanted to list ([D] **list**) only certain observations or summarize only those observations that met some criterion, we would use an if *exp* or an in *range* clause on the command.

In many problems, the desired subset of the data is not defined in terms of observation numbers (as specified with in *range*) but in terms of some logical condition. Then it is more useful to use the if *exp* qualifier. We could, of course, use if *exp* to express an in *range* condition. But the most common use of if *exp* involves the transformation of data or the specification of a statistical procedure for a subset of data identified by if *exp* as a logical condition. Here are some examples to illustrate these qualifiers:

```
. list state pop in 1/5
```

	state	pop
1.	Connecticut	3107.6
2.	Illinois	11426.5
3.	Indiana	5490.2
4.	Iowa	2913.8
5.	Kansas	2363.7

```
. list state pop medage if medage >= 32
```

	state	pop	medage
1.	Connecticut	3107.6	32.00
13.	New Jersey	7364.8	32.20
17.	Pennsylvania	11863.9	32.10

3.2.4   Missing data handling

Stata possesses 27 numeric missing value codes: the system missing value . and 26 others from .a through .z. They are treated as large positive values, and they sort in that order; plain . is the smallest missing value (see [U] **12.2.1 Missing values**). This allows qualifiers such as if *variable* <. to exclude all possible missing values.[1] To make your code as readable as possible, use the missing() ([D] **functions**) function described below.

Stata's standard practice for missing data handling is to omit those observations from any computation. For generate or replace, missing values are typically propagated so that any function of missing data is missing. In univariate statistical computations (such as summarize) computing a mean or standard deviation, only nonmissing cases are considered. For multivariate statistical commands, Stata generally practices *casewise deletion*, which is when an observation in which any variable is missing is deleted from the computation. The missing($x_1, x_2, \ldots, x_n$) function returns 1 if any of the arguments are missing, and 0 otherwise; that is, it provides the user with a casewise deletion indicator.

---

1. Before version 8, Stata user code often used qualifiers like if *variable* != . to rule out missing values. That is now dangerous practice because that qualifier will capture only the . missing data code. If any of the additional codes are present in the data (for instance, by virtue of having used Stat/Transfer to convert an SPSS or SAS dataset to Stata format), they will be handled properly only when if *variable* <. or if !missing(*variable*) is used.

Several Stata commands handle missing data in nonstandard ways. The functions `max()` and `min()` and the `egen` rowwise functions (`rowmax()`, `rowmean()`, `rowmin()`, `rowsd()`, and `rowtotal()`) all ignore missing values (see section 3.4). For example, `rowmean(x1,x2,x3)` will compute the mean of three, two, or one of the variables, returning missing only if all three variables' values are missing for that observation. The `egen` functions `rownonmiss()` and `rowmiss()` return, respectively, the number of nonmissing and missing elements in their varlists. Although `correlate` *varlist* ([R] **correlate**) uses casewise deletion to remove any observation containing missing values in any variable of the varlist from the computation of the correlation matrix, the alternative command `pwcorr` computes pairwise correlations using all the available data for each pair of variables.

We have discussed missing values in numeric variables, but Stata also provides for missing values in string variables. The empty, or null, string (`""`) is taken as missing. There is an important difference in Stata between a string variable containing one or more spaces and a string variable containing no spaces (although they will appear identical to the naked eye). This suggests that you should not include one or more spaces as a possible value of a string variable; take care if you do.

Recoding missing values: The mvdecode and mvencode commands

When importing data from another statistical package, spreadsheet, or database, differing notions of missing data codes can hinder the proper rendition of the data within Stata. Likewise, if the data are to be used in another program that does not use the `.` notation for missing data codes, there may be a need to use an alternative representation of Stata's missing data. The `mvdecode` and `mvencode` commands (see [D] **mvencode**) can be useful in those circumstances. The `mvdecode` command permits you to recode various numeric values to missing, as would be appropriate when missing data have been represented as -99, -999, 0.001, and so on. Stata's full set of 27 numeric missing data codes can be used, so that -9 can be mapped to `.a`, -99 can be mapped to `.b`, etc. The `mvencode` command provides the inverse function, allowing Stata's missing values to be revised to numeric form. Like `mvdecode`, `mvencode` can map each of the 27 numeric missing data codes to a different numeric value.

Many of the thorny details involved with the reliable transfer of missing data values between packages are handled competently by Stat/Transfer. This third-party application (remarketed by StataCorp) can handle the transfer of variable and value labels between major statistical packages and can create subsets of files' contents (e.g., only selected variables are translated into the target format); it is well worth the cost for those researchers who frequently import or export datasets.

3.2.5   String-to-numeric conversion and vice versa

Stata has two major kinds of variables: string and numeric. Quite commonly, a variable imported from an external source will be misclassified as string when it should be considered as numeric. For instance, if the first value read by insheet is NA, that variable will be classified as a string variable. Stata provides several methods for converting string variables to numeric.

First, if the variable has merely been misclassified as string, you can apply the brute force approach of the real() function, e.g., generate patid = real(patientid). This will create missing values for any observations that cannot be interpreted as numeric.

Second, a more subtle approach is given by the destring command, which can transform variables in place (with the replace option) and can be used with a varlist to apply the same transformation to an entire set of variables with one command. This is useful if there are several variables that require conversion. However, destring should be used only for variables that have genuine numeric content but happen to have been misclassified as string variables.

Third, if the variable truly has string content and you need a numeric equivalent, you can use the encode command. You should not apply encode to a string variable that has purely numeric content (for instance, one that has been misclassified as a string variable) because encode will attempt to create a value label for each distinct value of the variable. As an example, we create a numeric equivalent of the state variable:

```
. encode state, generate(stateid)
. describe state stateid

                  storage   display      value
variable name      type     format       label      variable label

state             str13     %-13s                    State
stateid           long      %13.0g       stateid     State
. list state stateid in 1/5
```

```
         state              stateid

 1.   Connecticut       Connecticut
 2.   Illinois             Illinois
 3.   Indiana               Indiana
 4.   Iowa                     Iowa
 5.   Kansas                 Kansas
```

Although stateid is numeric, it has automatically been given the value label of the values of state. To see the numeric values, use list with the nolabel option.

Numeric-to-string conversion

You may also need to generate the string equivalent of a numeric variable. Often it is easier to parse the contents of string variables and extract substrings that may have some particular significance. Such transformations can be applied to integer numeric variables by means of integer division and remainders, but these transformations are generally more cumbersome and error-prone. The limits to exact representation of

numeric values, such as integers, with many digits are circumvented by placing those
values in string form. A thorough discussion of these issues is given in Cox (2002c).

We discussed three methods for string-to-numeric conversion. For each method, the
inverse function or command is available for numeric-to-string conversion: the `string()`
function, `tostring`, and `decode`. The `string()` function is useful in allowing a numeric
display format ([D] **format**) to be used. This would allow, for instance, the creation
of a variable with leading zeros, which are integral in some ID-number schemes. The
`tostring` command provides a more comprehensive approach: it contains various safe-
guards to prevent the loss of information and can be used with a particular display
format. Like `destring`, `tostring` can be applied to a varlist to alter an entire set of
variables.

A common task is the restoration of leading zeros in a variable that has been transferred from a spreadsheet. For instance, U.S. zip (postal) codes and Social Security
numbers can start with zero. The `tostring` command is useful here. Say, for example,
that we have a variable, `zip`:

```
tostring zip, format(%05.0f) generate(zipstring)
```

The variable `zipstring` will contain strings of five-digit numbers with leading zeros
included, as specified by the `format()` option.

To illustrate `decode`, let's say that you have the `stateid` numeric variable defined
above (with its value label) in your dataset but you do not have the variable in string
form. You can create the string variable `statename` by using `decode`:

```
. decode stateid, generate(statename)
. list stateid statename in 1/5
```

	stateid	statename
1.	Connecticut	Connecticut
2.	Illinois	Illinois
3.	Indiana	Indiana
4.	Iowa	Iowa
5.	Kansas	Kansas

To use `decode`, the numeric variable to be decoded must have a value label.

Working with quoted strings

You may be aware that `display "this is a quoted string"` will display the contents
of that quoted string. What happens, though, if your string itself contains quotation
marks? Then you must resort to *compound double quotes*. A command such as

```
. display `"This is a "quoted" string."'
```

will properly display the string, with the inner quotation marks intact. If ordinary double quotes are used, Stata will produce an error message. Compound double quotes can often be used advantageously when there is any possibility that the contents of string variables might include quotation marks.

3.3   Functions for the generate command

The fundamental commands for data transformation are `generate` and `replace`. They function in the same way, but two rules govern their use. `generate` can be used only to create a *new* variable, one whose name is not currently in use. On the other hand, `replace` can be used only to revise the contents of an *existing* variable. Unlike other Stata commands whose names can be abbreviated, `replace` must be spelled out for safety's sake.

We illustrate the use of `generate` by creating a new variable in our dataset that measures the fraction of each state's population living in urban areas in 1980. We need only specify the appropriate formula, and Stata will automatically apply that formula to every observation that is specified by the `generate` command, using the rules of algebra. For instance, if the formula would result in a division by zero for a given state, the result for that state would be flagged as missing. We generate the fraction, `urbanized`, and use the `summarize` command to display its descriptive statistics:

```
. generate urbanized = popurb / pop
. summarize urbanized
```

Variable	Obs	Mean	Std. Dev.	Min	Max
urbanized	21	.6667691	.1500842	.3377319	.8903645

We see that the average state in this part of the United States is 66.7% urbanized, with that fraction ranging from 34% to 89%.

If the `urbanized` variable already existed, but we wanted to express it as a percentage rather than a decimal fraction, we must use `replace`:

```
. replace urbanized = 100 * urbanized
(21 real changes made)
. summarize urbanized
```

Variable	Obs	Mean	Std. Dev.	Min	Max
urbanized	21	66.67691	15.00843	33.77319	89.03645

`replace` reports the number of changes it made; here it changed all 21 observations.

The concern for efficiency of a do-file is first a concern for *human* efficiency. You should write the data transformations as a simple, succinct set of commands that can readily be audited and modified. You may find that there are several ways to create the same variable by using `generate` and `replace`. It is usually best to stick with the simplest and clearest form of these statements.

A variety of useful functions are located in Stata's programming functions category (`help programming functions` or [D] **functions**). For instance, several `replace` statements might themselves be replaced with one call to the `inlist()` or `inrange()` functions (see Cox [2006c]). The former will allow the specification of a variable and a list of values. It returns 1 for each observation if the variable matches one of the elements of the list, and 0 otherwise. The function can be applied to either numeric or string variables. For string variables, up to 10 string values can be specified in the list. For example,

```
. generate byte newengland = inlist(state, "Connecticut", "Maine",
> "Massachusetts", "New Hampshire", "Rhode Island", "Vermont")

. sort medage

. list state medage pop if newengland, sep(0)
```

	state	medage	pop
6.	Vermont	29.40	511.5
13.	New Hampshire	30.10	920.6
14.	Maine	30.40	1124.7
16.	Massachusetts	31.20	5737.0
17.	Rhode Island	31.80	947.2
19.	Connecticut	32.00	3107.6

The `inrange()` function allows the specification of a variable and an interval on the real line and returns 1 or 0 to indicate whether the variable's values fall within the interval (which can be open, i.e., one limit can be $\pm\infty$). For example,

```
. list state medage pop if inrange(pop, 5000, 9999), sep(0)
```

	state	medage	pop
2.	Michigan	28.80	9262.1
4.	Indiana	29.20	5490.2
16.	Massachusetts	31.20	5737.0
21.	New Jersey	32.20	7364.8

Several data transformations involve the use of *integer division*, that is, truncating the remainder. For instance, four-digit U.S. Standard Industrial Classification (SIC) codes 3211–3299 divided by 100 must each yield 32. This is accomplished with the `int()` function (defined in `help math functions`).[2] A common task involves extracting one or more digits from an integer code; for instance, the third and fourth digits of the codes above can be defined as

```
generate digit34 = SIC - int(SIC / 100) * 100
```

or

```
generate mod34 = mod(SIC,100)
```

---

2. Also see the discussion of the `floor()` and `ceil()` functions in section 3.3.3.

where the second construct makes use of the modulo (`mod()`) function (see Cox [2007d]). The third digit alone could be extracted with

```
generate digit3 = int((SIC - int(SIC / 100) * 100) / 10)
```

or

```
generate mod3 = (mod(SIC, 100) - mod(SIC, 10)) / 10
```

or even

```
generate sub3 = real(substr(string(SIC), 3, 1))
```

using the `string()` function to express `SIC` as a string, the `substr()` function to extract the desired piece of that string, and the `real()` function to convert the extracted string into numeric form.

As discussed in section 2.4, you should realize the limitations of this method in dealing with very long integers, such as U.S. Social Security numbers of nine digits or ID codes of 10 or 12 digits. The functions `maxbyte()`, `maxint()`, and `maxlong()` are useful here. An excellent discussion of these issues is given in Cox (2002c).

Lastly, we must mention one exceedingly useful function for `generate`: the `sum()` function, which produces cumulative or running sums. That capability is useful in the context of time-series data, where it can be used to convert a flow or other rate variable into a stock or other amount variable. If we have an initial capital stock value and a net investment series, the `sum()` of investment plus the initial capital stock defines the capital stock at each point in time. This function does not place the single sum of the series into the new variable. If that is what you want, use the `egen` function `total()`.

3.3.1   Using if exp with indicator variables

A key element of many empirical research projects is the *indicator variable*: a variable taking on the values $(0, 1)$ to indicate whether a particular condition is satisfied. These are also commonly known as *dummy variables* or *Boolean variables*. The creation of indicator variables is best accomplished by using a *Boolean condition*: an expression that evaluates to true or false for each observation. The `if` *exp* qualifier has an important role here as well. Using our dataset, it would be possible to generate indicator variables for small and large states with the following commands. As I note below, we must take care to handle potentially missing values by using the `missing()` function.

```
. generate smallpop = 0
. replace smallpop = 1 if pop <= 5000 & !missing(pop)
(13 real changes made)
. generate largepop = 0
. replace largepop = 1 if pop > 5000 & !missing(pop)
(8 real changes made)
```

```
. list state pop smallpop largepop, sep(0)
```

	state	pop	smallpop	largepop
1.	Connecticut	3107.6	1	0
2.	Illinois	11426.5	0	1
3.	Indiana	5490.2	0	1
4.	Iowa	2913.8	1	0
5.	Kansas	2363.7	1	0
6.	Maine	1124.7	1	0
7.	Massachusetts	5737.0	0	1
8.	Michigan	9262.1	0	1
9.	Minnesota	4076.0	1	0
10.	Missouri	4916.7	1	0
11.	N. Dakota	652.7	1	0
12.	Nebraska	1569.8	1	0
13.	New Hampshire	920.6	1	0
14.	New Jersey	7364.8	0	1
15.	New York	17558.1	0	1
16.	Ohio	10797.6	0	1
17.	Pennsylvania	11863.9	0	1
18.	Rhode Island	947.2	1	0
19.	S. Dakota	690.8	1	0
20.	Vermont	511.5	1	0
21.	Wisconsin	4705.8	1	0

This combination of `generate` and `replace` is required to define both the 0 and the 1 values. If you merely used `generate smallpop = 1 if pop <= 5000`, the variable `smallpop` would be set to missing, not zero, for all observations that did not satisfy the `if` condition. A simpler approach would merely use the Boolean conditions:

```
generate smallpop = (pop <= 5000)
generate largepop = (pop > 5000)
```

Although this latter approach is more succinct, it suffers from a potentially serious flaw: if any values of `pop` are missing, they will be coded as 1 in the variable `largepop` and 0 in the variable `smallpop` because all of Stata's missing value codes are represented as very large positive numbers. The solution requires an `if` *exp*, `if !missing(pop)`, added to the `generate` statements.[3]

```
generate smallpop = (pop <= 5000) if !missing(pop)
generate largepop = (pop > 5000) if !missing(pop)
```

If the data contain missing values, they should be taken into account by using `if` *exp* qualifiers. Even if the current data do not include missing values, it is good programming practice to allow for them.

The `if` *exp* in the example above *must* be used as shown. Placing the `!missing(pop)` inside the Boolean expression (e.g., `(pop > 5000 & !missing(pop))`) would mistak-

---

3. We could also apply this condition as `if pop < .`, taking advantage of the fact that all measurable values are less than the system missing value.

enly assign a value of 0 to `largepop` for missing values of `pop`. The `if` *exp* qualifier, on the other hand, will cause any missing values of `pop` to be properly reflected in `largepop`.[4]

3.3.2   The cond() function

You may often want to code a result variable as a if a condition is true and b if that condition is false. The `cond(x,a,b)` function provides this capability without requiring an `if` *exp* qualifier. x is the condition to be tested, a is the result when true, and b is the result when false. If you are familiar with the C or Perl programming languages, you will recognize the similarity to those languages' ternary conditional operator $x\,?\,a\!:b$.

Suppose that we want to separate states having a ratio of marriages to divorces (the net marriage rate) above and below 2.0. We define `netmarr2x` as taking on the values 1 and 2, and we attach value labels for ease of use. The variable can then be used in tabstat ([R] **tabstat**):

```
. generate netmarr2x = cond(marr/divr > 2.0, 1, 2)
. label define netmarr2xc 1 "marr > 2 divr" 2 "marr <= 2 divr"
. label values netmarr2x netmarr2xc
. tabstat pop medage, by(netmarr2x)
Summary statistics: mean
  by categories of: netmarr2x
```

netmarr2x	pop	medage
marr > 2 divr	5792.196	30.38333
marr <= 2 divr	4277.178	30.08889
Total	5142.903	30.25714

We see that states with a high net marriage rate are larger and have slightly older populations.

It is possible to nest the `cond()` function; that is, `cond()`'s second and third arguments can be additional `cond()` functions. However, this syntax can lead to rather unwieldy constructs with many parentheses. For a more positive view, see Kantor and Cox (2005).

---

4. An even simpler approach takes advantage of the mutually exclusive and exhaustive nature of these two indicator variables: `generate largepop = 1 - smallpop`. If missing values are properly handled in defining `smallpop`, the algebraic `generate` statement will propagate them in `largepop`, because any function of missing data produces missing data.

3.3.3 Recoding discrete and continuous variables

Stata's `recode` ([D] **recode**) command creates a new variable based on the coding of an existing categorical variable. If data transformations involve many similar statements, they are not being written efficiently. It is perfectly feasible to write many transformation statements, such as

```
replace newcode = 5 if oldcode == 2
replace newcode = 8 if oldcode == 3
replace newcode = 12 if inlist(oldcode, 5, 6, 7)
...
```

It is not a good idea to perform transformations this way, because the use of copy and paste steps to construct these similar statements is likely to lead to typing errors when constructing a sizable number of statements.

Stata's `recode` command usually produces more efficient and readable code.[5] For instance,

```
recode oldcode (2 = 5) (3 = 8) (5/7 = 12), gen(newcode)
```

will perform the above transformation. The equal sign should be taken to indicate assignment: *oldvalue(s)* → *newvalue*. Unlike the line-by-line approach above using `replace`, `recode` can be applied to an entire varlist. This is often handy where a questionnaire-based dataset contains several similar questions with the same coding. Use the `prefix()` option to define the variable name stub. Missing-data codes can be handled, all nonspecified values can be mapped to a single outcome, and value labels for the values of the newly created variables can be defined. In fact, `recode` can work in place, modifying the existing variables rather than creating new ones. As the documentation for `recode` suggests, that is probably not a good idea if any further modifications to the mapping arise.

Several recode functions are also available. The $recode(x, x_1, x_2, x_3, x_4, \ldots, x_n)$ function will code a continuous variable into intervals: $x \leq x_1, x_1 < x \leq x_2$, and so on. The resulting values are set equal to the threshold values. For example,

```
. generate medagebrack = recode(medage, 29, 30, 31, 32, 33)
. tabulate medagebrack
```

medagebrack	Freq.	Percent	Cum.
29	3	14.29	14.29
30	8	38.10	52.38
31	4	19.05	71.43
32	4	19.05	90.48
33	2	9.52	100.00
Total	21	100.00	

The `floor()` and `ceil()` functions can also be used to generate an integer variable (see Cox [2003c]). The `floor()` function always rounds down (to the floor), whereas

---

5. Sometimes, there is an algebraic expression that offers a one-line solution using `generate`.

the `ceil()` function always rounds up (to the ceiling). These functions can be used to implement top-coding or bottom-coding of values.

Formally, for value x, the `floor($x$)` function returns the integer n such that $n \leq x < n + 1$, while the `ceil($x$)` function returns the integer n such that $n - 1 < x \leq n$. For positive integers, `floor()` discards the decimal part. The `int()` function, on the other hand, returns the integer obtained by truncating x toward zero, so that `int($-3.14159$)` $= -3$. The `round($x$)` function returns the closest integer to x.[6]

Alternatively, to categorize x into a group defined by threshold values, the `irecode($x,x_1,x_2,x_3,x_4,\ldots,x_n$)` function will return the group number (where $x \leq x_1 \rightarrow 0$, $x_1 < x \leq x_2 \rightarrow 1$, and so on). To illustrate, let's create four population-size categories and present their descriptive statistics. Given that state populations in our sample ranged from about 0.5 million to 17.5 million in 1980, we split the distribution at 1, 4, and 8 million:

```
. generate size=irecode(pop, 1000, 4000, 8000, 20000)
. label define popsize 0 "<1m" 1 "1-4m" 2 "4-8m" 3 ">8m"
. label values size popsize
. tabstat pop, stat(mean min max) by(size)
Summary for variables: pop
      by categories of: size
```

size	mean	min	max
<1m	744.541	511.456	947.154
1-4m	2215.91	1124.66	3107.576
4-8m	5381.751	4075.97	7364.823
>8m	12181.64	9262.078	17558.07
Total	5142.903	511.456	17558.07

Rather than categorizing a continuous variable by using threshold values, we may want to group observations based on quantiles: quartiles, quintiles, deciles, or any other percentiles of their empirical distribution. We can readily create groupings of that sort with `xtile` (see [D] **pctile**):

```
. xtile medagequart = medage, nq(4)
. tabstat medage, stat(n mean min max) by(medagequart)
Summary for variables: medage
      by categories of: medagequart (4 quantiles of medage)
```

medagequart	N	mean	min	max
1	7	29.02857	28.3	29.4
2	4	29.875	29.7	30
3	5	30.54	30.1	31.2
4	5	32	31.8	32.2
Total	21	30.25714	28.3	32.2

---

6. The `round()` function also has a two-argument form; see **help math functions** for details.

3.4 Functions for the egen command

Although the functions available for use with `generate` or `replace` are limited to those listed in [D] **functions** (or, given subsequent additions, those listed in `help functions`), Stata's `egen` command provides an open-ended list of capabilities. Just as Stata's command set can be extended by placing additional `.ado` and `.sthlp` files on the ado-path, the functions that can be invoked with `egen` are those defined by ado-files whose names start with `_g` stored on the ado-path. A number of those functions are part of official Stata, as documented by [D] **egen** (or, given subsequent additions, by `help egen`). But your copy of Stata may include additional `egen` functions: either those you have written yourself, or those downloaded from the Statistical Software Components archive or from another Stata user's `net from...` site. In this section, I first discuss several official Stata functions, and then I present several useful additions developed by the Stata user community.

Although the syntax of an `egen` statement is similar to that of `generate`, several differences should be noted. Because only a subset of `egen` functions allow a `by varlist:` prefix or `by(varlist)` option, the documentation should be consulted to determine whether a particular function is "byable", in Stata parlance. Similarly, the explicit use of `_n` and `_N`, often useful in `generate` and `replace` commands (see section 3.5.1), is not compatible with `egen`. Because there is no way to specify that a variable created with a nonbyable `egen` function should use the logic of `replace`, it may be necessary to use a temporary variable as the `egen` result and use a subsequent `replace` to combine those values over by-groups.

Official egen functions

If you seek spreadsheet-like functionality in Stata's data transformations, you should become acquainted with the rowwise `egen` functions. Like the equivalent spreadsheet functions, the rowwise functions support the calculation of sums, averages, standard deviations, extrema, and counts across several Stata variables. Wildcards can be used; e.g., if you have the state-level U.S. Census variables `pop1890`, `pop1900`, ..., `pop2000`, you can use `egen nrCensus = rowmean(pop*)` to compute the average population of each state over those decennial censuses. As discussed in the treatment of missing values in section 3.2.4, the rowwise functions operate in the presence of missing values. The mean will be computed for all 50 states, although several were not part of the United States in 1890.

The number of nonmissing elements in the rowwise varlist can be computed by using `rownonmiss()` with `rowmiss()` as the complementary value. Other official rowwise functions include `rowmax()`, `rowmin()`, `rowtotal()`, and `rowsd()` (row standard deviation). The `rowfirst()` and `rowlast()` functions give the first and last nonmissing values in the varlist. You may find these useful if the variables refer to sequential items, for instance, wages earned per year over several years, with missing values when unemployed; `rowfirst()` would return the earliest wage observation, and `rowlast()`, the most recent.

Official `egen` also provides several statistical functions that compute a statistic for specified observations of a variable and place that constant value in each observation of the new variable. Because these functions generally allow the use of `by` *varlist:*, they can be used to compute statistics for each by-group of the data, as is discussed in section 3.5. This facilitates computing statistics for each household for individual-level data or for each industry for firm-level data. The `count()`, `mean()`, `min()`, `max()`, and `total()`[7,8] functions are especially useful in this context. To illustrate, let's `egen` the average population in each of the `size` groups defined above and express each state's population as a percentage of the average population in that group. Size category 0 includes the smallest states in our sample.

```
. bysort size: egen avgpop = mean(pop)
. generate popratio = 100 * pop / avgpop
. format popratio %7.2f
. list state pop avgpop popratio if size == 0, sep(0)
```

	state	pop	avgpop	popratio
1.	Rhode Island	947.2	744.541	127.21
2.	N. Dakota	652.7	744.541	87.67
3.	S. Dakota	690.8	744.541	92.78
4.	Vermont	511.5	744.541	68.69
5.	New Hampshire	920.6	744.541	123.65

Other functions in the same statistical category include `iqr()` (interquartile range), `kurt()` (kurtosis), `mad()` (median absolute deviation), `mdev()` (mean absolute deviation), `median()`, `mode()`, `pc()` (percentage or proportion of total), `pctile()`, `p(n)` (nth percentile), `rank()`, `sd()` (standard deviation), `skew()` (skewness), and `std()` (z score).

---

7. Before Stata version 9, the `egen sum()` function performed this task, but it was often confused with `generate`'s `sum()` function; hence, it was renamed.
8. An important note regarding `egen`'s `total()` function: it treats missing values as zeros so that the `total()` of a variable with all missing values is computed as zero rather than as missing. You can change this behavior with this function's `missing` option.

egen functions from the user community

The most comprehensive collection of additional `egen` functions is contained in Nicholas J. Cox's `egenmore` package, available by using the `ssc` command.[9] The `egenmore` package is a collection of many varied routines by Cox and others (including the author). Some of these routines extend the functionality of official `egen` routines, while others provide capabilities lacking in official Stata. Although some of the routines require Stata version 8.2 or newer, many will work with older versions of Stata.[10]

Several `egenmore` functions work with standard Stata dates, expressed in numeric form as discussed in section 2.4.2. The `egen bom()` and `eom()` functions create date variables corresponding to the first day or last day of a given calendar month. They can be used to generate the offset for any number of months (e.g., the last day of the third month hence). With the `work` option, the first (last) nonweekend day of the month can be specified (although there is no support for holidays). The companion functions `bomd()` and `eomd()` provide similar functionality using date variables; that is, they can be used to find the first (last) day of a month in which their date-variable argument falls. This is useful if you wish to aggregate observations by the calendar month in which they occurred.

Several `egenmore` functions extend `egen`'s statistical capabilities. The `corr()` function computes correlations (optionally, covariances) between two variables, `gmean()` and `hmean()` compute geometric and harmonic means, `rndint()` computes random integers from a specified uniform distribution, `semean()` computes the standard error of the mean, and `var()` computes the variance. The `filter()` function can apply any linear filter to data that have been declared time-series data by `tsset`, including panel data, for which the filter is applied separately to each panel.[11] The companion function `ewma()` can be used to apply an exponentially weighted moving average to time-series data.[12]

A function that has proved useful in data management is the `record()` function, meant to evoke "setting a record". We may want to compute the highest wage earned to date by each employee or the lowest stock price encountered to date. That value can be considered the record value and can be computed with this function. If our data contain annual wage rates for several employees over several years,

```
egen hiwage = record(wage), by(empid) order(year)
```

will compute for each employee (as specified with `by(empid)`) the highest wage earned to date, allowing ready evaluation of conditions under which wages have fallen because

---

9. The package is labeled `egenmore` because it further extends `egenodd`, which appeared in the *Stata Technical Bulletin* (Cox 1999, Cox 2000). Most of the `egenodd` functions now appear in official Stata, so they will not be further discussed here. I am grateful to Nicholas J. Cox for his thorough documentation of `egenmore` functionality in `help egenmore`.

10. Some of the `egenmore` routines provide limited date-and-time functionality for pre–Stata 10 users. As discussed in section 2.4.2, Stata now offers extensive support for timestamp variables, offering precision up to the millisecond.

11. Similar functionality is available from `tssmooth ma` ([TS] **tssmooth ma**).

12. Also see `tssmooth exponential` ([TS] **tssmooth exponential**).

of a job change, etc. Several other `egen` functions are available in `egenmore`, and a variety of other useful user-written functions are available in the Statistical Software Components archive.

In summary, several common data-management tasks can be expeditiously handled by `egen` functions. The open-ended nature of this command implies that new functions often become available, either through ado-file updates to official Stata or through contributions from the user community. The latter will generally be advertised on Statalist, with past messages accessible in the Statalist archives, and recent contributions are highlighted in `ssc new`.

3.5   Computation for by-groups

One of Stata's most useful features is the ability to transform variables or compute statistics over by-groups. By-groups are defined with the `by` *varlist*: prefix and are often useful in data transformations using `generate`, `replace`, and `egen`. Using by *varlist*: with one or more categorical variables, a command will be repeated automatically for each value of the `by` *varlist*:. However, it also has its limitations: by *varlist*: can execute only one command.[13]

3.5.1   Observation numbering: _n and _N

The observations in the dataset are numbered 1, 2, ..., 21 in the list on page 43. When you refer to an observation, you can do so with its observation number. The highest observation number, corresponding to the total number of observations, is known as _N, while the current observation number is known as _n. In some circumstances, the meanings of these two symbols will be altered. The observation numbers will change if a `sort` ([D] **sort**) command alters the order of the dataset in memory.

Under the control of a by-group, the meanings of _n and _N are altered. Within a by-group, _n is the current observation of the group, and _N is the last observation of the group. In this example, we `gsort` ([D] **gsort**) the state-level data by region and descending order of population. We then use `generate`'s running `sum()` function by `region:` to display the total population in each region that lives in the largest, two largest, three largest, ... states:

---

13. Also see section 7.2.1.

```
. gsort region -pop
. by region: generate totpop = sum(pop)
. list region state pop totpop, sepby(region)
```

	region	state	pop	totpop
1.	NE	New York	17558.1	17558.07
2.	NE	Pennsylvania	11863.9	29421.97
3.	NE	New Jersey	7364.8	36786.79
4.	NE	Massachusetts	5737.0	42523.83
5.	NE	Connecticut	3107.6	45631.4
6.	NE	Maine	1124.7	46756.06
7.	NE	Rhode Island	947.2	47703.22
8.	NE	New Hampshire	920.6	48623.83
9.	NE	Vermont	511.5	49135.28
10.	N Cntrl	Illinois	11426.5	11426.52
11.	N Cntrl	Ohio	10797.6	22224.15
12.	N Cntrl	Michigan	9262.1	31486.23
13.	N Cntrl	Indiana	5490.2	36976.45
14.	N Cntrl	Missouri	4916.7	41893.14
15.	N Cntrl	Wisconsin	4705.8	46598.9
16.	N Cntrl	Minnesota	4076.0	50674.87
17.	N Cntrl	Iowa	2913.8	53588.68
18.	N Cntrl	Kansas	2363.7	55952.36
19.	N Cntrl	Nebraska	1569.8	57522.18
20.	N Cntrl	S. Dakota	690.8	58212.95
21.	N Cntrl	N. Dakota	652.7	58865.67

We can use _n and _N in this context. They will be equal for the last observation of each by-group:

```
. by region: list region totpop if _n == _N
```

```
-> region = NE
```

	region	totpop
9.	NE	49135.28

```
-> region = N Cntrl
```

	region	totpop
12.	N Cntrl	58865.67

The computation of total population by region, stored as a new variable, also could have been performed with egen's total() function. We could calculate mean population (or any other statistic) over a by *varlist*: with more than one variable:

```
. generate popsize = smallpop + 2*largepop
. label variable popsize "Population size code"
. label define popsize 1 "<= 5 million" 2 "> 5 million", modify
. label values popsize popsize
. bysort region popsize: egen meanpop2 = mean(pop)
. list region popsize state pop meanpop2, sepby(region popsize)
```

	region	popsize	state	pop	meanpop2
1.	NE	<= 5 million	New Hampshire	920.6	1322.291
2.	NE	<= 5 million	Vermont	511.5	1322.291
3.	NE	<= 5 million	Maine	1124.7	1322.291
4.	NE	<= 5 million	Rhode Island	947.2	1322.291
5.	NE	<= 5 million	Connecticut	3107.6	1322.291
6.	NE	> 5 million	New York	17558.1	10630.96
7.	NE	> 5 million	Massachusetts	5737.0	10630.96
8.	NE	> 5 million	New Jersey	7364.8	10630.96
9.	NE	> 5 million	Pennsylvania	11863.9	10630.96
10.	N Cntrl	<= 5 million	S. Dakota	690.8	2736.153
11.	N Cntrl	<= 5 million	Missouri	4916.7	2736.153
12.	N Cntrl	<= 5 million	Minnesota	4076.0	2736.153
13.	N Cntrl	<= 5 million	Nebraska	1569.8	2736.153
14.	N Cntrl	<= 5 million	N. Dakota	652.7	2736.153
15.	N Cntrl	<= 5 million	Kansas	2363.7	2736.153
16.	N Cntrl	<= 5 million	Wisconsin	4705.8	2736.153
17.	N Cntrl	<= 5 million	Iowa	2913.8	2736.153
18.	N Cntrl	> 5 million	Indiana	5490.2	9244.112
19.	N Cntrl	> 5 million	Illinois	11426.5	9244.112
20.	N Cntrl	> 5 million	Ohio	10797.6	9244.112
21.	N Cntrl	> 5 million	Michigan	9262.1	9244.112

We can now compare each state's population with the average population of states of its size class (large or small) in its region.

Although egen's statistical functions can be handy, creating variables with constant values or constant values over by-groups in a large dataset will consume a great deal of Stata's available memory. If the constant values are needed only for a subsequent transformation, such as computing each state's population deviation from average size, and will not be used in later analyses, you should drop those variables at the earliest opportunity. Alternatively, consider other Stata commands that can provide this functionality. Ben Jann's center command (findit center) can transform a variable into deviation from mean form, and it works with by-groups.

Another important consideration is the interpreted nature of `egen` functions, which implies that they can be considerably slower than built-in functions or special-purpose commands. `egen` functions can be used to generate constant values, for example, total household income or average industry output, for each element of a by-group. If that is your objective—if you want to construct a dataset with one value per household or per industry—`egen` is an inefficient tool. Stata's `collapse` command is especially tailored to perform that function and will generate a single summary statistic for each by-group.

3.6   Local macros

One of the most important concepts for Stata do-file authors is the *local macro*. This entity does not have an exact equivalent in some statistical packages and matrix languages. If you are familiar with lower-level programming languages such as Fortran or C, you may find Stata's terminology for various objects rather confusing. In those languages, you refer to a variable with statements such as `x = 2`. Although you might have to declare `x` before its use—for instance, as `integer` or `float`—the notion of a variable in those languages refers to an entity that is assigned one value, either numeric or string. In contrast, the Stata variable refers to one column of the dataset, which contains _N values, one per observation.

So what corresponds to a Fortran or C variable in Stata's command language? Either a Stata *macro* or a *scalar*, to be discussed below. But that correspondence is not one to one, because a Stata macro can contain multiple elements. Stata's local macro is a container that can hold either one object—such as a number or a variable name—or a set of objects. A local macro can contain any combination of alphanumeric characters and can hold more than 8,600 characters in all versions of Stata (165,000 in Stata/IC; 1,000,000 in Stata/SE, Stata/MP). The Stata macro is really an alias that has both a name and a value. When its name is dereferenced, it returns its value. That operation can be carried out at any time. Alternatively, the macro's value can be easily modified with an additional command.

A macro can be either *local* or *global*, referring to its scope, which defines where its name will be recognized. A local macro is created in a do-file or an ado-file and ceases to exist when that do-file terminates, either normally or abnormally. A global macro exists for the duration of the Stata program or interactive session. There are some good reasons to use global macros, but like any global definition, they can have unintended consequences. I explain more about global macros in section 3.7.

The Stata command to define a local macro is `local` (see [P] **macro**). For example,

```
. local anxlevel None Mild Moderate Severe
. display "The defined levels of anxiety are: 'anxlevel'"
The defined levels of anxiety are: None Mild Moderate Severe
```

The first command names the local macro—as `anxlevel`—and then defines its value to be the list of four anxiety codes. To work with the value of the macro, we must dereference it. The expression `anxlevel` refers to the value of the macro. The macro's name is preceded by the left single-quote character (`) and followed by the right single-quote character (').[14] Stata uses different opening and closing quote characters to signify the beginning and end of a macroname because, as we shall see, macro references can be nested, one inside the other. If macros are nested, they are evaluated from the inside out; that is, `pid`year'' will first replace `year' with its value and then evaluate `pid*year*'.[15] To dereference the macro, the correct punctuation is vital. In the example's `display` statements, we must wrap the dereferenced macro in double quotes because `display` expects a double-quoted string argument or the value of a scalar expression, such as `display log(14)`.

Usually, the `local` statement is written without an equal sign (=). It is acceptable syntax to use an equal sign following the macro's name, but it is a bad idea to get in the habit of using it unless it is required. The equal sign causes the remainder of the expression to be evaluated rather than merely aliased to the macro's name. This is a common cause of head scratching, where a user will complain that "my do-file worked when I had eight regressors, but not when I had nine...." Defining a macro with an equal sign will cause evaluation of the remainder of the command as a numeric expression or as a string expression. A string expression cannot contain more than 244 characters, so a result longer than that will not be evaluated correctly.

When is it appropriate to use an equal sign in a `local` statement? It is appropriate whenever you *must* evaluate the macro's value. In this next example, we see a macro used as a counter that fails to do what we had in mind:[16]

```
. local count 0

. local anxlevel None Mild Moderate Severe

. foreach a of local anxlevel {
  2.     local count `count' + 1
  3.     display "Anxiety level `count': `a'"
  4. }
Anxiety level 0 + 1: None
Anxiety level 0 + 1 + 1: Mild
Anxiety level 0 + 1 + 1 + 1: Moderate
Anxiety level 0 + 1 + 1 + 1 + 1: Severe
```

Here we must use the equal sign to request evaluation rather than concatenation:

---

14. These characters are found in different places on keyboards for different languages. The right single quote is commonly known as the apostrophe.

15. For a thorough discussion of these issues, see Cox (2002a).

16. The `foreach` command is presented in section 7.3.

```
. local count 0

. local anxlevel None Mild Moderate Severe

. foreach a of local anxlevel {
  2.      local count = `count' + 1
  3.      display "Anxiety level `count': `a'"
  4. }
Anxiety level 1: None
Anxiety level 2: Mild
Anxiety level 3: Moderate
Anxiety level 4: Severe
```

The corrected example's `local` statement contains the name of the macro twice: once without punctuation, which defines its name, and again on the right-hand side of the equal sign, with its current value dereferenced by `count`. It is crucial to understand why the statement is written this way: we are redefining the macro in the first instance and referencing its current value in the second.

In contrast to this example, there are instances where we want to construct a macro within a loop, repeatedly redefining its value, and we *must* avoid the equal sign:

```
. local count 0

. local anxlevel None Mild Moderate Severe

. foreach a of local anxlevel {
  2.      local count = `count' + 1
  3.      local newlist `newlist' `count' `a'
  4. }
. display "`newlist'"
 1 None 2 Mild 3 Moderate 4 Severe
```

The `local newlist` statement introduces a new twist: it defines the local macro `newlist` as a string containing its own current content, space, *value_of_count*, space, *value_of_a*. The `foreach` ([P] **foreach**) statement defines the local macro `a` with the value of each anxiety level in turn. The first time through the loop, `newlist` does not exist, so how can we refer to its current value? Easy: Every Stata macro has an empty, or *null*, value unless it has explicitly been given a nonnull value. Thus it takes on the string " 1 None" the first time, and then the second time through, it concatenates that string with the new string " 2 Mild", and so on. Using the equal sign would be inappropriate in the `local newlist` statement because it would cause truncation of `newlist` at 244 characters. This would not cause trouble in this example, but it would be a serious problem if we had a longer list.

We can also use macro evaluation to generate macro values "on the fly". For instance, we could construct a loop (see section 7.3) that generates new variable names as local macros. Say that we have the variables v11, v12, . . . , v15, which should be renamed as x1971, x1972, . . . , x1975. We could do so with macro evaluation:

```
forvalues i = 11/15 {
        rename v`i'  = x`=1960 + `i''
}
```

The syntax '*=exp*' tells Stata to evaluate *exp* immediately. In this code fragment, Stata evaluates the expression 1960 + 'i' before evaluating the outer macro. The first time through the loop, i = 11 and the new variable name will be x1971.

From these examples, we might conclude that Stata's macros are useful in constructing lists, or as counters and loop indices. They are that, but they play a much larger role in Stata do-files and ado-files and in the return values of many Stata commands. Macros are one of the key elements of Stata's programming language that allow you to avoid repetitive commands and the retyping of computed results. Macros allow you to change the performance of your do-file by merely altering the contents of a local macro. In this manner, your do-file can be made general, and that set of Stata commands can be reused or adapted for use in similar tasks with minimal effort.

3.7   Global macros

Global macros are distinguished from local macros by their manner of creation (with the `global` statement; see [P] **macro**) and their means of reference. We obtain the value of the global macro `george` by typing `$george`, with the dollar sign taking the place of the punctuation surrounding the local macro's name when it is dereferenced. Global macros are often used to store items parametric to a program, such as a character string containing today's date that is to be embedded in all filenames created by the program, or the name of a default directory in which your datasets and do-files are to be accessed.

Unless there is an explicit need for a global macro—a symbol with *global scope*—it is usually preferable to use a local macro. It is easy to forget that a global symbol was defined in do-file *A*. By the time you run do-file *G* or *H* in that session of Stata, you may find that it does not behave as expected, because it picks up the value of the global symbol. Such problems are difficult to debug. Authors of Fortran or C programs have always been encouraged to keep definitions local unless they must be visible outside the module. That is good advice for Stata programmers as well.

3.8   Extended macro functions and macro list functions

Stata contains a versatile library of functions that can be applied to macros: the extended macro functions (`help extended_fcn`, or see [P] **macro**). These functions allow you to easily retrieve and manipulate the contents of macros. For instance,

```
. local anxlevel None Mild Moderate Severe

. local wds: word count `anxlevel'

. display "There are `wds' anxiety levels:"
There are 4 anxiety levels:

. forvalues i = 1/`wds' {
  2.          local wd: word `i' of `anxlevel'
  3.          display "Level `i' is `wd'"
  4. }
Level 1 is None
Level 2 is Mild
Level 3 is Moderate
Level 4 is Severe
```

In this example, we use the **word count** and **word** # **of** extended functions, both of which operate on strings. We do *not* enclose the macro's value (`anxlevel'`) in double quotes, because then it would be considered a single word.[17] This do-file will work for any definition of the list in **local anxlevel** without the need to define a separate **count** macro.

A wide variety of extended macro functions perform useful tasks, such as extracting the variable label or value label from a variable or determining its data type or display format; extracting the row or column names from a Stata matrix; or generating a list of the files in a particular directory that match a particular pattern (e.g., *.dta). The **subinstr()** function allows a particular pattern to be substituted in a macro, either the first time the pattern is encountered or in all instances.

Another useful set of functions support the manipulation of lists held in local macros. These functions are described in **help macro lists** and [P] **macro lists**. They can be used to identify the unique elements of a list or the duplicate entries; to sort a list; and to combine lists with Boolean operators such as AND, OR, etc. A set of handy list functions allows one list's contents to be subtracted from another, identifying the elements of list *A* that are not also found in list *B*. You can test lists for equality, defined for lists as containing the identical elements in the same order, or for weak equality, which does not consider ordering. Functions are available to produce the union or intersection of two lists and to **sort** the elements of a list in alphabetical order. The **uniq** list function returns the unique (distinct) elements of a list, while **dups** returns the duplicate elements. The **posof** function can be used to determine whether a particular entry exists in a list, and if so, in which position in the list. To consider the entire list, function *A* **in** *B* returns 1 if all elements in list *A* are found in *B*, and 0 otherwise. An excellent discussion of many of these issues can be found in Cox (2003a).

3.8.1 System parameters, settings, and constants: creturn

When using Stata macros, you may find the system parameters, settings, and constants available in **creturn** ([P] **creturn**) useful. For instance, you may want to embed today's

---

17. In this context, a word is a space-delimited token in the string. In the string `Stata was first released in 1985`, there are six words including "1985".

date in Stata output such as the title of a table or graph. The value c(current_date) will return that string, and c(current_time) will return the time. You can use c() return values to capture the current working directory (c(pwd)), the version of Stata that you are running (c(stata_version)), the name of the last file specified (c(filename)), and the date and time it was last saved (c(filedate)).

Several creturn values are constants: c(pi) is an accurate value of π (which can also be referenced as _pi), while c(alpha) (c(ALPHA)) returns a list of the lowercase (uppercase) letters of the alphabet. The list of months can be retrieved in full (or abbreviated 3-letter) form with c(Months) (c(Mons)), while the list of weekdays in full (or abbreviated) form is accessible in c(Weekdays) (c(Wdays)). These month and day lists are often useful when matching input data that may have been provided in that format; see Cox (2004e).

3.9   Scalars

In addition to Stata's variables and local and global macros, there are two additional entities related to every analysis command: scalars and matrices. Scalars, like macros, can hold either numeric or string values, but a scalar can hold only one value.[18] Most analysis commands return one or more results as numeric scalars. For instance, describe returns the scalars r(N) and r(k), corresponding to the number of observations and variables in the dataset. A scalar is also much more useful for storing one numeric result, such as the mean of a variable, rather than storing that value in a Stata variable containing _N copies of the same number. A scalar can be referred to in any subsequent Stata command by its name:

```
. use fem2, clear
. scalar root2 = sqrt(2.0)
. generate double rootage = age * root2
```

The distinction between a macro and a scalar appears when it is referenced. The macro must be dereferenced to refer to its value, while the scalar is merely named.

There is one important distinction between macros and Stata's scalars: the length of a string scalar is limited to the length of a string variable (244 bytes; see help limits), whereas a macro's length is, for most purposes, unlimited.[19]

Stata's scalars are typically used in a numeric context. When a numeric quantity is stored in a macro, it must be converted from its internal (binary) representation into a printable form. By storing the result of a computation—for instance, a variable's mean or standard deviation—in a scalar, no conversion of its value need take place. However, a scalar can appear only in an expression where a Stata variable or a numeric expression could be used. For instance, one cannot specify a scalar as part of an in

---

18. The length of a string scalar is limited to the length of a string variable (244 characters).
19. A macro is limited to 165,200 characters in Stata/IC and to over one million characters in Stata/SE and Stata/MP.

range qualifier because its value will not be extracted. It can be used in an `if` *exp* qualifier, because that contains a numeric expression. Most of Stata's statistical and estimation commands return various numeric results as scalars (see section 5.3).

Stata is capable of working with scalars of the same name as Stata variables. As the manual suggests, Stata will not become confused, but you well may. So, you should avoid using the same names for both entities; see Kolev (2006). I discuss Stata's *matrices* in section 3.10. Scalars and matrices share the same namespace, so you cannot have both a scalar named `gamma` and a matrix named `gamma`.

Stata's scalars play a useful role in do-files. By defining scalars at the beginning of the do-file and referring to them throughout the code, you make the do-file parametric. Doing this helps you avoid the difficulties of changing various constants in the do-file's statements everywhere they appear. You often may need to repeat a complex data transformation task for a different category. You may want to work with 18–24-year-old subjects rather than with 25–39-year-old subjects. Your do-files contain the qualifiers for minimum and maximum age throughout the program. If you define those age limits as scalars at the program's outset, the do-file becomes much simpler to modify and maintain.

We illustrate using the `fem2` dataset. These data are a modified version of the `fem` dataset on 118 female psychiatric patients presented in Rabe-Hesketh and Everitt (2006),[20] originally available in Hand et al. (1994). The `anxiety` measure is on a scale from 1=none to 4=severe, while `iq` is the intelligence quotient score. We use scalars to define segments of the IQ range for which we want to analyze `anxiety`.

```
. use fem2, clear
. scalar lb1 = 80
. scalar ub1 = 88
. scalar lb2 = 89
. scalar ub2 = 97
. scalar lb3 = 98
. scalar ub3 = 109
. forvalues i = 1/3 {
  2.        display _n "IQ " lb'i' " - " ub'i'
  3.        tabulate anxiety if inrange(iq, lb'i', ub'i')
  4. }
IQ 80 - 88
```

ANXIETY	Freq.	Percent	Cum.
1	3	12.00	12.00
2	13	52.00	64.00
3	7	28.00	92.00
4	2	8.00	100.00
Total	25	100.00	

---

20. I am grateful to Sophia Rabe-Hesketh and Brian Everitt for permission to use these data.

```
IQ 89 - 97
    ANXIETY |      Freq.      Percent        Cum.
------------+-----------------------------------
          1 |          5         7.35        7.35
          2 |         38        55.88       63.24
          3 |         24        35.29       98.53
          4 |          1         1.47      100.00
------------+-----------------------------------
      Total |         68       100.00
IQ 98 - 109
    ANXIETY |      Freq.      Percent        Cum.
------------+-----------------------------------
          2 |          9        75.00       75.00
          3 |          2        16.67       91.67
          4 |          1         8.33      100.00
------------+-----------------------------------
      Total |         12       100.00
```

In this example, the macros for `lb` and `ub` are dereferenced "on the fly" in the `display` and `tabulate` commands. If we had a more elaborate do-file in which we wanted to carry out a number of analyses for these three groups of patients based on IQ score, we need only refer to the scalars we have defined that delineate the groups.

3.10   Matrices

Stata has long provided a full-featured matrix language, which supports a broad range of matrix operations on real matrices, as described in [P] **matrix**. Stata also provides a dedicated matrix language, Mata, which operates in a separate environment within Stata, as is discussed in chapter 13.

Stata's estimation commands typically create both scalars and Stata matrices, in particular, the matrix `e(b)`, containing the set of estimated coefficients, and the matrix `e(V)`, containing the estimated variance–covariance matrix of the coefficients. These matrices can be manipulated by Stata's `matrix` commands and their contents used in later commands. Like all Stata estimation commands, `regress` ([R] **regress**) produces matrices `e(b)` and `e(V)` as the row vector of estimated coefficients (a $1 \times k$ matrix) and the estimated variance–covariance matrix of the coefficients (a $k \times k$ symmetric matrix), respectively. You can examine those matrices with the `matrix list` command or copy them for use in your do-file with the `matrix` statement. The `matrix beta = e(b)` command will create a matrix, `beta`, in your program as a copy of the last estimation command's coefficient vector:

```
. generate age2 = age^2

. regress weight age age2
```

Source	SS	df	MS		Number of obs =	107
					F(2, 104) =	10.86
Model	135.170028	2	67.5850139		Prob > F =	0.0001
Residual	647.10605	104	6.22217356		R-squared =	0.1728
					Adj R-squared =	0.1569
Total	782.276078	106	7.379963		Root MSE =	2.4944

weight	Coef.	Std. Err.	t	P>|t|	[95% Conf. Interval]	
age	.1652777	1.0152	0.16	0.871	-1.847902	2.178458
age2	.0009098	.0136163	0.07	0.947	-.0260919	.0279115
_cons	-5.918914	18.6551	-0.32	0.752	-42.91268	31.07485

```
. matrix b = e(b)

. matrix V = e(V)

. matrix list b

b[1,3]
             age         age2        _cons
y1     .16527767   .00090982   -5.9189145

. matrix list V

symmetric V[3,3]
              age         age2        _cons
  age    1.0306315
 age2   -.01380654     .0001854
_cons    -18.91199   .25270496   348.01273
```

When you use Stata's traditional `matrix` commands, the matrix size is limited. In Stata/IC, you cannot have more than 800 rows or 800 columns in a matrix.[21] This implies that many matrix tasks cannot be handled with traditional `matrix` commands in a straightforward manner. For instance, `mkmat` ([P] **matrix mkmat**) can create a Stata matrix from a varlist, but the number of observations that can be used is limited to 800 in Stata/IC. Beyond resorting to Mata, there are two points that should be made. First, Stata contains specialized operators such as `matrix accum` ([P] **matrix accum**) that can compute cross-product matrices from any number of observations. A regression of 10,000 observations on five variables (including a constant) involves a 5×5 cross-products matrix, regardless of N. Second, variations on this command, such as `matrix glsaccum`, `matrix vecaccum`, and `matrix opaccum`, generate other useful summarizations. In this sense, the limitation on matrix dimension is not binding.

The brute force approach is rarely appropriate when working with complex matrix expressions. For example, the seemingly unrelated regression estimator implemented by the `sureg` ([R] **sureg**) command is presented in textbooks as a generalized least-squares estimator of the form

$$\widehat{\beta} = (\mathbf{X}'\Omega^{-1}\mathbf{X})(\mathbf{X}'\Omega^{-1}\mathbf{y})$$

---

21. The limit in Stata/SE and Stata/MP of 11,000 rows or columns is much larger, but sizable matrices use a great deal of computer memory. As is discussed in chapter 13, Mata provides a more efficient solution in terms of its *views*; see `help mata st_view()` for details.

where **X** is block-diagonal in the individual equations' X matrices. If there are G equations, each with T observations, the Ω matrix is of the order $T \times G$ by $T \times G$. Given the algebra of partitioned matrices, every statistical package that performs seemingly unrelated regression takes advantage of the fact that this expression can be rewritten as the product of several terms, one per equation in the system. In that expression, each term is no more than a single equation's regression. A huge matrix computation can be simplified as a loop over the individual equations. Although you might be tempted to copy the matrix expression straight out of the textbook or journal article into code, that will often be an infeasible approach in Stata's traditional matrix commands or in Mata, and indeed in any matrix language limited by the computer's available memory. Some cleverness is often necessary when implementing complicated matrix expressions to reduce the problem to a workable size.

For those Stata users who are writing do-files, Stata matrices are likely to be useful in two particular contexts: that of saved results as described above and as a way of organizing information for presentation. References to matrix elements appear in square brackets. Because Stata does not have a vector data type, all Stata matrices have two subscripts, and both subscripts must be given in any reference. A range of rows or a range of columns can be specified in an expression; see [P] **matrix** for details. Stata's traditional matrices are distinctive in that their elements can be addressed both conventionally by their row and column *numbers* (counting from 1, not 0) and by their row and column *names*.

Stata's matrices are often useful devices for housekeeping purposes, such as the accumulation of results that are to be presented in tabular form. The `tabstat` command can generate descriptive statistics for a set of by-groups. Likewise, `statsmat` (written by Nicholas J. Cox and the author and available from `ssc`) can be used to generate a matrix of descriptive statistics for a set of variables or for a single variable over by-groups. A command I wrote with Joao Pedro Azevedo, `outtable`, can then be used to generate a LaTeX table; an example of its use is given in section 9.2. Michael Blasnik's `mat2txt` can be used to generate tab-delimited output. Stata matrices' row and column labels can be manipulated with `matrix rownames`, `matrix colnames`, and several extended macro functions described in section 3.8. This allows you to control the row and column headings on tabular output. Stata's traditional `matrix` operators make it possible to assemble a matrix from several submatrices. For instance, you can have one matrix for each country in a multicountry dataset.

In summary, judicious use of Stata's traditional `matrix` commands eases the burden of many housekeeping tasks and makes it feasible to update material in tabular form without retyping.

4 Cookbook: Do-file programming I

This cookbook chapter presents for Stata do-file programmers several recipes using the programming features described in the previous chapter. Each recipe poses a problem and a worked solution. Although you may not encounter this precise problem, you should be able to recognize its similarities to a task that you would like to automate in a do-file.

4.1 Tabulating a logical condition across a set of variables

The problem. When considering many related variables, you want to determine whether, for each observation, *all* variables satisfy a logical condition. Alternatively, you might want to know whether *any* satisfy that condition (for instance, taking on inappropriate values), or you might want to *count* how many of the variables satisfy the logical condition.[1]

The solution. This would seem to be a natural application of `egen`, because that command already contains a number of rowwise functions to perform computations across variables. For instance, the `anycount()` function counts the number of variables in its varlist whose values for each observation match those of an integer numlist, whereas the `rowmiss()` and `rownonmiss()` functions tabulate the number of missing and nonmissing values for each observation, respectively. The three tasks above are all satisfied by `egen` functions from Nicholas Cox's `egenmore` package: `rall()`, `rany()`, and `rcount()`, respectively. Why not use those functions, then?

Two reasons come to mind: First, recall that `egen` functions are interpreted code. Unlike the built-in functions accessed by `generate`, the logic of an `egen` function must be interpreted each time it is called. For a large dataset, the time penalty can be significant. Second, to use an `egen` function, you must remember that there is such a function, and you must remember its name. In addition to Stata's official `egen` functions, documented in the online help files, there are many user-written `egen` functions available, but you must track them down.

For these reasons, current good programming practice suggests that you should avoid `egen` function calls in instances where the performance penalty might be an issue. This is particularly important within an ado-file program but can apply to many do-files as well. Often you can implement the logic of an `egen` function with a few lines of Stata commands.

---

1. This recipe relies heavily on Nicholas J. Cox's `egenmore` help file.

To use the `egenmore rcount()` function, you must define the logical condition to be tested by using a specific syntax. For instance, imagine that we have a dataset of household observations, where the variables `child1-child12` contain the current age of each child (or missing values for nonexistent offspring).[2] We could use `rcount()` to determine the number of school-age children:

```
egen nschool = rcount(child1-child12), cond(@ > 5 & @ < 19)
```

where, as explained in `help egenmore`, the at sign (@) is a placeholder for each variable in *varlist* in turn. Alternatively, we could compute `nschool` with a `foreach` loop:[3]

```
generate nschool = 0
foreach v of varlist child1-child12 {
    replace nschool = nschool + inrange(`v', 6, 18)
}
```

The built-in `inrange()` function will execute more efficiently than the interpreted logic within `rcount()`. As a bonus, if we also wanted to compute an indicator variable signaling whether there are any school-age children in the household, we could do so within the same `foreach` loop:

```
generate nschool = 0
generate anyschool = 0
foreach v of varlist child1-child12 {
    replace nschool = nschool + inrange(`v', 6, 18)
    replace anyschool = max( anyschool, inrange(`v', 6, 18))
}
```

Here `anyschool` will remain at 0 for each observation unless one of the children's ages match the criteria specified in the `inrange` function. Once `anyschool` is switched to 1, it will remain so. An alternative (and more computationally efficient) way of writing this code takes advantage of the fact that `generate` is much faster than `replace`, because the latter command must keep track of the number of changes made in the variable.[4] Thus we could type

```
foreach v of varlist child1-child12 {
    local nschool "`nschool' + inrange(`v', 6, 18)"
}
generate byte nschool = `nschool'
generate byte anyschool = nschool > 0
```

In this variation, the local macro is built to include an `inrange()` clause for each of the possible 12 children. For a constructed dataset (`ex4.1.dta`) of one million observations, the `egen` code runs in 3.81 seconds, while the comparable code using `replace` and the `inrange()` function runs in 2.23 seconds.[5] By contrast, the last block of code, avoiding `replace`, runs in 0.78 seconds.

---

2. I assume that these variables are contiguous in the dataset (if not, we could use `order` to make them so).

3. The `foreach` command is presented in section 7.3.

4. I am grateful to a StataCorp reviewer for this suggestion.

5. Invoking `replace` twice to also define `anyschool` requires 4.73 seconds!

In summary, you may want to consider whether the convenience of an `egen` function is offset by its computational burden. Coding the logic in your do-file can be a more efficient approach.[6]

4.2 Computing summary statistics over groups

The problem. Your dataset has a hierarchical nature, where observations represent individuals who are also identified by their household ID code or represent records of individual patient visits that can be aggregated over the patient ID or over the clinic ID. In the latter case, you can define groups of observations belonging to a particular patient or to a particular clinic.

With this kind of hierarchical data structure, you may want to compute summary statistics for the groups. This can be performed readily in Stata with `tabstat`, but that command will only display a table of summary measures. Alternatively, you could use `collapse` to generate a dataset of aggregated values for a variety of summary statistics, or you could use `contract` to generate a collapsed dataset of frequencies. However, you may find that these options do not fit the bill.

What if you want to juxtapose the summary statistics for each aggregate unit with the individual observations to compute one or more variables for each record? For instance, you might have repeated-measures data for a physician's patients measuring their height, weight, and blood pressure at the time of each office visit. You might want to flag observations where their weight is above their median weight, or where their blood pressure is above the 75th percentile of their repeated measurements.

The solution. Computations such as these can be done with a judicious use of by-groups (see section 3.5). For instance,

```
by patientid: egen medwt = median(weight)
by patientid: egen bp75 = pctile(bp), p(75)
```

We have stressed that you should avoid using variables to store constant values (which would occur if you omitted the `by patientid:` prefix). But here we are storing a separate constant for each `patientid`. You can now compute indicators for weight, blood pressure, and at-risk status by using the `byte` data type for these binary variables:

```
generate byte highwt = weight > medwt & !missing(weight, medwt)
generate byte highbp = bp > bp75 & !missing(bp, bp75)
generate byte atrisk = highwt & highbp
```

If you need to calculate a sum for each group (`patientid` here), you can use the `total()` function for `egen`. Alternatively, to improve computational efficiency, you could use

---

6. As I discuss in chapter 13, Mata functions may also prove useful in reducing the computational burden involved with tasks like these.

```
by patientid: generate atriskvisits = sum(atrisk)
by patientid: generate n_atrisk = atriskvisits if _n == _N
gsort -n_atrisk
list patientid n_atrisk if inrange(n_atrisk, 1, .)
```

This sequence of commands uses the `sum()` function from `generate`, which is a *running sum*. Its value when `_n == _N` is the total for that `patientid`. We store that value as `n_atrisk` and sort it in descending order with `gsort`.[7] The `list` command then prints one record per `patientid` for those patients with at least one instance of `atrisk` in their repeated measures.

4.3 Computing the extreme values of a sequence

The problem. Let's assume that you have hierarchical data, such as observations of individual patient visits to a clinic. In the previous recipe, we described how summary statistics for each patient could be calculated. These include extrema, for instance, the highest weight ever recorded for each patient or the lowest serum cholesterol reading. What you may need, however, is the *record* to date for those variables: the maximum (minimum) value observed so far in the sequence. This is a "record" value in the context of setting a record, for instance, maximum points scored per game or minimum time recorded for the 100-yard dash. How might you compute these values for hierarchical data?[8]

The solution. First, let's consider a single sequence (that is, data for a single patient in our example above). You might be tempted to think that this is a case where looping over observations will be essential—and you would be wrong! We exploit the fact that Stata's `generate` and `replace` commands respect Stata's sort order (see Newson [2004]). We need record only the first observation's value, and then we can use `replace` to generate the "record high":

```
sort visitdate
generate maxwt = weight in 1
replace maxwt = max(maxwt[_n - 1], weight) in 2/l
```

Usually, you need not worry about missing values, because the `max()` function is smart enough to ignore them unless it is asked to compare missing with missing. If we want to calculate a similar measure for each `patientid` in the dataset, we use the same mechanism:

```
sort patientid visitdate
by patientid: generate minchol = serumchol if _n == 1
by patientid: replace minchol = min(minchol[_n - 1], serumchol) if _n > 1
```

---

7. The `gsort` command is presented in section 3.5.1.
8. This recipe relies heavily on the response to the Stata frequently asked question "How do I calculate the maximum or minimum seen so far in a sequence?"
   (http://www.stata.com/support/faqs/data/sequence2.html), written by Nicholas J. Cox.

With repeated-measures data, we cannot refer to observations 1, 2, etc., because those are absolute references to the entire dataset. Under the control of a by-group, the _n and _N values are redefined to refer to the observations in that by-group, allowing us to refer to _n in the `generate` command and the prior observation in that by-group with a [_n - 1] subscript.

4.4   Computing the length of spells

The problem. Assume that you have ordered data (for instance, a time series of measurements) and you would like to examine *spells* in the data. These might be periods during which a qualitative condition is unchanged, as signaled by an indicator variable. As examples, consider the sequence of periods during which a patient's cholesterol remains above the recommended level or a worker remains unemployed or a released offender stays clear of the law. Alternatively, spells might signal repeated values of a measured variable, such as the number of years that a given team has been ranked first in its league. Our concern with spells can involve identifying their existence and measuring their duration. This discussion of these issues relies heavily on Cox (2007b). I am grateful to Nick Cox for his cogent exposition.

The solution. One solution to this problem involves using a ready-made Stata command, `tsspell`, written by Nicholas J. Cox. This command can handle any aspect of our investigation. It does require that the underlying data be defined as a Stata time series with `tsset`. This makes it less than ideal if your data are ordered but not evenly spaced, such as patient visits to their physician, which can be irregularly timed.[9] Another issue arises, though: that raised in section 4.1 with respect to `egen`. The `tsspell` program is fairly complicated interpreted code, which may impose a computational penalty when applied to a very large dataset. You may need only one simple feature of the program for your analysis. Thus you may want to consider analyzing the spells in do-file code, which is much simpler than the invocation of `tsspell`. As in section 4.1, you can generally avoid explicit looping over observations, and you will want to do so whenever possible.

Assume that you have a variable denoting the ordering of the data (which might be a Stata date or date-and-time variable, but need not be) and that the data have been sorted on that variable. The variable of interest is `employer`, which takes on the values A, B, C, ..., or missing for periods of unemployment. You want to identify the beginning of each spell with an indicator variable. How do we know that a spell has begun? The condition

```
generate byte beginspell = employer != employer[_n-1]
```

will suffice to define the start of each new spell (using the `byte` data type to define this indicator variable). Of course, the data can be *left-censored* in the sense that we do not start observing the employee's job history on his or her date of hire. But the fact that `employer[_n-1]` is missing for period 1 does not matter, because it will be captured as

---

9. As Cox points out (Cox 2007b, 250), the ordered data may not have a time dimension at all, but may refer to spatial orientation.

the start of the first spell. What about spells of unemployment? If they are coded as a
missing value of `employer`, they will be considered spells as well.

First, consider some fictitious data on an employee. She is first observed working
for firm A in 1987, and then she is laid off in 1990. After a spell of unemployment, she
is hired by firm B in 1992, and so on.

```
. use ex_employee, clear
. generate byte beginspell = employer != employer[_n-1]
. list, sepby(employer) noobs
```

year	employer	wage	begins~l
1987	A	8.25	1
1988	A	8.50	0
1989	A	8.75	0
1990		.	1
1991		.	0
1992	B	7.82	1
1993	B	7.98	0
1994	B	8.12	0
1995	B	8.40	0
1996	B	8.52	0
1997	C	9.00	1
1998	A	9.25	1
1999		.	1
2000		.	0
2001	D	10.18	1
2002	D	10.37	0
2003	E	11.00	1
2004	E	11.40	0
2005	E	11.80	0
2006		.	1

`beginspell` properly flags each change in employment status, including entry into unemployment. If we wanted to flag only spells of unemployment, we could do so with

```
. generate byte beginunemp = missing(employer) &
> ((employer != employer[_n-1]) | (_n == 1))
```

which would properly identify the years in which unemployment spells commenced as
1990, 1999, and 2006.[10]

---

10. Recognize that these data are also *right-censored* in that we observe the beginning of an unemployment spell in 2006, but we do not know its duration.

With an indicator variable flagging the start of a spell, we can compute how many changes in employment status this employee has faced, because the count of that indicator variable provides that information. We can also use this notion to tag each spell as separate:

```
. generate spellnr = sum(beginspell)
. list, sepby(employer) noobs
```

year	employer	wage	begins~l	beginu~p	spellnr
1987	A	8.25	1	0	1
1988	A	8.50	0	0	1
1989	A	8.75	0	0	1
1990	.		1	1	2
1991	.		0	0	2
1992	B	7.82	1	0	3
1993	B	7.98	0	0	3
1994	B	8.12	0	0	3
1995	B	8.40	0	0	3
1996	B	8.52	0	0	3
1997	C	9.00	1	0	4
1998	A	9.25	1	0	5
1999	.		1	1	6
2000	.		0	0	6
2001	D	10.18	1	0	7
2002	D	10.37	0	0	7
2003	E	11.00	1	0	8
2004	E	11.40	0	0	8
2005	E	11.80	0	0	8
2006	.		1	1	9

We have observed nine spells for this employee, and the first and last of these are censored.

What if we now want to calculate the average wage paid by each employer?

```
sort spellnr
by spellnr: egen meanwage = mean(wage)
```

Or what if we want to calculate the duration of employment with each employer (the length of each employment spell)?

```
by spellnr: gen length = _N if !missing(employer)
```

Here we are taking advantage of the fact that the time variable is an evenly spaced time series. If we had unequally spaced data, we would want to use Stata's date functions to compute the duration of each spell.

This example may seem not all that useful because it refers to a single employee's employment history. However, all the techniques we have illustrated work equally well when applied in the context of panel or longitudinal data, as long as they can be placed on a time-series calendar. If we add an `id` variable to these data and `xtset id year`, we can reproduce all the results above merely by using the `by id:` prefix. In the last three examples, we must sort by both `id` and `spellnr`. For example,

```
sort id spellnr
by id spellnr: egen meanwage = mean(wage)
```

is now required to compute the mean wage for each spell of each employee in a panel context.

Several additional aspects of spells may be of interest. Returning to the single employee's data, we may want to flag only employment spells at least three years long. Using the `length` variable, we can generate such an indicator, as in

```
. sort spellnr
. by spellnr: gen length = _N if !missing(employer)
(5 missing values generated)
. generate byte longspell = (length >= 3 & !missing(length))
```

. list year employer length longspell, sepby(employer) noobs

year	employer	length	longsp~l
1987	A	3	1
1988	A	3	1
1989	A	3	1
1990		.	0
1991		.	0
1992	B	5	1
1993	B	5	1
1994	B	5	1
1995	B	5	1
1996	B	5	1
1997	C	1	0
1998	A	1	0
1999		.	0
2000		.	0
2001	D	2	0
2002	D	2	0
2003	E	3	1
2004	E	3	1
2005	E	3	1
2006		.	0

For more details on the handling of spells, see Cox (2007b).

4.5   Summarizing group characteristics over observations

The problem. Let's say your dataset has a hierarchical nature, such as observations representing individuals who are also identified by their household ID code or group identifier. You would like to compute summary measures for each household (group) and attach them to the individual records.[11] Here is some fictitious data:

---

11. This recipe relies heavily on the response to Stata frequently asked question "How do I create variables summarizing for each individual properties of the other members of a group?" (http://www.stata.com/support/faqs/data/members.html), written by Nicholas J. Cox.

```
. use household_id, clear
. list fam person female age, sepby(fam) noobs
```

fam	person	female	age
1	3	1	14
1	2	1	16
1	1	1	36
2	5	0	10
2	4	1	12
2	3	0	14
2	2	1	42
2	1	0	45
3	6	1	3
3	5	1	7
3	4	1	9
3	3	0	11
3	2	1	36
3	1	0	39

For simplicity, we have created data with no missing values. Some of the statements below would have to be "bulletproofed" if there was the possibility of missing values for some individuals.

The solution. For some calculations of interest, we have already addressed this issue in section 4.2. We can readily calculate, for instance, the total household income for each family by using the `egen total()` function with a `by:` prefix. We could do the same with a logical condition: for instance, how many adults (at least 18 years old) are there in each family?

```
. sort fam
. by fam: egen nadult = total(age >= 18)
```

How many siblings does each child have?

```
. by fam: egen nchild = total(age <= 17) if age <= 17
. generate sibs = nchild - 1
```

The `if` clause `age <= 17` defines `nchild` only for children's records. We can extend this logic to count the number of male children:

```
. by fam: egen nboys = total(age <= 17 & !female) if age <= 17
```

Or we can count, for each child, how many brothers he or she has who are also children:

```
. by fam: egen nbros = total(age <= 17 & !female) if age <= 17
. replace nbros = nbros - (age <= 17 & !female)
```

where the subtraction in the `replace` statement states that you cannot be your own brother. Which children have older siblings who are also children?

```
. sort fam (age)
. by fam: generate byte older = (age <= 17 & age[_n+1] <= 17) * (age < age[_n+1])
```

How old is each child's oldest sibling (missing if the child is the oldest child)?

```
. by fam: egen oldestchild = max(age) if age<=17
. generate oldestsibling = cond(age < oldestchild & age <= 17, oldestchild, .)

. list fam female age nadult sibs nboys nbros older oldestsibling, sepby(fam)
> noobs
```

fam	female	age	nadult	sibs	nboys	nbros	older	oldest~g
1	1	14	1	1	0	0	1	16
1	1	16	1	1	0	0	0	.
1	1	36	1	.	.	.	0	.
2	0	10	2	2	2	1	1	14
2	1	12	2	2	2	2	1	14
2	0	14	2	2	2	1	0	.
2	1	42	2	.	.	.	0	.
2	0	45	2	.	.	.	0	.
3	1	3	2	3	1	1	1	11
3	1	7	2	3	1	1	1	11
3	1	9	2	3	1	1	1	11
3	0	11	2	3	1	0	0	.
3	1	36	2	.	.	.	0	.
3	0	39	2	.	.	.	0	.

Many tricks of this sort can be used to evaluate group characteristics for each member of a group, and most of them allow us to avoid explicit looping over observations as one might do in many programming languages. With judicious use of logical conditions and the `by:` prefix or a `by()` option, you can write efficient and succinct code to perform these tasks.

4.6   Using global macros to set up your environment

The problem. Suppose you are working on a large research project where the same datasets and Stata do-files can be accessed on several different computers, some Linux-based and some Windows-based. The do-files have references to datasets to be read and written, which must be adjusted when the material is accessed from a different computer system.

The solution. Global macros (see section 3.7) can be used to deal with these issues. Set up a master do-file for the project that defines the base directory to be accessed on a Windows system:

```
global BASEDIR "c:/documents/project226"
global SHAREDIR "c:/documents/datastore"
global USER "gould"
cd "$BASEDIR"
do job1
```

On a Linux or Mac OS X machine, your do-file will instead be

```
global BASEDIR "/Users/baum/projects/226"
global SHAREDIR "/Users/baum/research/data"
global USER "cfb"
cd "$BASEDIR"
do job1
```

In both file specifications, I use the forward slash (/) rather than the backslash usually used in a Windows/DOS environment. Stata properly handles the slash for Windows, and by using the forward slash, you avoid the difficulty that arises with the backslash's special meaning to Stata's interpreter.

With this logic, the master do-file, when executed, sets the system-specific directory to the value of the global macro BASEDIR and uses cd to make that the current directory. The master do-file then calls do-file job1.do. Within that do-file, you may need to reference files that are outside the project directory, for instance, files shared with other research projects. By defining the global macro SHAREDIR, your do-files can refer to files in that shared directory with syntax like

```
use "$SHAREDIR/datafile1.dta"
```

The quotation marks in the cd and use statements will deal with the case where spaces can occur within the file specification. It is best to avoid spaces within directory names, but they are an unfortunate fact of life on many systems.

You may also want to embed additional information in a log file produced by do-files of the research project, for instance, the identity of the user who ran the job. You can do that by placing statements such as

```
display "Analysis run by $USER from $BASEDIR at `c(current-date)'
`c(current-time)'"
```

in each do-file to be executed. See section 4.8 for more on c(current_date) and c(current_time).

In summary, global macros can be useful in establishing your Stata environment, particularly if you routinely work on more than one computer or you share files related to a research project with co-workers.

4.7   List manipulation with extended macro functions

The problem. You have a file containing four countries' gross domestic product values, and you would like to regress each in turn on the other three countries' values. To do this, you would like to be able to manipulate lists of items. The macro list functions, described in section 3.8, are available for this task.

The solution. We use the gdp4cty file, which is in Stata's *long format*[12] in this example. The levelsof ([P] **levelsof**) command assembles a list of the cty values defined for gdp. That command displays the list and stores it in the local macro gdplist. We then apply reshape wide to place the data in the *wide format*, with a separate gross domestic product variable for each country. A foreach loop builds up the macro allcty with those four variable names.

```
. use gdp4cty, clear

. keep cty gdp date

. levelsof cty, local(gdplist)
`"DE"' `"FR"' `"UK"' `"US"'

. reshape wide gdp, i(date) j(cty) string
(note: j = DE FR UK US)

Data                          long   ->   wide
```

	long	->	wide
Number of obs.	400	->	100
Number of variables	3	->	5
j variable (4 values)	cty	->	(dropped)
xij variables:			
	gdp	->	gdpDE gdpFR ... gdpUS

```
. describe

Contains data
  obs:            100
  vars:             5
  size:         2,800 (99.9% of memory free)
```

variable name	storage type	display format	value label	variable label
date	float	%tq		
gdpDE	float	%9.0g		DE gdp
gdpFR	float	%9.0g		FR gdp
gdpUK	float	%9.0g		UK gdp
gdpUS	float	%9.0g		US gdp

```
Sorted by:   date

. foreach c of local gdplist {
  2.          local allcty "`allcty' gdp`c'"
  3. }
```

We can then use one of the macro list functions (help macro lists). The minus list function (−) gives us the first list with all the elements of the second list removed.[13] In a foreach loop over countries, we produce the macro allbut as the other three variable names, and we use this macro in the regress command.

---

12. For a discussion of long format, wide format, and the reshape command, see section 5.5.
13. You may also find the OR (|), AND (&), and list sort functions useful.

```
. foreach c of local allcty {
  2.              local allbut: list allcty - c
  3.              qui regress 'c' 'allbut' date
  4.              display _newline "Dependent variable 'c':"
  5.              mat list e(b)
  6. }
Dependent variable gdpDE:

e(b)[1,5]
            gdpFR        gdpUK        gdpUS         date        _cons
y1     .0605496    .10816974    .09276896   -.82440282   445.32199

Dependent variable gdpFR:

e(b)[1,5]
            gdpDE        gdpUK        gdpUS         date        _cons
y1    .05673715   -.16807133   -.08535245    -1.474414    678.2173

Dependent variable gdpUK:

e(b)[1,5]
            gdpDE        gdpFR        gdpUS         date        _cons
y1    .08791943   -.14578622   -.17140193   -.37449064   621.80752

Dependent variable gdpUS:

e(b)[1,5]
            gdpDE        gdpFR        gdpUK         date        _cons
y1    .06297068   -.06182947   -.14314374    -.0741045   622.66212
```

4.8   Using creturn values to document your work

The problem. Suppose you would like to document aspects of the environment in the
log file produced by your do-file.

The solution. The items you need to capture are available in the set of `creturn` values.
For instance, you might include these lines in your do-file:

```
local date 'c(current_date)'
local time 'c(current_time)'
local vers 'c(stata_version)'
local mem  'c(memory)'
local flav = cond('c(MP)', "MP", cond('c(SE)', "SE", "IC"))
local cwd 'c(pwd)'
display _newline "Run 'date' at 'time' on Stata/'flav' version 'vers',
>     memory = 'mem' bytes"
display _newline "Current working directory: 'cwd'"
```

Running this do-file produces the following:

```
. do mydofile

. local date 'c(current_date)'

. local time 'c(current_time)'

. local vers 'c(stata_version)'

. local mem  'c(memory)'

. local flav = cond('c(MP)', "MP", cond('c(SE)', "SE", "IC"))

. local cwd 'c(pwd)'
```

```
. display _newline "Run 'date' at 'time' on Stata/'flav' version 'vers',
> memory  = 'mem' bytes"
Run 5 Sep 2008 at 21:49:58 on Stata/MP version 10.1, memory = 307200000 bytes
. display _newline "Current working directory: 'cwd'"
Current working directory: /Users/baum/doc/ITSP/dof.8824
```

If you had already loaded a dataset into memory, you could display additional characteristics of the data. For example,

```
. sysuse auto, clear
(1978 Automobile Data)
. display _newline "Datafile: 'c(filename)' (N='c(N)', k='c(k)')  as of 'c(file
> date)'"
Datafile: /Applications/Stata/ado/base/a/auto.dta (N=74, k=12)  as of 13 Apr 20
> 07 17:45
```

By embedding this information in the log file, you can capture details of how and when the do-file was run, which can be particularly useful in a shared environment, such as where several research assistants are performing various tasks to update the data.

5 Do-file programming: Validation, results, and data management

5.1 Introduction

This chapter discusses three topics: data validation, reusing computed results, and commands for data management. Data validation tasks involve automating the process of ensuring that your data are sensible. I discuss the reuse of computed results to stress that you never need to copy values from Stata output and reenter them. That practice is error-prone and potentially irreproducible, whereas a do-file that automatically makes use of earlier computations is consistently reliable. Likewise, results of estimation can be automatically formatted for presentation in several formats, reducing the need for retyping and for judgments about precision.

The last four sections of this chapter focus on reorganizing datasets. The `reshape` command allows an existing dataset to be altered in form, whereas the `append` and `merge` commands combine the dataset in memory with one or more external Stata-format files. Several lesser-known but useful data-management commands are also presented.

5.1.1 What you should learn from this chapter

- How to effectively validate data through scripted data checking
- How to access and reuse computed results in `return` and `ereturn` lists
- How to save and tabulate estimated results
- The mechanics of reorganizing datasets with `reshape`
- Procedures for combining data with `append` and `merge`

5.2 Data validation: The assert, count, and duplicates commands

The first step in effective data management should always be *sanity checking*. Do all values of the raw data make sense? Are there any coding errors that are apparent in the range of data values? Are there values of numeric variables that should properly be coded as some sort of missing data, as discussed in section 3.2.4? A recommended practice for data management containing an audit trail involves the creation of a do-file that reads the raw data, applies several checks to ensure that data values are appropriate, and writes the initial Stata binary data file.[1] This data file should not be modified in later programs or interactive analysis. Each program that uses the file and creates additional variables, subsets, or merges of the data should save the resulting modified file under a new name. Each step in the data validation and transformation process can then be documented and re-executed if the need arises. Even if the raw data are provided in Stata binary format from an official source, you should assume that coding errors may be present.

This methodology, involving the creation of an audit trail of all operations applied to the data, should begin at the start of data management. Statalist postings often contain phrases such as "I did the original data transformations (or merges) in Excel, and now I need to. . . ." Even if you are more familiar with a spreadsheet syntax than with the Stata commands needed to replicate that syntax, using Stata is greatly preferable because its operations on the data can be precisely documented and replicated. Consider two research assistants starting with the same set of 12 spreadsheets, instructed to construct one spreadsheet by performing some complicated append or merge processes with copy and paste. What is the probability that the two research assistants will produce identical results? Many experienced researchers would agree that it is likely to be less than one.

The proposed solution would be to export the 12 spreadsheets to text format, and then read them into Stata by using a do-file that loops over the `.txt` or `.csv` files, applies the same transformations to each one, and performs the appropriate `append` or `merge` operation. That do-file, once properly constructed, will produce a reproducible result. The do-file can be easily modified to perform a similar task, such as handling 12 spreadsheets containing cost elements rather than revenues. Comments should be added to the do-file documenting its purpose, dates of creation/modification, and creator/modifier. You can either place an asterisk (`*`) or double forward slashes (`//`) at the beginning of each comment line, or you can use the block comment syntax (`/*` to begin a comment, `*/` to end it) to add several lines of comments to a do-file. Although it will take some time to learn how to use Stata's programming features to set up such a systematic process of data management, it will prove invaluable in any context when questions arise about the data or when a modified version of the original data appears.

The data checks recommended above start with the elementary use of `describe` and `summarize`, which provide some useful information about the data that have been imported (typically by using `insheet`, `infile`, or `infix`). Let's consider a version of the `census2a` dataset that has been altered to illustrate data validation:

---

1. An intriguing extension of this notion, `ckvar`, is described in Rising (2007).

```
. use census2b, clear
(Version of census2a for data validation purposes)

. describe

Contains data from census2b.dta
  obs:             50                          Version of census2a for data
>                                                validation purposes
  vars:             5                          23 Sep 2004 15:49
  size:          2,050 (99.9% of memory free)

              storage  display   value
variable name   type   format    label    variable label

state          str14    %14s
region         str7     %9s
pop            float    %9.0g
medage         float    %9.0g
drate          float    %9.0g

Sorted by:
```

The log displays the data types of the five variables. The first two are string variables (of maximum length 14 and 7 characters, respectively) while the other three are float variables. Here all data types appear to be appropriate.

Let's now consider descriptive statistics for the numeric variables in these data:

```
. summarize pop-drate
    Variable |       Obs        Mean    Std. Dev.       Min        Max

         pop |        49     4392737      4832522         -9   2.37e+07
      medage |        50       35.32     41.25901       24.2        321
       drate |        50       104.3     145.2496         40       1107
```

Several anomalies are revealed for the numeric variables. Population data appear to be missing for one state, which is unlikely for properly organized census data. Furthermore, population takes on a negative value for at least one state, indicating some coding errors. We use the knowledge that the values of U.S. states' populations in recent decades should be greater than several hundred thousand but no more than about 30 million. In your own research projects, you will have similar subject-matter knowledge that helps you define sensible ranges of values.

For instance, a maximum age value of 321 would suggest that Ponce de Leon is alive and well. Because the drate (death rate) variable has a mean of 104 per 100,000, a value of 10 times that amount suggests a coding error.

You may also find the codebook ([D] **codebook**) command useful in displaying information about each variable and highlighting any unusual values:

```
. codebook
```

---

state   (unlabeled)

```
              type:  string (str14), but longest is str13
      unique values:  49                          missing "":  0/50
           examples:  "Georgia"
                      "Maryland"
                      "Nevada"
                      "S. Carolina"

            warning:  variable has embedded blanks
```

---

region   (unlabeled)

```
              type:  string (str7)
      unique values:  4                           missing "":  2/50
          tabulation:  Freq.  Value
                          2   ""
                         12   "N Cntrl"
                          9   "NE"
                         16   "South"
                         11   "West"

            warning:  variable has embedded blanks
```

---

pop   (unlabeled)

```
              type:  numeric (float)
             range:  [-9,23667902]               units:  1
      unique values:  49                       missing .:  1/50

              mean:  4.4e+06
          std. dev:  4.8e+06

       percentiles:        10%       25%       50%       75%       90%
                        511456    947154   3.0e+06   5.3e+06   1.1e+07
```

---

medage   (unlabeled)

```
              type:  numeric (float)
             range:  [24.2,321]                  units:  .1
      unique values:  37                       missing .:  0/50

              mean:  35.32
          std. dev:  41.259

       percentiles:        10%       25%       50%       75%       90%
                          27.5      28.7     29.75      30.2     31.85
```

```
drate                                                                 (unlabeled)

                   type:   numeric (float)
                  range:   [40,1107]                      units:   1
          unique values:   30                        missing .:   0/50

                   mean:        104.3
               std. dev:       145.25

            percentiles:          10%       25%      50%      75%      90%
                                  68.5       79     85.5       93       98
```

Rather than just using the Data Editor to visually scan for the problems sprinkled throughout this small illustrative dataset, we are interested in data-validation techniques that can be applied to datasets with thousands of observations. We use assert to apply sanity checks for these three variables, and in the event of failure, we list the offending observations. If all checks are passed, this do-file should run without error:

```
use census2b, clear
                         // check pop
list if  !inrange(pop, 300000, 3e7)
assert inrange(pop, 300000, 3e7)
                         // check medage
list if   !inrange(medage, 20, 50)
assert  inrange(medage, 20, 50)
                         // check drate
list if   !inrange(drate, 10, 104+145)
assert  inrange(drate, 10, 104+145)
```

The first list command considers that population should be at least 300,000 and that it should be less than 30 million (3.0×10^7), and it lists the observations (if any) that fall outside that range by using the unary *not* operator (!) and the inrange() function. By reversing the logical conditions in the list command, we construct the assertion that all cases have valid and nonmissing (< .) values for pop. Likewise, we assert that each state's median age should be between 20 and 50 years. Finally, we assert that the death rate should be at least 10 per 100,000 and less than approximately $\widehat{\mu}+\widehat{\sigma}$ from that variable's descriptive statistics. Let's run the data-validation do-file:

```
. use census2b, clear
(Version of census2a for data validation purposes)
.
                         // check pop
. list if   !inrange(pop, 300000, 3e7)
```

	state	region	pop	medage	drate
4.	Arkansas	South	-9	30.6	99
10.	Georgia	South	.	28.7	81
15.	Iowa	N Cntrl	0	30	90

```
. assert inrange(pop, 300000, 3e7)
3 contradictions in 50 observations
assertion is false
r(9);

end of do-file
r(9);
```

As we would expect, the do-file fails to run to completion because the first **assert** locates three erroneous values of **pop**. We should now correct these entries and rerun the do-file until it executes without error. In these data, the numeric variables are continuous. In a dataset with categorical variables, you could use the **inlist()** function to validate that one of a set of particular integer values was present in each categorical variable.

This little example could be expanded to a lengthy do-file (or a set of nested do-files) that checks each of several hundred variables. The same logic could be used: the do-file should exit without error if all assertions are satisfied. In other words, no news is good news.

You can use **tabulate** ([R] **tabulate oneway**) to check the values of string variables in your dataset. In the **census2b** dataset, we will want to use **region** as an identifier variable in later analysis, expecting that each state is classified in one of four U.S. regions.

```
. use census2b, clear
(Version of census2a for data validation purposes)
. list state if region == ""
```

```
. tabulate region

     region |      Freq.     Percent        Cum.
------------+-----------------------------------
     N Cntrl |         12       25.00       25.00
          NE |          9       18.75       43.75
       South |         16       33.33       77.08
        West |         11       22.92      100.00
------------+-----------------------------------
       Total |         48      100.00
. assert !missing(region)
2 contradictions in 50 observations
assertion is false
r(9);

end of do-file
r(9);
```

The tabulation reveals that only 48 states have `region` defined. The assertion that we should have no missing values of `region` fails, and a list of values where the variable equals string missing (the null string) identifies Alaska and Hawaii as the misclassified entries.

Validating data with `tabulate` ([R] **tabulate twoway**) can be used to good advantage by generating cross-tabulations. Consider, for instance, a dataset of medical questionnaire respondents in which we construct a two-way table of `gender` and the number of completed pregnancies, `NCPregnancy`. Not only should the latter variable have a lower bound of zero and a sensible upper bound, but also its cross-tabulation with `gender=="Male"` should yield zero values unless some females have been misclassified as males. We could check this with

```
. assert NCPregnancy = 0 if gender == "Male"
```

You can use `duplicates` ([D] **duplicates**) to check variables that should take on distinct values. This command can handle much more complex cases in which a combination of variables must be unique (or a so-called *primary key*, in database terminology),[2] but we will apply it only to the variable `state`:

```
. use census2b, clear
(Version of census2a for data validation purposes)
. duplicates list state
Duplicates in terms of state
```

obs:	state
16	Kansas
17	Kansas

```
. duplicates report state
Duplicates in terms of state
```

copies	observations	surplus
1	48	0
2	2	1

```
. assert r(unique_value) == r(N)
assertion is false
r(9);

end of do-file

r(9);
```

---

2. As an example, U.S. senators' surnames may not be unique, but the combination of surname and state code almost surely will be unique.

The return item, r(unique_value), is set equal to the number of unique observations found. If that value falls short of the number of observations, r(N), duplicates exist. The identification of duplicates in this supposedly unique identifier implies that the dataset must be corrected before its further use. The duplicates command could also be applied to numeric variables to detect the same condition.

In summary, several sound data-management principles can improve the quality of analysis conducted with Stata. You should bring the data into Stata for manipulation as early in the analysis process as possible. You should construct a well-documented do-file to validate the data, ensuring that variables that should be complete are complete, that unique identifiers are unique, and that only sensible values are present in every variable. That do-file will run to completion without error if and only if all sanity checks are passed. Last, the validated and, if necessary, corrected file should not be modified in later analysis. Subsequent data transformations or merges should create new files rather than overwrite the original contents of the validated file. Strict adherence to these principles, although time consuming in the moment, will ultimately save a good deal of your time and provide those responsible for the research with the assurance that the data have been generated in a reproducible and well-documented fashion.

5.3   Reusing computed results: The return and ereturn commands

Each of Stata's commands stores its results. Sometimes this is done noisily, as when a nonzero return code (help _rc) is accompanied by an error message, but usually this is done silently. You may not be aware of the breadth and usefulness of the results made available for further use by Stata commands. Using stored results can greatly simplify your work with Stata because a do-file can be constructed to use the results of a previous statement in a computation, title, graph label, or even a conditional statement.

We must distinguish between *r-class* and *e-class* commands. Each Stata command belongs to a class that may be r, e, or, less commonly, s. This applies both to commands that are built in (such as summarize) and to the 80% of official Stata commands that are implemented in the ado-file language.[3] The e-class commands are *estimation* commands, which return e(b) and e(V)—the estimated parameter vector and its variance–covariance matrix, respectively—to the calling program; e-class commands also return other information (help ereturn). Almost all other official Stata commands are r-class commands, which return results to the calling program (help return). Let's deal first with the simpler case of r-class commands.

Virtually every Stata command, including those that you might not think of as generating results, places items in the saved results, which can be displayed by typing

---

3. If this distinction interests you, which will report that a command is either built in (i.e., compiled C or Mata code) or located in a particular ado-file on your hard disk. If it is stored in an ado-file, the viewsource ([P] viewsource) command can be used to view its code.

return list.[4] For instance, consider `describe` applied to a dataset of United Nations peacekeeping missions:

```
. use un, clear

. describe

Contains data from un.dta
  obs:            58
  vars:           15                          13 Jun 2006 10:52
  size:        3,886 (99.9% of memory free)
```

variable name	storage type	display format	value label	variable label
name	str10	%10s		
mistype	float	%9.0g		
contype	float	%9.0g		
sevviol	float	%9.0g		
area	float	%9.0g		
loctype	float	%9.0g		
addloc	float	%9.0g		
borders	float	%9.0g		
primact	float	%9.0g		
spinv	float	%9.0g		
duration	float	%9.0g		
troop	float	%9.0g		
expend	float	%9.0g		
deaths	float	%9.0g		
completed	byte	%8.0g		

```
Sorted by:  duration

. return list

scalars:
              r(changed) =  0
             r(widthmax) =  18074
               r(k_max) =  2048
               r(N_max) =  14562
                r(width) =  63
                    r(k) =  15
                    r(N) =  58

. local sb: sortedby

. display "dataset sorted by: `sb'"
dataset sorted by: duration
```

The saved results for the `describe` command contain items of a single type: scalars, as described in section 3.9. `r(N)` and `r(k)` contain the number of observations and variables present in the dataset in memory. `r(changed)` is an indicator variable that will be set to 1 as soon as a change is made to the contents of the data. We also demonstrate here how information about the dataset's sort order can be retrieved by using one of the extended macro functions discussed in section 3.8. Any of the scalars defined in the saved results can be used in a following statement without displaying the saved results. A subsequent r-class command will replace the contents of the saved results with its return values, so if you want to use any of these items, you should save them to local macros or scalars. For a more practical example, consider `summarize`:

---

4. Significant exceptions are `generate`, `replace`, and `egen`.

```
. summarize troop, detail
                              troop

        Percentiles      Smallest
 1%           2                2
 5%          28               15
10%          50               28        Obs                    49
25%         189               36        Sum of Wgt.            49

50%        1187                         Mean             4270.102
                            Largest     Std. Dev.        7624.842
75%        5522              18500
90%       15522              19828      Variance         5.81e+07
95%       19828              25945      Skewness         2.943846
99%       39922              39922      Kurtosis         12.32439
. scalar iqr = r(p75) - r(p25)

. display "IQR = " iqr
IQR = 5333

. scalar semean = r(sd) / sqrt(r(N))

. display "Mean = " r(mean) " S.E. = " semean
Mean = 4270.102 S.E. = 1089.2632
```

The `detail` option displays the full range of results available—here all in the form of scalars—after the `summarize` command. We compute the interquartile range of the summarized variable and its standard error of mean as scalars and display those quantities. We often need the mean of a variable for further computations but do not wish to display the results of `summarize`. Here the `meanonly` option of `summarize` both suppresses output and calculates the variance or standard deviation of the series. The scalars `r(N)`, `r(mean)`, `r(min)`, and `r(max)` are still available (see Cox [2007e]).

When working with time-series or panel data, it is often useful to know whether the data have been `xtset`, and if so, which variable is serving as the panel variable and (if defined) as the calendar variable.[5] For example, consider this extract from the National Longitudinal Surveys:

---

5. Panel data need not refer to time series for each individual; see [XT] **xtset**. The `tsset` command can also be used to declare data as a time series of cross-sections.

```
. use unionT, clear
(NLS Women 14-24 in 1968)

. xtset
      panel variable:  idcode (unbalanced)
       time variable:  year, 1970 to 1988, but with gaps
               delta:  1 unit

. return list

scalars:
             r(tdelta) =  1
               r(tmax) =  1988
               r(tmin) =  1970
               r(imax) =  5159
               r(imin) =  1

macros:
            r(panelvar) : "idcode"
             r(timevar) : "year"
               r(unit1) : "."
               r(tsfmt) : "%8.0g"
             r(tdeltas) : "1 unit"
               r(tmaxs) : "1988"
               r(tmins) : "1970"
            r(balanced) : "unbalanced"
```

In this example, the returned scalars include the first and last time periods in this panel dataset (the years 1970 and 1988) and the range of the idcode variable, which is designated as r(panelvar). The macros also include the time-series calendar variable r(timevar) and the range of that variable in a form that can be readily manipulated, for instance, for graph titles. The value r(tdelta) indicates that the interval between time measurements is one time unit, or one year (see r(tdeltas)). The designation of the dataset as unbalanced implies that there are differing numbers of time-series observations available for different units (individuals).

Several statistical commands are r-class because they do not fit a model. The correlate command will return one estimated correlation coefficient, the correlation of the first two variables, regardless of the number of variables in the command's varlist. However, the command also returns the full correlation matrix as matrix r(C), which can be copied for further use. The ttest ([R] ttest) command is also r-class, and we can access its saved results to retrieve all the quantities it computes:

(Continued on next page)

```
. ttest age, by(union)
Two-sample t test with equal variances
```

Group	Obs	Mean	Std. Err.	Std. Dev.	[95% Conf. Interval]	
0	20389	30.32302	.0456294	6.515427	30.23358	30.41245
1	5811	30.81535	.0837133	6.381461	30.65124	30.97946
combined	26200	30.43221	.0400895	6.489056	30.35364	30.51079
diff		-.4923329	.0964498		-.6813797	-.3032861

```
    diff = mean(0) - mean(1)                                  t =  -5.1046
Ho: diff = 0                               degrees of freedom =    26198

    Ha: diff < 0              Ha: diff != 0              Ha: diff > 0
 Pr(T < t) = 0.0000      Pr(|T| > |t|) = 0.0000       Pr(T > t) = 1.0000
. return list

scalars:
                r(sd) =  6.489056317184585
              r(sd_2) =  6.381460760617624
              r(sd_1) =  6.515426706403353
                r(se) =  .0964497730823021
               r(p_u) =  .9999998330518228
               r(p_l) =  1.66948177204e-07
                 r(p) =  3.33896354408e-07
                 r(t) =  -5.10455202650672
              r(df_t) =  26198
              r(mu_2) =  30.81535019790053
               r(N_2) =  5811
              r(mu_1) =  30.32301731325715
               r(N_1) =  20389
```

The saved results contain scalars representing each of the displayed values from `ttest` except the total number of observations (which can be computed as `r(N_1) + r(N_2)`), the standard errors of the group means, and the confidence-interval limits.

5.3.1   The ereturn list command

An even broader array of information is provided after any e-class (estimation) command, as displayed by `ereturn list` ([P] **ereturn**). Most e-class commands return four types of Stata objects: scalars, such as `e(N)`, summarizing the estimation process; macros, providing such information as the name of the response variable (`e(depvar)`) and the estimation method (`e(model)`); matrices `e(b)` and `e(V)`, as described in section 5.3; and a Stata pseudovariable, `e(sample)`.[6] For example, consider a simple regression on the United Nations peacekeeping dataset:

---

6. Although [U] **18.10.2 Saving results in e()** and the `ereturn list` output describe `e(sample)` as a function, it is perhaps better considered a variable, albeit one that does not appear in the dataset.

```
. use un, clear
. regress deaths duration troop
```

Source	SS	df	MS
Model	103376.04	2	51688.0199
Residual	60815.7935	39	1559.37932
Total	164191.833	41	4004.67886

```
Number of obs =      42
F(  2,     39) =   33.15
Prob > F       =  0.0000
R-squared      =  0.6296
Adj R-squared  =  0.6106
Root MSE       =  39.489
```

deaths	Coef.	Std. Err.	t	P>|t|	[95% Conf. Interval]	
duration	.137803	.0405977	3.39	0.002	.0556864	.2199197
troop	.0059166	.0007648	7.74	0.000	.0043697	.0074635
_cons	-6.556093	8.179624	-0.80	0.428	-23.10094	9.988759

```
. ereturn list
scalars:
                  e(N) =  42
               e(df_m) =  2
               e(df_r) =  39
                  e(F) =  33.1465341558571
                 e(r2) =  .6296052472842496
               e(rmse) =  39.48897720443585
                e(mss) =  103376.0398278876
                e(rss) =  60815.7935054457
               e(r2_a) =  .6106106445808778
                 e(ll) =  -212.4320571229659
               e(ll_0) =  -233.2889619227027

macros:
            e(cmdline) : "regress deaths duration troop"
              e(title) : "Linear regression"
                e(vce) : "ols"
             e(depvar) : "deaths"
                e(cmd) : "regress"
         e(properties) : "b V"
            e(predict) : "regres_p"
              e(model) : "ols"
          e(estat_cmd) : "regress_estat"

matrices:
                 e(b) :  1 x 3
                 e(V) :  3 x 3

functions:
             e(sample)
```

Two particularly useful scalars in this list are e(df_m) and e(df_r), the model and residual degrees of freedom, respectively. These show the numerator and denominator degrees of freedom for e(F). The e(rmse) scalar allows for retrieval of the Root MSE of the equation. Two of the scalars do not appear in the printed output: e(ll) and e(ll_0), the likelihood function evaluated for the estimated model and for the null model, respectively.[7,8] Although the name of the response variable is available in macro

---

7. For ordinary least-squares regression with a constant term, the null model is that considered by the ANOVA *F* statistic: the intercept-only model with all slope coefficients constrained to zero.

8. These likelihood values can be displayed with the **estimates stats** ([R] **estimates**) command following estimation.

e(depvar), the names of the regressors are not shown here. They can be retrieved from the matrix e(b), as we now illustrate. Because the estimated parameters are returned in a $1 \times k$ row vector, the variable names are column names of that matrix:

```
. local regressors: colnames e(b)
. display "Regressors: 'regressors'"
Regressors: duration troop _cons
```

Another result displayed above should be noted: e(sample), listed as a function rather than a scalar, macro, or matrix. The e(sample) pseudovariable returns 1 if an observation was included in the estimation sample, and 0 otherwise. The regress command honors any if *exp* and in *range* qualifiers and then practices casewise deletion to remove any observations with missing values across the set (y, X). Thus the observations actually used in generating the regression estimates may be fewer than those specified in the regress command. A subsequent command such as summarize *varlist* if *exp* (or in *range*) will not necessarily provide the descriptive statistics of the observations on X that entered the regression unless all regressors and the y variable have the same pattern of missing values. But the set of observations actually used in estimation can easily be determined with the qualifier if e(sample):

```
. summarize regressors if e(sample)
```

This command will yield the appropriate summary statistics from the regression sample. It can be retained for later use by placing it in a new variable:

```
. generate byte reg1sample = e(sample)
```

where we use the byte data type to save memory by saving e(sample) in an indicator (0,1) variable.

The estat ([R] estat) command can be used to display several items after any estimation command. Some of those items (ic, summarize, and vce) are common to all estimation commands, whereas others depend upon the specific estimation command that precedes estat. For our fitted regression model,

```
. estat summarize
 Estimation sample regress              Number of obs =      42

    Variable |       Mean     Std. Dev.       Min         Max
-------------+-----------------------------------------------
      deaths |   34.16667     63.28253         0         234
    duration |   90.04762     152.8856         2         641
       troop |     4785.5     8116.061         2       39922
```

produces summary statistics, computed over the estimation sample, for the response variable and all regressors from the previous regress command.

In the next example, we use the matrix list command to display the coefficient matrix generated by our regression: e(b), the k-element row vector of estimated coefficients. Like all Stata matrices, this array bears row and column labels, so an element can be addressed by either its row and column numbers[9] or its row and column names.

```
. matrix list e(b)
e(b)[1,3]
        duration       troop       _cons
y1    .13780304    .00591661   -6.5560927
```

The `estat` command can be used to display the estimated variance–covariance matrix (VCE) by using `estat vce`.[10] This command provides a number of options to control the display of the matrix. For example,

```
. estat vce
Covariance matrix of coefficients of regress model
        e(V) |    duration        troop        _cons
    ---------+---------------------------------------
    duration |    .00164817
       troop |    3.504e-06    5.849e-07
       _cons |   -.16518387   -.00311436    66.906254
```

The diagonal elements of the VCE are the squares of the estimated standard errors of the respective coefficients.

Many official Stata commands, as well as many user-written routines, use the information available from `ereturn list`. How can a command like `estat ovtest` ([R] **regress postestimation**) compute the necessary quantities after `regress`? Because it can retrieve all relevant information, the names of the regressors, the name of the dependent variable, and the net effect of all `if` *exp* and `in` *range* qualifiers (from `e(sample)`) from the results left behind as e-class scalars, macros, matrices, or functions by the e-class command. Any do-file you write can perform the same magic if you use `ereturn list` to find the names of each quantity left behind for your use and store the results you need in local macros or scalars immediately after the e-class command. As noted before, retaining scalars as scalars is preferable to maintain full precision. You should not store scalar quantities in Stata variables unless there is good reason to do so.

The e-class commands can be followed by any of the `estimates` package of commands, described in the next section. Estimates can be saved in memory, combined in tabular form, and saved to disk for use in a later session.

5.4   Storing, saving, and using estimated results

The `estimates` package of commands makes it easy to work with different sets of estimation results. You store a set of results in memory with the `estimates store` *name* command, which stores them under *name* and, optionally, a descriptive title. Up to

---

9. Stata matrices' rows and columns are numbered starting with 1.
10. Prior to Stata version 9, the `vce` command provided this functionality.

300 sets of estimates can be stored in memory.[11] Stored estimates can be reviewed with `estimates replay` *namelist*, where *namelist* can refer to one or several sets of estimates in memory. Most usefully, estimates in memory can be combined in tabular form with `estimates table` or with user-written programs such as `estout` or `outreg2` (available from the Statistical Software Components archive). A set of estimates can also permanently be saved to a disk file with `estimates save` *filename* and read from disk with `estimates use` *filename*. At that point, you can review those prior estimates, tabulate them, or even apply postestimation commands such as `test` ([R] **test**) or `lincom` ([R] **lincom**) as long as these commands do not depend on the original dataset.

To organize several equations' estimates into a tabular form for scrutiny or publication, you can use `estimates table`. You specify that a table is to be produced containing several sets of results. If you do not specify, `estimates table` assumes that you mean only the *active* set of results. To juxtapose several sets of estimates, you give their names in a namelist or specify all available sets with `*` or `_all`. Stata automatically handles alignment of the coefficients into the appropriate rows of a table. Options allow for the addition of estimated standard errors (`se`), t-values (`t`), p-values (`p`), or significance stars (`star`). Each of these quantities can be given its own display `format` if the default is not appropriate, so the coefficients, standard errors, t- and p-values need not be rounded by hand. Variable labels can be displayed in place of variable names with the `label` option. You can also choose to present coefficients in exponentiated form with the `eform` option. The order of coefficients in the table can be controlled by the `keep()` option rather than relying on the order in which they appear in the list of estimates' contents. Certain parameter estimates can be removed from the coefficient table with `drop()`. Any result left in `e()` can be added to the table with the `stats()` option; this option can also be used to add several additional criteria, such as the Akaike information criterion and Bayesian information criterion.

Let's consider an example using several specifications from a model of air quality in several U.S. cities:

```
. use airquality, clear
. quietly regress so2 temp manuf pop
. estimates store model1
. quietly regress so2 temp pop wind
. estimates store model2
. quietly regress so2 temp wind precip days
. estimates store model3
. quietly regress so2 temp manuf pop wind precip days
. estimates store model4
```

---

11. See `help limits`; the limit of 20 sets quoted in [R] **estimates** is outdated.

```
. estimates table model1 model2 model3 model4, stat(r2_a rmse)
> b(%7.3f) se(%6.3g) p(%4.3f)
```

Variable	model1	model2	model3	model4
temp	-0.587	-1.504	-1.854	-1.268
	.371	.43	.861	.621
	0.122	0.001	0.038	0.049
manuf	0.071			0.065
	.0161			.0157
	0.000			0.000
pop	-0.047	0.020		-0.039
	.0154	.0051		.0151
	0.004	0.000		0.014
wind		-2.858	-1.685	-3.181
		2.22	2.58	1.82
		0.206	0.518	0.089
precip			0.539	0.512
			.525	.363
			0.312	0.167
days			0.006	-0.052
			.236	.162
			0.980	0.750
_cons	58.196	128.501	128.862	111.728
	20.5	36.7	67.7	47.3
	0.007	0.001	0.065	0.024
r2_a	0.581	0.386	0.172	0.611
rmse	15.191	18.393	21.356	14.636

legend: b/se/p

After fitting and storing four different models of SO_2 (sulphur dioxide) concentration, we use **estimates table** to present the coefficients, estimated standard errors, and *p*-values in tabular form. The **stat()** option adds summary statistics from the **e()** results.

```
. estimates table model4 model1 model3 model2, stat(r2_a rmse ll)
> b(%7.3g) star label title("Models of sulphur dioxide concentration")
Models of sulphur dioxide concentration
```

Variable	model4	model1	model3	model2
Mean temperature	-1.27*	-.587	-1.85*	-1.5**
Mfg. workers, 000	.0649***	.0712***		
Population	-.0393*	-.0466**		.0203***
Mean wind speed	-3.18		-1.69	-2.86
Mean precipitation	.512		.539	
Mean days quality=poor	-.0521		.006	
Constant	112*	58.2**	129	129**
r2_a	.611	.581	.172	.386
rmse	14.6	15.2	21.4	18.4
ll	-164	-168	-181	-175

legend: * p<0.05; ** p<0.01; *** p<0.001

We suppress the standard errors and display significance stars for the estimates while displaying variable labels (rather than names) with the `label` option. We add the log-likelihood value for each model by using the `stat()` option. The `estimates` command can be used after any Stata estimation command, including multiple-equation commands.

We can also execute postestimation commands on the stored estimates by typing `estimates for` *namelist*. In the four models above, let's test the hypothesis that the effect of mean temperature on SO_2 concentration is -1.6, a hypothetical value from other researchers' studies of this relationship.

```
. estimates for model1 model2 model3 model4: test temp = -1.6
```

```
Model model1
```
```
 ( 1)   temp = -1.6

       F(  1,     37) =     7.45
            Prob > F =    0.0096
```

```
Model model2
```
```
 ( 1)   temp = -1.6

       F(  1,     37) =     0.05
            Prob > F =    0.8236
```

```
Model model3
```
```
 ( 1)   temp = -1.6

       F(  1,     36) =     0.09
            Prob > F =    0.7697
```

```
Model model4
```
```
 ( 1)   temp = -1.6

       F(  1,     34) =     0.29
            Prob > F =    0.5964
```

We find that in only one case—that of `model1`—can the hypothesis be rejected by the data. We need not have the original data in memory to perform these tests. We can reproduce the results from `model1`:

```
. estimates replay model1
```

```
Model model1
```

Source	SS	df	MS		Number of obs =	41
					F(3, 37) =	19.50
Model	13499.2473	3	4499.7491		Prob > F =	0.0000
Residual	8538.65513	37	230.774463		R-squared =	0.6125
					Adj R-squared =	0.5811
Total	22037.9024	40	550.947561		Root MSE =	15.191

so2	Coef.	Std. Err.	t	P>|t|	[95% Conf. Interval]	
temp	-.5871451	.3710077	-1.58	0.122	-1.338878	.1645878
manuf	.0712252	.0160601	4.43	0.000	.0386842	.1037661
pop	-.0466475	.0153719	-3.03	0.004	-.0777939	-.0155011
_cons	58.19593	20.48789	2.84	0.007	16.68352	99.70835

The four sets of estimates also can be documented with `estimates notes`[12] and saved
to disk with `estimates save` for later use or exchange with another researcher:

```
. forvalues i = 1/4 {
  2.          estimates restore model'i'
  3.          estimates notes: from file 'c(filename)' saved 'c(filedate)'
  4.          estimates save so2_model'i', replace
  5. }
(results model1 are active now)
file so2_model1.ster saved
(results model2 are active now)
file so2_model2.ster saved
(results model3 are active now)
file so2_model3.ster saved
(results model4 are active now)
file so2_model4.ster saved
```

In a later Stata session (or after emailing these `.ster` files to another Stata user), we can
retrieve any of these saved estimates and work with them. We `clear` and `estimates`
`clear` to illustrate that neither data nor previously stored estimates are needed.

```
. clear

. estimates clear

. estimates describe using so2_model3
  Estimation results saved on 30jul2008 19:45, produced by
      . regress so2 temp wind precip days
  Notes:
    1.  from file airquality.dta saved 30 Jun 2007 08:41
. estimates use so2_model3

. estimates store so2_model3

. estimates table *
```

Variable	so2_model3
temp	-1.8539796
wind	-1.6852669
precip	.5385879
days	.00600175
_cons	128.86154

---

12. See section 3.8.1 regarding the use of `c()` return values.

We must `store` the estimates to make them accessible to `estimates table`, although we could replay them with the original estimation command (`regress`) without taking that step. Note, however, that variable labels are not accessible without the original data in memory.

5.4.1   Generating publication-quality tables from stored estimates

Ben Jann's `estout` package is a full-featured solution to preparing publication-quality tables in various output formats (Jann 2005; 2007). This routine, which Jann describes as a wrapper for `estimates table`, reformats stored estimates in a variety of formats, combines summary statistics from model estimation, and produces output in several formats (such as Stata Markup and Control Language for display in the Viewer window, tab-delimited or comma-separated values [for word processors or spreadsheets], LATEX, rich text format [`.rtf`], and HTML). A companion program in that package, `estadd`, allows for the addition of user-specified statistics to the `e()` arrays accessible by `estimates`. A simplified version of `estout` is available as `esttab`, and the utility command `ests to` stores estimates without the need to name them. `ests to` can also be used as a prefix command (see section 7.2). Complete documentation, including many examples using the `estout` package, is available at http://repec.org/bocode/e/estout.

As an example,

```
. use airquality, clear

. eststo clear

. eststo: quietly regress so2 temp manuf pop
(est1 stored)

. eststo: quietly regress so2 temp pop wind
(est2 stored)

. eststo: quietly regress so2 temp wind precip days
(est3 stored)

. eststo: quietly regress so2 temp manuf pop wind precip days
(est4 stored)

. esttab using esttab_example.tex, ar2 label se nostar nodepvars brackets
> nomtitles title("Models of sulphur dioxide concentration") booktabs
> alignment(D{.}{.}{-1}) replace
(output written to esttab_example.tex)
```

produces a formatted LATEX table, displayed as table 5.1.

Table 5.1. Models of sulphur dioxide concentration

	(1)	(2)	(3)	(4)
Mean temperature	−0.587	−1.504	−1.854	−1.268
	[0.371]	[0.430]	[0.861]	[0.621]
Mfg. workers, 000	0.0712			0.0649
	[0.0161]			[0.0157]
Population	−0.0466	0.0203		−0.0393
	[0.0154]	[0.00514]		[0.0151]
Mean wind speed		−2.858	−1.685	−3.181
		[2.219]	[2.582]	[1.815]
Mean precipitation			0.539	0.512
			[0.525]	[0.363]
Mean days quality=poor			0.00600	−0.0521
			[0.236]	[0.162]
Constant	58.20	128.5	128.9	111.7
	[20.49]	[36.73]	[67.69]	[47.32]
Observations	41	41	41	41
Adjusted R^2	0.581	0.386	0.172	0.611

Standard errors are in brackets.

These useful programs are available by typing `ssc`, as is an alternative: Roy Wada's `outreg2`, which also has the facility to work with stored estimates. We illustrate the production of a LaTeX table here, but these programs can also generate a rich text format (`.rtf`) table or one in tab-delimited form, which can be imported into standard office software.

5.5 Reorganizing datasets with the reshape command

When data have more than one identifier per record, they can be organized in different ways. For instance, it is common to find online displays or downloadable spreadsheets of data for individual units—for instance, U.S. states—with the unit's name as the row label and the year as the column label. If these data were brought into Stata in this format, they would be in the *wide form*, with the same measurement (population) for different years denoted as separate Stata variables. For example,

```
. use NEstates, clear
. keep if inlist(year,1990,1995,2000)
. drop dpi*
. keep year popCT popMA popRI
. order year
. xpose, clear
```

```
. drop in 1
. gen str state = "CT" in 1
. replace state = "MA" in 2
. replace state = "RI" in 3
. rename v1 pop1990
. rename v2 pop1995
. rename v3 pop2000
. order state
. list, noobs
```

state	pop1990	pop1995	pop2000
CT	3291967	3324144	3411750
MA	6022639	6141445	6362076
RI	1005995	1017002	1050664

There are several Stata commands—such as **egen** rowwise functions—that work effectively on data stored in the wide form. The wide form can also be a useful form of data organization for producing graphs.

Alternatively, we can imagine stacking each year's population figures from this display into one variable, **pop**. In this format, known in Stata as the *long form*, each datum is identified by two variables: the state name and the year to which it pertains.

```
. reshape long pop, i(state) j(year)
. list, noobs sepby(state)
```

state	year	pop
CT	1990	3291967
CT	1995	3324144
CT	2000	3411750
MA	1990	6022639
MA	1995	6141445
MA	2000	6362076
RI	1990	1005995
RI	1995	1017002
RI	2000	1050664

This data structure is required for many of Stata's statistical commands, such as the **xt** package of panel-data commands. As many knowledgeable Stata users have pointed out, the long form is also useful for data management using by-groups and the computation of statistics at the individual level, often implemented with the **collapse** command.

Inevitably, you will acquire data (either raw data or Stata datasets) that are stored in either the wide or the long form, and you will find that translation to the other format is necessary to carry out your analysis. The solution to this problem is Stata's **reshape** command. **reshape** is an immensely powerful tool for reformulating a dataset in memory without recourse to external files. In statistical packages lacking a data-reshape feature, common practice entails writing the data to one or more external text files and reading it back in. With the proper use of `reshape`, this is not necessary in Stata. But `reshape` requires that the data to be reshaped are labeled in such a way that they can be handled by the mechanical rules that the command applies. In situations beyond the simple application of `reshape`, it may require some experimentation to construct the appropriate command syntax. This is all the more reason for enshrining that code in a do-file, because some day you are likely to come upon a similar application for `reshape`.

In the rest of this section, we will work with several variations on a longitudinal (panel) dataset of U.S. school district characteristics. Let's consider the original form of this dataset:

```
. use mathpnl_long, clear
(modified mathpnl.dta from Wooldridge (2000))

. describe
Contains data from mathpnl_long.dta
  obs:         2,200                          modified mathpnl.dta from
                                                Wooldridge (2000)
  vars:            7                          28 Jun 2007 09:41
  size:       57,200 (99.5% of memory free)
```

variable name	storage type	display format	value label	variable label
distid	float	%9.0g		district identifier
expp	int	%9.0g		expenditure per pupil
revpp	int	%9.0g		revenue per pupil
avgsal	float	%9.0g		average teacher salary
math4score	float	%9.0g		% satisfactory, 4th grade math
math7score	float	%9.0g		% satisfactory, 7th grade math
year	int	%9.0g		year

```
Sorted by:  distid  year
```

We see that the dataset contains 2,200 observations, identified by the numeric `distid` school district identifier and `year`.

```
. xtset
       panel variable:  distid (strongly balanced)
        time variable:  year, 1992 to 1998
               delta:  2 units
```

```
. tabulate year
```

year	Freq.	Percent	Cum.
1992	550	25.00	25.00
1994	550	25.00	50.00
1996	550	25.00	75.00
1998	550	25.00	100.00
Total	2,200	100.00	

The dataset is in the long form, as `xtset` shows, and is strongly balanced; that is, each of the 550 districts appears in the sample for each of the four years. The `delta: 2 units` qualifier indicates that the time series are biennial, or once every two years.

Imagine that we did not have this dataset available, but instead had its wide form equivalent, perhaps acquired from a spreadsheet on school district expenditures. The wide-form dataset looks like the following:

```
. use mathpnl_wide, clear
(modified mathpnl.dta from Wooldridge (2000))

. describe

Contains data from mathpnl_wide.dta
  obs:            550                          modified mathpnl.dta from
                                                 Wooldridge (2000)
  vars:            21                          28 Jun 2007 09:41
  size:        39,600 (99.6% of memory free)
```

variable name	storage type	display format	value label	variable label
distid	float	%9.0g		district identifier
expp1992	int	%9.0g		1992 expp
revpp1992	int	%9.0g		1992 revpp
avgsal1992	float	%9.0g		1992 avgsal
math4score1992	float	%9.0g		1992 math4score
math7score1992	float	%9.0g		1992 math7score
expp1994	int	%9.0g		1994 expp
revpp1994	int	%9.0g		1994 revpp
avgsal1994	float	%9.0g		1994 avgsal
math4score1994	float	%9.0g		1994 math4score
math7score1994	float	%9.0g		1994 math7score
expp1996	int	%9.0g		1996 expp
revpp1996	int	%9.0g		1996 revpp
avgsal1996	float	%9.0g		1996 avgsal
math4score1996	float	%9.0g		1996 math4score
math7score1996	float	%9.0g		1996 math7score
expp1998	int	%9.0g		1998 expp
revpp1998	int	%9.0g		1998 revpp
avgsal1998	float	%9.0g		1998 avgsal
math4score1998	float	%9.0g		1998 math4score
math7score1998	float	%9.0g		1998 math7score

```
Sorted by: distid
```

We see that there are now 550 observations, one per school district, with separate variables containing the five measurements for each year. Those Stata variables are systematically named, which is an important consideration. It means that we can readily use wildcards, for instance, `summarize expp*`, to compute descriptive statistics for all expenditure-per-pupil variables. It also eases our task of reshaping the dataset.

Let's say that we want to convert this wide-form dataset into its long-form equivalent. The `reshape` command works with the notion of $x_{i,j}$ data. Its syntax lists the variables to be stacked up and specifies the i and j variables, where the i variable indexes the rows and the j variable indexes the columns in the existing form of the data.

```
. reshape long expp revpp avgsal math4score math7score, i(distid) j(year)
(note: j = 1992 1994 1996 1998)
Data                            wide   ->   long

Number of obs.                   550   ->    2200
Number of variables               21   ->       7
j variable (4 values)                   ->   year
xij variables:
        expp1992 expp1994 ... expp1998   ->   expp
      revpp1992 revpp1994 ... revpp1998   ->   revpp
   avgsal1992 avgsal1994 ... avgsal1998   ->   avgsal
math4score1992 math4score1994 ... math4score1998->math4score
math7score1992 math7score1994 ... math7score1998->math7score
```

We use `reshape long` because the data are in the wide form and we want to place them in the long form. We provide the variable names to be stacked without their common suffixes (here the `year` embedded in their wide-form variable name). The `i` variable is `distid` and the `j` variable is `year`. Together, these variables uniquely identify each measurement. Stata's description of `reshape` speaks of i defining a unique observation and j defining a subobservation logically related to that observation. Any additional variables that do not vary over j are not specified in the `reshape` statement, because they will be automatically replicated for each j. After this command, you will have a dataset that is identical to the long-form dataset displayed above:

(Continued on next page)

```
. describe
Contains data
   obs:         2,200                        modified mathpnl.dta from
                                             Wooldridge (2000)
   vars:            7
   size:       57,200 (99.5% of memory free)

                   storage   display    value
variable name      type      format     label     variable label

distid             float     %9.0g                district identifier
year               int       %9.0g
expp               int       %9.0g
revpp              int       %9.0g
avgsal             float     %9.0g
math4score         float     %9.0g
math7score         float     %9.0g

Sorted by:   distid  year
     Note:   dataset has changed since last saved
```

What if you wanted to reverse the process and translate the data from the long to the wide form?

```
. reshape wide expp revpp avgsal math4score math7score, i(distid) j(year)
(note: j = 1992 1994 1996 1998)
Data                                 long   ->   wide

Number of obs.                       2200   ->       550
Number of variables                     7   ->        21
j variable (4 values)                year   ->   (dropped)
xij variables:
                                     expp   ->   expp1992 expp1994 ... expp1998
                                    revpp   ->   revpp1992 revpp1994 ... revpp1998
                                   avgsal   ->   avgsal1992 avgsal1994 ...
> avgsal1998
                               math4score   ->   math4score1992 math4score1994 ...
> math4score1998
                               math7score   ->   math7score1992 math7score1994 ...
> math7score1998
```

You can see that this command is identical to the first **reshape** command, with the exception that now you **reshape wide**, designating the target form. It reproduces the wide-form dataset described above.

This example highlights the importance of having appropriate variable names for **reshape**. If our wide-form dataset contained the **expp1992**, **Expen94**, **xpend_96**, and **expstu1998** variables, there would be no way to specify the common stub labeling the choices. However, one common case can be handled without the renaming of variables. Say that we have the variables **exp92pp**, **exp94pp**, **exp96pp**, and **exp98pp**. The command

```
reshape long exp@pp, i(distid) j(year)
```

will deal with that case, with the @ as a placeholder for the location of the j component of the variable name. In more difficult cases where the repeated use of `rename` ([D] **rename**) may be tedious, `renvars` (Cox and Weesie 2001; 2005) may be useful.

This discussion has only scratched the surface of `reshape`'s capabilities. There is no substitute for experimentation with this command after a careful perusal of `help reshape`, because it is one of the most complicated elements of Stata. For more guidance with `reshape` usage, see the following chapter and the response to the Stata frequently asked question "I am having problems with `reshape` command. Can you give further guidance?"[13].

5.6   Combining datasets

You may be aware that Stata can work with only one dataset at a time. How, then, do you combine datasets in Stata? First, it is important to understand that at least one of the datasets to be combined must already have been saved in Stata format. Second, you should realize that each of Stata's commands for combining datasets provides a certain functionality that should not be confused with the functionalities of other commands. For instance, consider the `append` command with two stylized datasets:

$$
\text{Dataset 1:} \begin{pmatrix} \text{id} & \text{var1} & \text{var2} \\ 112 & \vdots & \vdots \\ 216 & \vdots & \vdots \\ 449 & \vdots & \vdots \end{pmatrix}
$$

$$
\text{Dataset 2:} \begin{pmatrix} \text{id} & \text{var1} & \text{var2} \\ 126 & \vdots & \vdots \\ 309 & \vdots & \vdots \\ 421 & \vdots & \vdots \\ 604 & \vdots & \vdots \end{pmatrix}
$$

---

13. This response was written by Nicholas J. Cox and is available at
    http://www.stata.com/support/faqs/data/reshape3.html.

These two datasets contain the same variables, as they must for `append` to sensibly combine them. If Dataset 2 contained idcode, `Var1`, and `Var2`, the two datasets could not sensibly be appended without renaming the variables.[14] Appending these two datasets with common variable names creates a single dataset containing all the observations:

$$
\text{Combined:}
\begin{pmatrix}
\text{id} & \text{var1} & \text{var2} \\
112 & \vdots & \vdots \\
216 & \vdots & \vdots \\
449 & \vdots & \vdots \\
126 & \vdots & \vdots \\
309 & \vdots & \vdots \\
421 & \vdots & \vdots \\
604 & \vdots & \vdots
\end{pmatrix}
$$

The rule for `append`, then, is that if datasets are to be combined, they should share the same variable names and data types (string versus numeric). In the above example, if `var1` in Dataset 1 was a `float` variable while that variable in Dataset 2 was a `string` variable, they could not be `append`ed. It is permissible to `append` two datasets with differing variable names in the sense that Dataset 2 could also contain an additional variable or variables (for example, `var3`, and `var4`). The values of those variables in the observations coming from Dataset 1 would then be set to missing.

While `append` combines datasets by adding observations to the existing variables, the other key command, `merge`, combines variables for the existing observations. Consider these two stylized datasets:

$$
\text{Dataset 1:}
\begin{pmatrix}
\text{id} & \text{var1} & \text{var2} \\
112 & \vdots & \vdots \\
216 & \vdots & \vdots \\
449 & \vdots & \vdots
\end{pmatrix}
$$

$$
\text{Dataset 3:}
\begin{pmatrix}
\text{id} & \text{var22} & \text{var44} & \text{var46} \\
112 & \vdots & \vdots & \vdots \\
216 & \vdots & \vdots & \vdots \\
449 & \vdots & \vdots & \vdots
\end{pmatrix}
$$

---

14. In Stata, var1 and Var1 are two separate variables.

We can `merge` these datasets on the common *merge key* (here the `id` variable):

$$
\text{combined} : \begin{pmatrix}
\text{id} & \text{var1} & \text{var2} & \text{var22} & \text{var44} & \text{var46} \\
112 & \vdots & \vdots & \vdots & \vdots & \vdots \\
216 & \vdots & \vdots & \vdots & \vdots & \vdots \\
449 & \vdots & \vdots & \vdots & \vdots & \vdots
\end{pmatrix}
$$

The rule for `merge`, then, is that if datasets are to be combined on one or more merge keys, they must each have one or more variables with a common name and data type (string versus numeric). In the example above, each dataset must have a variable named `id`. That variable can be either numeric or string, as long as that characteristic of the merge key variables matches across the datasets to be merged. Of course, we need not have exactly the same observations in each dataset. If Dataset 3 contained observations with additional `id` values, those observations would be merged with missing values for `var1` and `var2`.

As we shall see in section 5.8, we have illustrated the simplest kind of merge: the *one-to-one merge*. Stata supports several other types of merges. The key concept should be clear: the `merge` command combines datasets "horizontally", adding variables' values to existing observations. With these concepts in mind, let's consider these commands in the context of our prior example.

5.7   Combining datasets with the append command

Data files downloaded from their providers often contain only one time period's information, or data about only one entity. A research project may require that you download several separate datasets and combine them for statistical analysis. In the school district example from section 5.5, imagine that we could only access separate data files for each of the four years: 1992, 1994, 1996, and 1998. Combining these datasets into one dataset would be a task for `append`. However, there are two caveats. First, it is possible that the individual datasets lack a year variable. It would, after all, be a constant value within the 1992 dataset. But if we append these datasets, we will lose the information about which observation pertains to which year. If individual datasets lack a dataset identifier, we must add one. That is as simple as typing `generate int year = 1992` and saving those data.[15] Second, as discussed previously, we must ensure that the variable names across the individual datasets are identical. If necessary, `rename` or `renvars` (Cox and Weesie 2001; 2005) should be used to ensure that variable names match.

---

15. The same caveat applies if individual datasets contain information for a single entity. For instance, if one file contains time series of Connecticut infant mortality data and a second contains that variable for Massachusetts, we must ensure that each file contains a state identifier variable of some sort before combining them.

You can then use **append** to combine the four school district datasets:

```
. use mathpnl1992, clear
(modified mathpnl.dta from Wooldridge (2000))

. append using mathpnl1994

. append using mathpnl1996

. append using mathpnl1998

. label data ""

. save mathpnl_appended, replace
file mathpnl_appended.dta saved

. describe

Contains data from mathpnl_appended.dta
  obs:          2,200
  vars:            14                           30 Jul 2008 19:45
  size:       118,800 (99.9% of memory free)
```

variable name	storage type	display format	value label	variable label
distid	float	%9.0g		district identifier
lunch	float	%9.0g		% eligible for free lunch
enrol	float	%9.0g		school enrollment
expp	int	%9.0g		expenditure per pupil
revpp	int	%9.0g		revenue per pupil
avgsal	float	%9.0g		average teacher salary
drop	float	%9.0g		high school dropout rate, %
grad	float	%9.0g		high school grad. rate, %
math4score	float	%9.0g		% satisfactory, 4th grade math
math7score	float	%9.0g		% satisfactory, 7th grade math
year	int	%9.0g		year
staff	float	%9.0g		staff per 1000 students
cpi	float	%9.0g		consumer price index
rexpp	float	%9.0g		real spending per pupil, 1997$

```
Sorted by:
```

The combined dataset contains all four years' data in the long form, with each observation identified by **distid** and **year**, identical to the long-form dataset described previously.

5.8   Combining datasets with the merge command

The long-form dataset we constructed above is useful if you want to add aggregate-level information to individual records. For instance, imagine that you have downloaded a separate file identifying the state in which each school district appears:

```
. use mathpnl_long, clear

. keep distid year

. keep if year == 1992

. generate int state = runiform() * 50 + 1

. drop year
```

```
. label var state "State code"
. label data ""
. sort distid
. save mathpnl_state, replace
. describe
Contains data from mathpnl_state.dta
  obs:           550
  vars:            2                          30 Jul 2008 19:45
  size:        5,500 (99.9% of memory free)

              storage  display    value
variable name   type   format     label      variable label

distid          float  %9.0g                 district identifier
state           int    %8.0g                 State code

Sorted by:  distid
```

Say you want to compute descriptive statistics and generate graphs for our school district data separately for certain states. With the aid of this auxiliary dataset, you can easily code each of the 550 districts' four time-series observations with the appropriate state code:

```
. use mathpnl_long, clear
(modified mathpnl.dta from Wooldridge (2000))
. merge distid using mathpnl_state, sort uniqusing
variable distid does not uniquely identify observations in the master data
. tab _merge
     _merge |      Freq.     Percent        Cum.

          3 |      2,200      100.00      100.00

      Total |      2,200      100.00
. assert _merge == 3
. drop _merge
. save mathpnl_longs, replace
file mathpnl_longs.dta saved
```

This use of **merge** is known as a *one-to-many* match-merge, where the state code for each school district is added to the individual records of each of the districts. The **distid** variable is the merge key. Unless both the master file (**mathpnlL.dta**) and the using file (**mathpnl_state.dta**) are sorted by the merge key, the **sort** option should be specified.

By default, **merge** creates a new variable, **_merge**, that takes on integer values for each observation of 1 if that observation was found only in the master dataset, 2 if it was found only in the using dataset, or 3 if it was found in both datasets. Here we expect **tab _merge** to reveal that all values equal 3, as it does. We also use the **uniqusing** option to ensure that there are no duplicate values of the district ID in the **using** file. That file should uniquely map each district to a single state code, so a duplicate value of **distid** must be a data-entry error. If the same **distid** mistakenly appears on two records in the `using` file, asserting `uniqusing` will cause `merge` to fail. Absent of such errors, each of the district's observations (four in number, one for each year) will be mapped to one state code.

This strategy will also work if you have a comprehensive list of school districts in the using file. Consider that the 550 school districts in `mathpnl_long.dta` are a random sample of U.S. school districts and that the information you have located identifying districts by state lists all U.S. school districts. The one-to-many merge still works perfectly well. After the merge, `_merge` will indicate that each `distid` in the using file that did not match a `distid` in our sample now has `_merge=2`. Those observations in the `merge`d file should be deleted (with `drop if _merge == 2`) before it is saved.

Alternatively, if `_merge` reveals that any observation has `_merge=1`, then this indicates that some of our `distid` values are not being successfully matched in the `using` file. This probably indicates errors in one of the files and should be corrected. The `assert _merge == 3` in the commands above will ensure that the do-file aborts if there are unmatched records in the master file.[16]

In your particular application, you may find that `_merge` values of 1 or 2 are appropriate. The key notion is that you should always tabulate `_merge` and consider whether the results of the merge are sensible in the context of your work. It is an excellent idea to use the `uniqmaster`, `uniqusing`, or `unique` options of the `merge` command whenever those conditions should logically be satisfied in your data, and one of those options is required if the `sort` option is also used.

In comparison to a lengthy and complicated do-file using a set of `replace` statements, the `merge` technique is far better. By merely altering the contents of the using file, we can correct any difficulties that appear in the one-to-many merge. Furthermore, if we had several school district–specific variables to be added to the individual observations, these variables would all be handled with one `merge` command. We might want to merge aggregate time-series information instead. For example, if we wanted to compute real expenditures per pupil, we would match on the merge key `year` to incorporate a price deflator (varying by year, but not by school district) in the `merge`d file. This technique proves exceedingly useful when working with individual data and panel data where we have aggregate information to be combined with the individual-level data.

5.8.1   The dangers of many-to-many merges

Earlier, we discussed using `merge` to attach aggregate characteristics to individual records using a one-to-many merge. There are good reasons to use a one-to-many merge, as we did above with school district–level characteristics, or its inverse: a many-to-one merge, which would essentially reverse the roles of the master and using datasets. The `merge` command can also be used to combine datasets with a one-to-one match-merge, as we illustrated in section 5.6. This would be appropriate if we had two or more datasets whose observations pertained to the same units: e.g., U.S. state population fig-

---

16. The do-file will also abort if there are unmatched records in the using file. If this is not desired, you could validate the merge using `assert _merge >= 2`.

ures from the 1990 and 2000 Censuses. Then you would want to assert _merge = 3 if you know that each dataset should contain the same observations, and you should use the unique option of merge.

There is a great danger in stumbling into the *many-to-many* merge, the alternative to the one-to-many or one-to-one merge. This problem arises when there are multiple observations in both datasets for some values of the merge key variable(s). The result of match-merging two datasets that both have more than one value of the merge key variable(s) is unpredictable, because it depends on the sort order of the datasets. Repeated execution of the same do-file will most likely result in a different number of cases in the result dataset without any error indication. There is no unique outcome for a many-to-many merge; when it is encountered, it usually results from a coding error in one of the files. The duplicates command (see section 5.2) is useful in tracking down such errors. To prevent such difficulties when using merge, you should specify either the uniqmaster or the uniqusing option in a match-merge. If no uniq... option is used, observations may be matched inappropriately.

In a one-to-one match-merge, the unique option should be used because it implies both uniqmaster and uniqusing, asserting that the merge key should be unique in both datasets.[17] Imagine that you have four separate school district data files, within which each variable is labeled with its year (e.g., expp1992 and revpp1992). You would not want to append these datasets because their variable names differ. But you could generate a wide-form dataset from the individual datasets with merge:

```
. use mathpnl1992m, clear
. forvalues y=1994(2)1998 {
  2.          merge distid using mathpnl`y'm, sort unique
  3.          assert _merge == 3
  4.          drop _merge
  5. }
. save mathpnlm, replace
file mathpnlm.dta saved
```

We have used the unique option to specify that each of the four datasets in the merge contains unique values of distid.

5.9   Other data-management commands

Stata provides several data-management commands less commonly used than append and merge. In this section, we discuss the fillin ([D] **fillin**), cross ([D] **cross**), stack ([D] **stack**), separate ([D] **separate**), joinby ([D] **joinby**), and xpose ([D] **xpose**) commands.

---

17. Those familiar with relational database-management systems, such as structured query language (SQL), will recognize that the concept of uniqueness refers to the merge key serving as a valid and unique primary key for the dataset.

5.9.1   The fillin command

We spoke earlier of a balanced panel: a longitudinal dataset in which there are N cross-sectional units, each with T time-series observations. Many panel datasets are unbalanced, with differing numbers of time-series observations available for different units. This does not create problems for Stata's panel-data (xt) commands, but it can cause difficulty if the data are to be exported to other software, particularly matrix languages that expect the dataset to contain $N \times T$ observations. The fillin command deals with this issue by adding observations with missing data so that all interactions of its varlist exist, thus making a complete rectangularization of the varlist.[18] See Cox (2005d) for more details.

The fillin command will work in any context, not only that of panel data. For instance, we might have a list of patients' ages and the average cholesterol level for male and female patients of that age. That dataset might not contain every combination of age and gender. If we want to rectangularize it, we would specify age gender in the varlist.

5.9.2   The cross command

The cross command forms every pairwise combination of the data in memory with the data in the using dataset. Imagine that you have a list of racial categories as the sole variable in memory: race, coded White, Black, Asian, and Hispanic. The file cancers.dta contains a list of the 10 most common forms of cancer: Lung, Colon, Breast, Cervical, Prostate, etc., coded as the variable cancertype. If you type

```
cross using cancers
```

you will create a dataset with 40 observations in which each of the racial categories is matched with each of the cancer categories in turn. See Franklin (2006) for more details.

5.9.3   The stack command

The stack command allows you to vertically stack the variables in its varlist.[19] For instance,

```
stack x1 x2 x3 x4, into(z1 z2)
```

---

18. The tsfill command ([TS] **tsfill**) is a specialized version of fillin. When tsfill is executed with its full option, a panel dataset will be balanced by the addition of missing observations as needed.

19. In linear algebra, this is the vec() operation, available in Stata's matrix language as the vec() function (help vec()) and in Mata (help mata vec()).

will create z1 as the "stack" of x1 and x3, and z2 as the "stack" of x2 and x4. Variables can be repeated in the varlist to juxtapose a single variable against others. For example,

```
stack x1 x2 x1 x3 x1 x4, into(z1 z2)
```

will create two new variables, one containing three copies of x1 and the other containing the stacked form of x2, x3, and x4.

When might you use stack? It is often a useful tool when preparing data for graphics where some form of the graph ([G] **graph twoway scatter**) command expects a single variable. For example,

```
. use mathpnl_wide, clear
(modified mathpnl.dta from Wooldridge (2000))

. preserve

. stack expp1992 avgsal1992 expp1998 avgsal1998, into(expp sal) wide clear

. scatter expp1992 expp1998 sal, m(oh x) scheme(s2mono)
> title("School District Expenditures vs. Average Salaries")
> xtitle("Average salaries") ytitle("Expenditures")
> legend(position(5) ring(0) label(1 "1992") label(2 "1998"))

. restore
```

Using the wide-form school district data, we use stack to juxtapose expenditures per pupil in 1992 and 1998 with average teacher salaries in those two years, and we produce a scatterplot, as figure 5.1 illustrates.

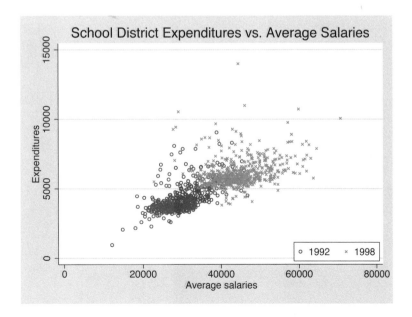

Figure 5.1. Superimposed scatterplots

stack clears the dataset in memory (automatically if its clear option is specified), retaining only those variables listed in the into() option unless the wide option is specified. With the wide option, the variables in the varlist are retained as well. In any case, a new variable, _stack, is created that identifies the groups. If you do not want to disrupt the dataset in memory, use preserve and restore as illustrated above and as described in section 2.6.

5.9.4   The separate command

The separate command can convert one variable into several new variables, either on the basis of a true-or-false Boolean expression (for two groups only) or in terms of a variable given in by(). In the latter case, the number of distinct values of the variable in by() (which can be of numeric or string type) will determine the number of new variables created.

As an example, consider the long-form school district dataset, and imagine that you want to create distinct expenditure-per-pupil variables for each year:

```
. use mathpnl_long, clear
(modified mathpnl.dta from Wooldridge (2000))

. separate expp, by(year) veryshortlabel
```

variable name	storage type	display format	value label	variable label
expp1992	int	%9.0g		1992
expp1994	int	%9.0g		1994
expp1996	int	%9.0g		1996
expp1998	int	%9.0g		1998

```
. summarize expp*
```

Variable	Obs	Mean	Std. Dev.	Min	Max
expp	2200	5205.833	1255.723	946	13982
expp1992	550	4181.165	933.6362	946	9041
expp1994	550	4752.56	993.0163	1147	10461
expp1996	550	5737.953	981.382	2729	10449
expp1998	550	6151.653	1028.371	3811	13982

We make use of the undocumented veryshortlabel option, which is a more austere variant of the shortlabel option; see Cox (2005e).

The separate command can also be useful in creating variables that correspond to a qualitative identifier, for instance, different variables for each ethnic group's blood pressure measurements.

5.9.5   The joinby command

The `joinby` command creates a new dataset by forming all possible pairwise combinations of the two datasets, given a merge key. It is similar in that sense to `cross`, which does not use a merge key. In many instances where a researcher considers using `joinby`, they probably want to do a `merge`.[20]

One instance in which `joinby` might come in handy is that involving a many-to-many merge, as discussed in section 5.8.1. Let's say that you have one dataset containing firms' subsidiaries, all identified by their `firmid` number. (Many large multinational corporations consist of several component firms, usually corresponding to the country in which they are incorporated). You have a second file that contains, for each `firmid`, a set of product codes. You want to construct a dataset in which each subsidiary is matched with each product sold by the parent firm. The variable `firmid` appears with duplicate entries in both datasets. If you use `joinby`,

```
use subsidiaries, clear
joinby firmid using products
```

you will create the desired data structure.

By default, if a value of `firmid` appears in one dataset but not the other, its observations will be dropped.[21] The merge key for `joinby` need not be a single variable. As in the example above, if both subsidiaries and product records are coded by country, you might use

```
joinby firmid country using products
```

to produce a dataset that will contain subsidiary–product combinations for each country in which that product is sold.

5.9.6   The xpose command

Another Stata data-management command is capable of making radical changes to the organization of the data: `xpose`. This is the *transpose* command, which turns observations into variables and vice versa. This functionality will be familiar to those who have used spreadsheets or matrix languages. It is rarely useful in Stata, because applying `xpose` will usually destroy the contents of string variables. If all variables in the dataset are numeric, this command can be useful.

Rather than using `xpose`, you should consider whether the raw data might be read in with the `byvariable()` option of `infile`. If there is truly a need to transpose the data, it was probably not created sensibly in the first place.

---

20. Those familiar with relational database-management systems will recognize `joinby` as the structured query language *outer join*, which is a technique to be avoided in most database tasks.

21. The `unmatched` option reverses this behavior.

6 Cookbook: Do-file programming II

This cookbook chapter presents for Stata do-file programmers several recipes using the programming features described in the previous chapter. Each recipe poses a problem and a worked solution. Although you may not encounter this precise problem, you should be able to recognize its similarities to a task that you would like to automate in a do-file.

6.1 Efficiently defining group characteristics and subsets

The problem. Say that your cross-sectional dataset contains a record for each patient who has been treated at one of several clinics. You want to associate each patient's clinic with a location code (for urban clinics, the Standard Metropolitan Statistical Area [SMSA]). The SMSA identifier is not on the patient's record, but it is available to you. How do you get this associated information on each patient's record without manual editing? One cumbersome technique (perhaps familiar to users of other statistical packages) involves writing a long sequence of statements with if *exp* clauses. Thankfully, there is a better way.[1]

The solution. Let's presume that we have a Stata dataset, `patient`, containing the individual's details as well as `clinicid`, the clinic ID. Assume that it can be dealt with as an integer. If it were a string code, that could easily be handled as well.

Create a text file, `clinics.raw`, containing two columns: the clinic ID (`clinicid`) and the SMSA Federal Information Processing Standards Code (`smsa`).[2] For instance,

```
12367   1120
12467   1120
12892   1120
13211   1200
14012   4560
 . . .    . . .
23435   5400
29617   8000
32156   9240
```

---

1. This recipe is adapted from a response to the Stata frequently asked question "How do you define group characteristics in your data in order to create subsets?"
 (http://www.stata.com/support/faqs/data/characteristics.html), written by Christopher F. Baum.
2. The Federal Information Processing Standards Code is a four-digit integer assigned to each SMSA. To see a list of SMSAs and their codes from 1960, visit
 http://www.census.gov/population/estimates/metro-city/60mfips.txt.

where SMSA codes 1120, 1200, 4560, 5400, 8000, and 9240 refer to the Massachusetts SMSAs of Boston, Brockton, Lowell, New Bedford, Springfield–Chicopee–Holyoke, and Worcester, respectively.

Read the file into Stata with `infile clinicid smsa using clinics`, and `save` the file as Stata dataset `clinic_char`. Now `use` the patient file and type the commands

```
merge clinicid using clinic_char, sort uniqusing
tab _merge
```

Use the `uniqusing` option, as discussed in section 5.8, to ensure that the `clinic_char` dataset has one record per clinic. After the merge is performed, you should find that all patients now have an `smsa` variable defined. If there are missing values in `smsa`, list the `clinicid`s that are missing for that variable and verify that they correspond to nonurban locations. When you are satisfied that the merge has worked properly, type

```
drop _merge
```

You have performed a one-to-many merge, attaching the same SMSA identifier to all patients who have been treated at clinics in that SMSA. You can now use the `smsa` variable to attach SMSA-specific information to each patient record with `merge`.

Unlike an approach depending on a long list of conditional statements, such as

```
replace smsa=1120 if inlist(clinicid,12367,12467,12892,...)
```

this approach leads you to create a Stata dataset containing your clinic ID numbers so that you can easily see whether you have a particular code in your list. This approach would be especially useful if you revise the list for a new set of clinics.

6.1.1   Using a complicated criterion to select a subset of observations

As Nicholas J. Cox has pointed out in a response to a Stata frequently asked question,[3] this approach can also be fruitfully applied if you need to work with a subset of observations that satisfies a complicated criterion. This might be best defined in terms of an indicator variable that specifies the criterion (or its complement). The same approach can be used for both. Construct a file containing the identifiers that define the criterion (in the example above, it would be the clinic ID to be included in the analysis). Merge that file with your dataset and examine the `_merge` variable. That variable will take on the value 1, 2, or 3, with a value of 3 indicating that the observation falls within the subset. You can then define the desired indicator:

```
generate byte subset1 = _merge == 3
drop _merge
regress ... if subset1
```

---

3. The question is titled "How do I select a subset of observations using a complicated criterion?", and Cox's response is available at http://www.stata.com/support/faqs/data/selectid.html.

Using this approach, any number of subsets can be easily constructed and maintained, avoiding the need for complicated conditional statements.

6.2   Applying reshape repeatedly

The problem. Are your data the wrong shape?[4] That is, are they not organized in the structure that you need to conduct the analysis you have in mind? Data sources often provide the data in a structure suitable for presentation but very clumsy for statistical analysis.

The solution. One of the key data-management tools Stata provides is `reshape`. If you need to modify the structure of your data, you should surely be familiar with `reshape` and its two functions: `reshape wide` and `reshape long`. Sometimes, you may have to apply `reshape` twice to solve a particularly knotty data-management problem.

As an example, consider this question posed on Statalist by an individual who has a dataset in the wide form:

country	tradeflow	Yr1990	Yr1991
Armenia	imports	105	120
Armenia	exports	90	100
Bolivia	imports	200	230
Bolivia	exports	80	115
Colombia	imports	100	105
Colombia	exports	70	71

He would like to reshape the data into the long form:

country	year	imports	exports
Armenia	1990	105	90
Armenia	1991	120	100
Bolivia	1990	200	80
Bolivia	1991	230	115
Colombia	1990	100	70
Colombia	1991	105	71

We must exchange the roles of years and tradeflows in the original data to arrive at the desired structure, suitable for analysis as `xt` data. This can be handled by two successive applications of `reshape`:

---

4. This recipe is adapted from Stata tip 45 (Baum and Cox 2007). I am grateful to Nicholas J. Cox for his contributions to this Stata tip.

```
. clear

. input str8 country str7 tradeflow Yr1990 Yr1991

         country   tradeflow      Yr1990       Yr1991
  1. Armenia imports 105 120
  2. Armenia exports 90 100
  3. Bolivia imports 200 230
  4. Bolivia exports 80 115
  5. Colombia imports 100 105
  6. Colombia exports 70 71
  7. end

. reshape long Yr, i(country tradeflow)
(note: j = 1990 1991)
Data                                    wide   ->   long
─────────────────────────────────────────────────────────────
Number of obs.                             6   ->     12
Number of variables                        4   ->      4
j variable (2 values)                          ->    _j
xij variables:
                             Yr1990 Yr1991   ->   Yr
─────────────────────────────────────────────────────────────
```

This transformation swings the data into long form, with each observation identified by country, tradeflow, and the new variable _j, taking on the values of year. We now perform reshape wide to make imports and exports into separate variables:

```
. rename _j year

. reshape wide Yr, i(country year) j(tradeflow) string
(note: j = exports imports)
Data                                    long   ->   wide
─────────────────────────────────────────────────────────────
Number of obs.                            12   ->      6
Number of variables                        4   ->      4
j variable (2 values)               tradeflow   ->   (dropped)
xij variables:
                                         Yr   ->   Yrexports Yrimports
─────────────────────────────────────────────────────────────
```

Transforming the data to wide form once again, the i() option contains country and year because those are the desired identifiers on each observation of the target dataset. We specify that tradeflow is the j() variable for reshape, indicating that it is a string variable. The data now have the desired structure.

Although we have illustrated this double-reshape transformation with only a few countries, years, and variables, the technique generalizes to any number of each.

As another example of successive applications of reshape, consider the World Bank's World Development Indicators dataset.[5] Their extract program generates a comma-separated–values (CSV) database, readable by Excel or Stata, but the structure of those data hinders analysis as panel data. For a recent year, the header line of the CSV file is

---

5. Available for purchase at http://econ.worldbank.org.

```
"Series code","Country Code","Country Name","1960","1961","1962","1963",
"1964","1965","1966","1967","1968","1969","1970","1971","1972","1973",
"1974","1975","1976","1977","1978","1979","1980","1981","1982","1983",
"1984","1985","1986","1987","1988","1989","1990","1991","1992","1993",
"1994","1995","1996","1997","1998","1999","2000","2001","2002","2003","2004"
```

That is, each row of the CSV file contains a variable and country combination, with the columns representing the elements of the time series.[6]

Our target dataset structure is that appropriate for panel-data modeling with the variables as columns and rows labeled by country and year. Two applications of `reshape` will again be needed to reach the target format. We first `insheet` the data and transform the triliteral country code into a numeric code with the country codes as labels:

```
. insheet using wdiex.raw, comma names
. encode countrycode, generate(cc)
. drop countrycode
```

We then must deal with the time-series variables being named v4–v48, because the header line provided invalid Stata variable names (numeric values) for those columns. We use `rename`, as described in section 3.6, to change v4 to d1960, v5 to d1961, and so on. We use a technique for macro expansion, involving the equal sign, by which an algebraic expression can be evaluated within a macro. Here the target variable name contains the string 1960, 1961, ..., 2004:

```
. forvalues i=4/48 {
         rename v'i' d'=1956+'i''
. }
```

We now are ready to carry out the first `reshape`. We want to identify the rows of the reshaped dataset by both the country code (`cc`) and the variable name (`seriescode`). The `reshape long` command will transform a fragment of the World Development Indicators dataset containing two series and four countries:

```
. reshape long d, i(cc seriescode) j(year)
(note: j = 1960 1961 1962 1963 1964 1965 1966 1967 1968 1969 1970 1971 1972
> 1973 1974 1975 1976 1977 1978 1979 1980 1981 1982 1983 1984 1985 1986 1987
> 1988 1989 1990 1991 1992 1993 1994 1995 1996 1997 1998 1999 2000 2001 2002
> 2003 2004)
```

Data		wide	->	long
Number of obs.		7	->	315
Number of variables		48	->	5
j variable (45 values)			->	year
xij variables:				
	d1960 d1961 ... d2004		->	d

---

6. A variation occasionally encountered will resemble this structure, but with time periods in reverse chronological order. The solution given can be used to deal with this problem as well.

```
. list in 1/15
```

	cc	seriesc~e	year	countryname	d
1.	AFG	adjnetsav	1960	Afghanistan	.
2.	AFG	adjnetsav	1961	Afghanistan	.
3.	AFG	adjnetsav	1962	Afghanistan	.
4.	AFG	adjnetsav	1963	Afghanistan	.
5.	AFG	adjnetsav	1964	Afghanistan	.
6.	AFG	adjnetsav	1965	Afghanistan	.
7.	AFG	adjnetsav	1966	Afghanistan	.
8.	AFG	adjnetsav	1967	Afghanistan	.
9.	AFG	adjnetsav	1968	Afghanistan	.
10.	AFG	adjnetsav	1969	Afghanistan	.
11.	AFG	adjnetsav	1970	Afghanistan	-2.97129
12.	AFG	adjnetsav	1971	Afghanistan	-5.54518
13.	AFG	adjnetsav	1972	Afghanistan	-2.40726
14.	AFG	adjnetsav	1973	Afghanistan	-.188281
15.	AFG	adjnetsav	1974	Afghanistan	1.39753

The rows of the data are now labeled by year, but one problem remains: all the variables for a given country are stacked vertically. To unstack the variables and put them in shape for `xtreg` ([XT] **xtreg**), we must carry out a second **reshape** to spread the variables across the columns, specifying `cc` and `year` as the *i* variables and *j* as `seriescode`. Because that variable has string content, we use the **string** option.

```
. reshape wide d, i(cc year) j(seriescode) string
(note: j = adjnetsav adjsavC02)
Data                              long    ->    wide

Number of obs.                     315    ->     180
Number of variables                  5    ->       5
j variable (2 values)       seriescode    ->    (dropped)
xij variables:
                                     d    ->    dadjnetsav dadjsavC02

. order cc countryname

. tsset cc year
       panel variable:  cc (strongly balanced)
        time variable:  year, 1960 to 2004
```

After this transformation, the data are now in shape for **xt** modeling, tabulation, or graphics.

As illustrated here, the **reshape** command can transform even the most inconvenient data structure into the structure needed for your research. It may take more than one application of **reshape** to get to where you need to be, but **reshape** can do the job.

6.3 Handling time-series data effectively

The problem. Daily data are often generated by nondaily processes: for instance, the financial markets are closed on weekends and many holidays.[7] Stata's time-series date schemes ([U] **24 Dealing with dates and times**) allow for daily (and intradaily) data, but gaps in time series can be problematic. A model that uses lags or differences will lose multiple observations every time a gap appears, discarding many of the original data points. Analysis of "business-daily" data often proceeds by assuming that Monday follows Friday, and so on. At the same time, we usually want data to be placed on Stata's time-series calendar so that useful tools such as the `tsline` ([TS] **tsline**) graph will work and label data points with readable dates.

The solution. In a 2006 Stata Users Group presentation in Boston, David Drukker spoke to this point. His solution was to generate two date variables, one containing the actual calendar dates and another numbering successive available observations consecutively. The former variable (`caldate`) is `tsset` when the calendar dates are to be used, while the latter variable (`seqdate`) is `tsset` when statistical analyses are to be performed.

To illustrate, I downloaded the daily data on the three-month U.S. Treasury Bill rate with Drukker's `freduse` command (Drukker 2006) and retained the August 2005–present data for analysis.[8]

```
. clear
. freduse DTB3
(14263 observations read)
. rename daten caldate
. tsset caldate
        time variable:  caldate, 04jan1954 to 03sep2008, but with gaps
                delta:  1 day
. keep if tin(1aug2005,)
(13455 observations deleted)
. label var caldate date
. tsline DTB3
```

These data do not contain observations for weekends and are missing for U.S. holidays. You may not want to drop the observations containing missing data, though, because we may have complete data for other variables. For instance, exchange rate data are available every day. If there were no missing data in our series—only missing observations—you could use Drukker's suggestion and `generate seqdate = _n`. Because there are observations for which `DTB3` is missing, you must follow a more complex route:

---

7. This recipe is adapted from Stata tip 40 (Baum 2007). I am grateful to David M. Drukker for his contributions to this Stata tip.

8. If you run this example, you will retrieve a longer time series and produce different output than what is shown.

```
. quietly generate byte notmiss = !missing(DTB3)
. quietly generate seqdate = cond(notmiss,sum(notmiss),.)
. tsset seqdate
        time variable:  seqdate, 1 to 776
                delta:  1 unit
```

The variable `seqdate` is created as the sequential day number for every nonmissing day and is itself missing when DTB3 is missing. This allows us to use this variable in `tsset` and then to use time-series operators (see [U] **11.4 varlists**) in `generate` or estimation commands such as `regress`. You may want to display the transformed data (or results from estimation, such as predicted values) on a time-series graph. We can view the data graphically with `tsline`. To do this, we just revert to the other `tsset` declaration:

```
. quietly generate dDTB3 = D.DTB3
. label var dDTB3 "Daily change in 3-mo Treasury rate"
. tsset caldate
        time variable:  caldate, 01aug2005 to 03sep2008, but with gaps
                delta:  1 day
. tsline dDTB3, yline(0)
```

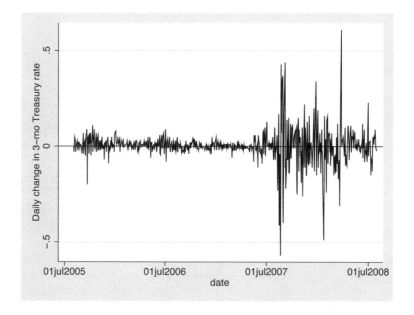

If you retain both the `caldate` and `seqdate` variables in our saved dataset, you will always be able to view these data either on a time-series calendar or as a sequential series.

In my research, I need to know how many calendar days separate each observed point (1 for Thursday–Friday but 3 for Friday–Monday) and then sum the dDTB3 by month, weighting each observation by the square root of the days of separation:

```
. tsset seqdate
        time variable:  seqdate, 1 to 776
                delta:  1 unit
. quietly generate dcal = D.caldate if !missing(seqdate)
. quietly generate month = mofd(caldate) if !missing(seqdate)
. format %tm month
. sort month (seqdate)
. quietly by month: generate adjchange = sum(dDTB3/sqrt(dcal))
. quietly by month: generate sumchange = adjchange if _n==_N & !missing(month)
. list month sumchange if !missing(sumchange), sep(0) noobs
```

month	sumchange
2005m8	-.003812
2005m9	-.0810769
2005m10	.2424316
2005m11	-.063453
2005m12	.096188
2006m1	.2769615
2006m2	.099641
2006m3	.0142265
2006m4	.0938675
2006m5	.0350555
2006m6	.0327906
2006m7	.0304485
2006m8	-.083812
2006m9	-.123094
2006m10	.1338675
2006m11	-.0428446
2006m12	-.015
2007m1	.0523205
2007m2	.0023205
2007m3	-.1142265
2007m4	-.0973205
2007m5	-.2245855
2007m6	.012376
2007m7	.0089786
2007m8	-.9057735
2007m9	-.4818504
2007m10	-.0882864
2007m11	-.9123026
2007m12	-.029279
2008m1	-1.177312
2008m2	-.2422094
2008m3	-.6493376
2008m4	-.000718
2008m5	.377735
2008m6	-.1321539
2008m7	-.201188
2008m8	-.0783419
2008m9	-.02

6.4   reshape to perform rowwise computation

The problem. Suppose, in the context of a panel (longitudinal) dataset, some attributes are stored in separate Stata variables. We want to perform some computations across those Stata variables.

The solution. In this example, taken from a question on Statalist, the user has a panel of elections identified by U.S. state and year with a set of variables listing names of candidates (given names only in this fictitious example):

```
. use cb5a, clear
. list
```

	state	year	cand1	cand2	cand3
1.	TX	2001	Tom	Dick	Harry
2.	TX	2005	Dick	Jane	Harry
3.	MA	2002	John	Jim	Jack
4.	MA	2003	Jim	Jill	Joan
5.	MA	2005	John	Jill	Jim

We want to compute the number of candidates who have stood for election in each state over the available years. Several candidates have stood more than once. This problem is solved by placing the data in Stata's long form with `reshape long`. The unique row identifier is constructed from the state and year variables with the `concat()` function from `egen`:

```
. egen rowid = concat(state year)
. reshape long cand, i(rowid) j(candnr)
(note: j = 1 2 3)
Data                                wide   ->   long
─────────────────────────────────────────────────────
Number of obs.                         5   ->      15
Number of variables                    6   ->       5
j variable (3 values)                      ->   candnr
xij variables:
                   cand1 cand2 cand3   ->   cand
─────────────────────────────────────────────────────
```

```
. list
```

	rowid	candnr	state	year	cand
1.	MA2002	1	MA	2002	John
2.	MA2002	2	MA	2002	Jim
3.	MA2002	3	MA	2002	Jack
4.	MA2003	1	MA	2003	Jim
5.	MA2003	2	MA	2003	Jill
6.	MA2003	3	MA	2003	Joan
7.	MA2005	1	MA	2005	John
8.	MA2005	2	MA	2005	Jill
9.	MA2005	3	MA	2005	Jim
10.	TX2001	1	TX	2001	Tom
11.	TX2001	2	TX	2001	Dick
12.	TX2001	3	TX	2001	Harry
13.	TX2005	1	TX	2005	Dick
14.	TX2005	2	TX	2005	Jane
15.	TX2005	3	TX	2005	Harry

```
. save cb5along, replace
file cb5along.dta saved
```

Alternatively, we could have created the `rowid` variable with

```
generate rowid = state + string(year)
```

The `concat()` function has the advantage that it will perform any needed string conversions "on the fly".

We now use by-groups and the `egen total()` function to produce the desired state-level tally. We return the data, now containing this state-level statistic, to the wide form:

```
. bysort state cand: generate byte last=(_n == _N)
. egen totcand = total(last), by(state)
. drop last
. quietly reshape wide
. list, sepby(state)
```

	rowid	cand1	cand2	cand3	state	year	totcand
1.	MA2002	John	Jim	Jack	MA	2002	5
2.	MA2003	Jim	Jill	Joan	MA	2003	5
3.	MA2005	John	Jill	Jim	MA	2005	5
4.	TX2001	Tom	Dick	Harry	TX	2001	4
5.	TX2005	Dick	Jane	Harry	TX	2005	4

If only a table of states and the number of distinct candidates was needed, the `drop` and `egen` statements could be replaced with `tabstat` applied to the `last` variable:

```
. reshape long cand, i(rowid) j(candnr)
. bysort state cand: generate byte last=(_n == _N)
. tabstat last, stat(sum) nototal by(state)
Summary for variables: last
      by categories of: state

state |       sum
------+----------
   MA |         5
   TX |         4
------+----------
```

As an alternative, we could use the `tag()` function from `egen`, which tags one observation in each distinct group defined by the `egen` varlist. Tagged observations are flagged by 1, while others have a value of 0. Thus we could calculate the desired total for each state:

```
. use cb5along, clear
. egen indivcount = tag(state cand)
. egen totcand = total(indivcount), by(state)
. tabstat totcand, nototal by(state)
Summary for variables: totcand
      by categories of: state

state |      mean
------+----------
   MA |         5
   TX |         4
------+----------
```

Although we illustrate this concept of using `reshape long` in the context of panel data, it is equally useful in many cross-section datasets. Many applications that involve the use of rowwise operators in a spreadsheet can be better handled in Stata by transforming the data to the long form and transforming it back to wide form if desired.

6.5 Adding computed statistics to presentation-quality tables

The problem. Assume you need to produce presentation-quality tables from estimation output that include quantities not available in the saved results. Stata's `mfx` ([R] **mfx**) command can compute many measures of marginal effects, and its computed quantities can be stored with the saved estimates via `estimates store`. However, there is no way to access those stored quantities with `estimates table`.

The solution. A solution to this problem, removing the need for manual editing of the tables, is available from Ben Jann's `estout` package, described in section 5.4.1. First, the quantities to be retrieved from the estimates must be made into additional scalars and stored with the estimates by using estadd. Then those scalars can be requested in the esttab output as additional statistics and can be given appropriate labels.

In the example below, we estimate two forms of an equation from the airquality dataset (Rabe-Hesketh and Everitt 2006) using ordinary least squares (regress) and instrumental variables (ivregress 2sls). The variable of interest is so2, the sulphur dioxide concentration in each city's atmosphere. This pollutant is modeled as a function of the city's population, number of manufacturing establishments, and average wind speed. In the instrumental-variables estimation, we consider population as possibly endogenous, and instrument it with average temperature and mean precipitation.

After each estimation, we use mfx compute, eyex to compute the elasticities of so2 with respect to each regressor.[9] Because mfx is an e-class command, its results are accessible: here in matrices e(Xmfx_eyex) and e(Xmfx_se_eyex) for the elasticity-point estimates and their standard errors, respectively. The [1,1] element of each matrix refers to pop, because it is the first regressor. We store that element of the matrix as a scalar, nu for the elasticity and nuse for its standard error, before storing the estimates.

```
. use airquality, clear
. // estimate with OLS
. qui regress so2 pop manuf wind
. mfx compute, eyex
Elasticities after regress
      y  = Fitted values (predict)
         =    30.04878
```

variable	ey/ex	Std. Err.	z	P>|z|	[	95% C.I.	]	X
pop	-1.158096	.30621	-3.78	0.000	-1.75826	-.55793		608.61
manuf	1.291947	.25205	5.13	0.000	.797928	1.78597		463.098
wind	-.4346578	.55942	-0.78	0.437	-1.53111	.661794		9.4439

```
. matrix nu =  e(Xmfx_eyex)
. estadd scalar nu = nu[1, 1]
. matrix nuse = e(Xmfx_se_eyex)
. estadd scalar nuse = nuse[1, 1]
. estimates store one
. // reestimate using IV
. qui ivregress 2sls so2 (pop = temp precip) manuf wind
```

---

9. Elasticities are discussed in section 9.5.

```
. mfx compute, eyex

Elasticities after ivregress
     y  = Fitted values (predict)
        =    30.04878
```

variable	ey/ex	Std. Err.	z	P>|z|	[	95% C.I.	]	X
pop	-2.599056	.83744	-3.10	0.002	-4.24041	-.957699		608.61
manuf	2.37258	.64775	3.66	0.000	1.10301	3.64215		463.098
wind	-.5755015	.68932	-0.83	0.404	-1.92655	.775546		9.4439

```
. matrix nu =  e(Xmfx_eyex)
. estadd scalar nu = nu[1, 1]
. matrix nuse = e(Xmfx_se_eyex)
. estadd scalar nuse = nuse[1, 1]
. estimates store two
```

We now are ready to use esttab to present the estimates. The stat() option allows us to name the nu and nuse scalars as added statistics and to give them appropriate labels.

```
. esttab one two, mtitles("OLS" "IV") stat(nu nuse, labels("Elas.(pop)" "SE"))
```

	(1) OLS	(2) IV
pop	-0.0572*** (-3.97)	-0.128** (-3.26)
manuf	0.0838*** (5.63)	0.154*** (3.93)
wind	-1.383 (-0.78)	-1.831 (-0.84)
_cons	39.09* (2.32)	54.15* (2.47)
Elas.(pop)	-1.158	-2.599
SE	0.306	0.837

```
t statistics in parentheses
* p<0.05, ** p<0.01, *** p<0.001
```

Using additional options with esttab, you could create a table in LaTeX, HTML, CSV, rich text, or tab-delimited format.

6.5.1   Presenting marginal effects rather than coefficients

When presenting the results from limited dependent variable estimation commands such as probit ([R] probit) or logit ([R] logit), it is usual practice to display the effects on the probability of observing unity rather than to display the raw coefficient estimates. The latter estimates reflect $\partial I/\partial X_j$, that is, the change in the *index variable*, or *latent variable*, resulting from a change in X_j. Stata can readily compute the quantities $\partial \Pr(y = 1)/\partial X_j$ with `mfx`, but how can you present them in a publication-quality table for a set of equations?

The `margin` option for Ben Jann's `esttab` and `estout` commands provides this capability. With that option specified, display of the raw coefficient estimates is suppressed, while the desired marginal effects and their standard errors are displayed instead.

As an example, let's generate an indicator variable for cities with sulphur dioxide concentration above and below the median level, and estimate two forms of a probit model:

```
. qui summarize so2, detail
. generate hiso2 = (so2 > r(p50)) & !missing(so2)
. summarize precip, meanonly
. generate hiprecip = (precip > r(mean)) & !missing(precip)
. qui probit hiso2 pop manuf
. mfx compute
Marginal effects after probit
     y  = Pr(hiso2) (predict)
        =  .50262647
```

variable	dy/dx	Std. Err.	z	P>|z|	[95% C.I.]	X
pop	−.0008596	.00055	−1.57	0.115	−.00193 .000211	608.61
manuf	.001419	.00068	2.08	0.037	.000084 .002754	463.098

```
. estimates store three
. qui probit hiso2 pop manuf hiprecip
. mfx compute
Marginal effects after probit
     y  = Pr(hiso2) (predict)
        =  .503741
```

variable	dy/dx	Std. Err.	z	P>|z|	[95% C.I.]	X
pop	−.0009268	.00056	−1.67	0.095	−.002016 .000163	608.61
manuf	.0015382	.00071	2.17	0.030	.000151 .002925	463.098
hiprecip*	.1562071	.17248	0.91	0.365	−.181851 .494265	.560976

```
(*) dy/dx is for discrete change of dummy variable from 0 to 1
. estimates store four
```

It is straightforward to present the marginal effects by using `esttab`:

(Continued on next page)

```
. esttab three four, margin
```

	(1) hiso2	(2) hiso2
pop	-0.000860 (-1.57)	-0.000927 (-1.67)
manuf	0.00142* (2.08)	0.00154* (2.17)
hiprecip (d)		0.156 (0.91)
N	41	41

```
Marginal effects; t statistics in parentheses
 (d) for discrete change of dummy variable from 0 to 1
* p<0.05, ** p<0.01, *** p<0.001
```

As in the prior example, this tabular output could be presented in any of several formats, including the default Stata Markup and Control Language.

6.6   Generating time-series data at a lower frequency

The problem. Assume your data are on a time-series calendar but you want to express them at a lower time-series frequency. For instance, you may have monthly data, but you would like to present them as quarterly data. You might want to retain only one observation per quarter (such as the quarter-end value), sum the values over the quarter, or average the values over the quarter, depending on the type of data.

The solution. To solve this problem, I will make use of a handy tool, `tsmktim` (Baum and Wiggins 2000), that will compute a time-series calendar variable given the first observation's value.[10] The `tsmktim` command also automatically applies `tsset` with the appropriate frequency so that time-series operators and functions can be used. It is your responsibility to ensure that the data are correctly sorted in time order and have no missing time periods before invoking `tsmktim`.

We access the `air2` dataset of airline passenger boardings. This dataset contains the variable `time` taking on the values 1949.0, 1949.083, 1949.167, etc., but no proper Stata calendar variable. We generate such a variable, `ym`, with `tsmktim`. The `start(1949m1)` option specifies that these data are to be placed on a monthly calendar. To compute the month of each observation, we use the Stata `month()` and `dofm()` date functions. The last month of each quarter can be found with the `mod()` function; quarter-end month numbers are evenly divisible by 3, providing the indicator variable `eoq`.

---

10. The most recent version of `tsmktim` is available from the Statistical Software Components archive.

```
. use air2, clear
(TIMESLAB: Airline passengers)

. * put a time-series calendar on the data
. tsmktim ym, start(1949m1)
        time variable:  ym, 1949m1 to 1960m12
                delta:  1 month

. * get the month number
. generate mnr = month(dofm(ym))

. * find the end-of-quarter months
. generate eoq = (mod(mnr, 3) == 0)

. list ym mnr eoq air in 1/16, sep(4) noobs
```

ym	mnr	eoq	air
1949m1	1	0	112
1949m2	2	0	118
1949m3	3	1	132
1949m4	4	0	129
1949m5	5	0	121
1949m6	6	1	135
1949m7	7	0	148
1949m8	8	0	148
1949m9	9	1	136
1949m10	10	0	119
1949m11	11	0	104
1949m12	12	1	118
1950m1	1	0	115
1950m2	2	0	126
1950m3	3	1	141
1950m4	4	0	135

We now can consider each of the proposed tasks. To retain only the quarter-end observations, we need to `keep` only the values indicated by `eoq` and use `tsmktim` on the new series to define it as quarterly data:

```
. // keep only last month of quarter
. preserve

. keep if eoq
(96 observations deleted)

. tsmktim yq, start(1949q1)
        time variable:  yq, 1949q1 to 1960q4
                delta:  1 quarter
```

(Continued on next page)

```
. list yq air in 1/16, sep(4) noobs
```

yq	air
1949q1	132
1949q2	135
1949q3	136
1949q4	118
1950q1	141
1950q2	149
1950q3	158
1950q4	140
1951q1	178
1951q2	178
1951q3	184
1951q4	166
1952q1	193
1952q2	218
1952q3	209
1952q4	194

To sum the values over the quarter, we must compute the quarter associated with each month in Stata's calendar system by using the `qofd()` function. We can then use `collapse` to create the new dataset of quarterly values, each of which is the sum of the monthly values of `air`. We again use `tsmktim` to generate the new quarterly calendar variable.

```
. // sum the values over the quarter
. restore, preserve
. generate qtr = qofd(dofm(ym))
. format qtr %tq
. collapse (sum) airsum=air, by(qtr)
. tsmktim yq, start(1949q1)
        time variable:  yq, 1949q1 to 1960q4
                delta:  1 quarter
```

```
. list yq airsum in 1/16, sep(4) noobs
```

yq	airsum
1949q1	362
1949q2	385
1949q3	432
1949q4	341
1950q1	382
1950q2	409
1950q3	498
1950q4	387
1951q1	473
1951q2	513
1951q3	582
1951q4	474
1952q1	544
1952q2	582
1952q3	681
1952q4	557

If instead we wanted the average values over each quarter, we would merely apply collapse, which computes the mean as a default:

```
. // average the values over the quarter
. restore, preserve
. generate qtr = qofd(dofm(ym))
. format qtr %tq
. collapse airavg=air, by(qtr)
. tsmktim yq, start(1949q1)
        time variable:  yq, 1949q1 to 1960q4
                delta:  1 quarter
```

(*Continued on next page*)

```
. list yq airavg in 1/16, sep(4) noobs
```

yq	airavg
1949q1	120.667
1949q2	128.333
1949q3	144
1949q4	113.667
1950q1	127.333
1950q2	136.333
1950q3	166
1950q4	129
1951q1	157.667
1951q2	171
1951q3	194
1951q4	158
1952q1	181.333
1952q2	194
1952q3	227
1952q4	185.667

Because these tasks with time-series data are encountered commonly, a variation on the official `collapse` command is available in the `tscollap` routine (Baum 2000).[11] The syntax of `tscollap` mirrors that of `collapse`, with one additional mandatory option: the frequency to which the time series are to be collapsed. With `tscollap`, `to(q)` specifies that the monthly data are to be collapsed to quarterly. Alternatively, the `generate()` option can be used to name the time-series calendar variable that will be created. To illustrate,

```
. restore

. tsset ym
        time variable:  ym, 1949m1 to 1960m12
                delta:  1 month
. tscollap (last) aireoq=air (sum) airsum=air (mean) airavg=air, to(q) gen(yq)


Converting from M to Q

        time variable:  yq, 1949q1 to 1960q4
                delta:  1 quarter
```

---

11. The most recent version of `tscollap` is available from the Statistical Software Components archive.

```
. list in 1/16, sep(4) noobs
```

aireoq	airsum	airavg	yq
132	362	120.667	1949q1
135	385	128.333	1949q2
136	432	144	1949q3
118	341	113.667	1949q4
141	382	127.333	1950q1
149	409	136.333	1950q2
158	498	166	1950q3
140	387	129	1950q4
178	473	157.667	1951q1
178	513	171	1951q2
184	582	194	1951q3
166	474	158	1951q4
193	544	181.333	1952q1
218	582	194	1952q2
209	681	227	1952q3
194	557	185.667	1952q4

Here we see that all three operations on the `air` time series can be performed by a single invocation of `tscollap`.

For another recipe involving generating lower-frequency measures from time-series data, see section 8.3.

7 Do-file programming: Prefixes, loops, and lists

7.1 Introduction

Stata's facility for efficient data management, statistical analysis, and graphics is largely based on the ability to program repetitive tasks, avoid manual effort, and reduce errors. In developing your use of these techniques, you should always consider how you can get the computer to do what it does best. Any research project involves repetitive tasks of some sort, and a modest amount of time taken automating those tasks will greatly reduce your workload and likelihood of errors.

In this chapter, I discuss three programming features specifically applicable to repetitive tasks: Stata's prefix commands, loop commands, and list constructs.

7.1.1 What you should learn from this chapter

- How to use the by and xi prefix commands effectively
- How to use statsby and rolling to collect statistics
- How to perform Monte Carlo simulations
- How to compute bootstrap and jackknife estimates of precision
- How to perform loops with forvalues and foreach

7.2 Prefix commands

Stata's prefix commands perform repetitive tasks without explicit specification of the range of values over which the tasks are to be performed. This is of particular value because the distribution of values can be discontinuous or form no particular pattern. These commands often will serve your needs in performing tasks with minimal programming, but they also have limitations. For instance, the by ([D] **by**) prefix repetitively executes a single command but cannot perform multiple commands. In terms of ease of use, it is a good strategy to consider whether a prefix will serve your needs before turning to more complicated programming constructs.

7.2.1   The by prefix

Many Stata users are already familiar with the by prefix, as discussed in section 3.5. The syntax

> by *varlist* [, sort] : *command*

specifies that *command* is to be repeated for each distinct value of *varlist*. Consider the simplest case, where *varlist* contains one variable: an integer (categorical) variable if numeric or a string variable. If the data are not sorted by that variable (or, in general terms, by *varlist*) before using by, the sort option or the form bysort *varlist*: should be used. *command* will be repeated for each distinct value of the variable regardless of how many values that entails. Repetitions will follow the sorting order of the variable, whether string or numeric.

If *varlist* contains multiple variables, the same principle applies. The data must be sorted by *varlist*, or bysort can be used; *command* will be repeated for each distinct combination of values in *varlist*. Consider the bpress dataset containing blood pressure measurements of individuals categorized by gender (with indicator variable sex, 0=male) and one of three agegrp codes (1=30–45, 2=46–59, 3=60+).

```
. use bpress, clear
(fictional blood-pressure data)

. bysort sex agegrp: summarize bp
```

-> sex = Male, agegrp = 30-45

Variable	Obs	Mean	Std. Dev.	Min	Max
bp	40	119.95	12.53498	95	146

-> sex = Male, agegrp = 46-59

Variable	Obs	Mean	Std. Dev.	Min	Max
bp	40	128.15	13.90914	102	155

-> sex = Male, agegrp = 60+

Variable	Obs	Mean	Std. Dev.	Min	Max
bp	40	134.075	10.27665	116	155

-> sex = Female, agegrp = 30-45

Variable	Obs	Mean	Std. Dev.	Min	Max
bp	40	116.05	9.483995	101	138

-> sex = Female, agegrp = 46-59

Variable	Obs	Mean	Std. Dev.	Min	Max
bp	40	117.725	10.10201	97	140

-> sex = Female, agegrp = 60+

Variable	Obs	Mean	Std. Dev.	Min	Max
bp	40	127.475	12.02985	108	155

The `summarize` command will be executed for each combination of `sex` and `agegrp`. If a combination did not exist in the data (for instance, if there were no elderly males in our sample), it will merely be skipped.

Another way that you might approach this problem uses the `egen` function `group()`. This function creates a variable that takes on the values 1, 2, etc., for the groups formed by *varlist*. To create groups by gender and age group, we could use

```
. egen group = group(sex agegrp), label
. by group, sort: summarize bp
```

The `label` option causes the groups defined by `egen` to be labeled with the elements of each group so that the `summarize` headers will read

```
-> group = Male 30-45
```

and so on. For additional pointers on using the `by` prefix, see Cox (2002b).

The **by** prefix option **rc0** (return code zero), specified before the colon, is often useful. You might specify *command* to be an estimation command that would fail if insufficient observations existed in a particular by-group. This would normally abort the do-file, which might be appropriate if you expected that all by-groups could be successfully processed. Often, though, you know that there are some infeasible by-groups. Rather than having to program for that eventuality, you can specify **rc0** to force **by** to continue through the remaining by-groups.

7.2.2   The xi prefix

Although **by** repeats a command based on the distinct values of *varlist*, we often want to produce indicator (binary, or dummy) variables for each of those distinct values. The **xi** ([R] **xi**) prefix serves two roles. It can be used by itself to produce a set of indicator variables from one or more terms. More commonly, it is used in conjunction with another command in which those terms play a role. To understand its function, we must first define the terms that **xi** comprehends.

In the simplest form, a term can be just **i.** *varname*, where *varname* refers to a numeric or string categorical variable.[1] For instance, by typing

```
. xi i.agegrp
```

you are asking Stata to create two indicator variables, omitting the lowest value (which Stata refers to as *naturally coded* indicators). By default, the indicator variables will be named **_Iagegrp2** and **_Iagegrp3**, and their variable labels will contain the name of the categorical variable and its level. Using the **prefix(**string**)** option, you can specify the prefix to be something other than **_I**. You can refer to the set of indicator variables with a wildcard. For example,

```
. regress bp _I*
```

will include both **agegrp** indicator variables in the regression. When you use **xi**, all previously created variables with the same prefix (for example, **_I**) will be automatically dropped.

For a single term (such as **i.agegrp**), similar functionality is available by using the command

tabulate *varname*, **generate(**prefix**)**

with the distinction that **tabulate** will generate a full set of indicator variables.

More commonly, **xi** can be used as a prefix, preceding another Stata command to produce the needed indicator variables "on the fly":

```
xi: regress bp i.agegrp
```

---

1. **I.** *varname* (with a capital I) can also be used.

More than one term can be included in either format of `xi`. For example,

```
xi: regress income i.agegrp i.sex
```

would include two **agegrp** dummies and one **sex** dummy (denoting females) in the regression:

```
. xi: regress bp i.agegrp i.sex
i.agegrp          _Iagegrp_1-3       (naturally coded; _Iagegrp_1 omitted)
i.sex             _Isex_0-1          (naturally coded; _Isex_0 omitted)

      Source |       SS       df       MS              Number of obs =     240
-------------+------------------------------           F(  3,   236) =   23.98
       Model |  9559.19583      3  3186.39861           Prob > F      =  0.0000
    Residual |     31353.6    236  132.854237           R-squared     =  0.2336
-------------+------------------------------           Adj R-squared =  0.2239
       Total |  40912.7958    239  171.183246           Root MSE      =  11.526

----------------------------------------------------------------------------
          bp |      Coef.   Std. Err.      t    P>|t|     [95% Conf. Interval]
-------------+--------------------------------------------------------------
  _Iagegrp_2 |     4.9375   1.822459     2.71   0.007     1.347134    8.527866
  _Iagegrp_3 |     12.775   1.822459     7.01   0.000     9.184634    16.36537
      _Isex_1 |     -6.975   1.488031    -4.69   0.000    -9.906521   -4.043479
        _cons |   121.4875   1.488031    81.64   0.000      118.556     124.419
----------------------------------------------------------------------------
```

The advantages of `xi` are clearly evident when interactions of variables are to be used in a model. In the example above, you may want to allow **agegrp** and **sex** to have nonindependent effects on **bp**. This implies that two interaction terms must also be created, one for each combination of included **agegrp** and **sex** terms. This can be achieved by typing

```
xi: regress bp i.agegrp*i.sex
```

This model will include both the *main effects* of the qualitative factors **agegrp** and **sex** as well as the *interaction effects* of those two factors. The interaction terms will be named to specify their elements: for example, _IageXsex_2_1 is the interaction (**X**) of **agegrp** = 2 and **sex** = 1. The variable name **agegrp** has been shortened to ensure that the variable names obey length limits.

(Continued on next page)

```
. xi: regress bp i.agegrp*i.sex
i.agegrp          _Iagegrp_1-3         (naturally coded; _Iagegrp_1 omitted)
i.sex             _Isex_0-1            (naturally coded; _Isex_0 omitted)
i.age~p*i.sex     _IageXsex_#_#        (coded as above)
```

Source	SS	df	MS		
Model	9989.17083	5	1997.83417		
Residual	30923.625	234	132.152244		
Total	40912.7958	239	171.183246		

Number of obs = 240
F(5, 234) = 15.12
Prob > F = 0.0000
R-squared = 0.2442
Adj R-squared = 0.2280
Root MSE = 11.496

bp	Coef.	Std. Err.	t	P>|t|	[95% Conf. Interval]
_Iagegrp_2	8.2	2.570528	3.19	0.002	3.135666 13.26433
_Iagegrp_3	14.125	2.570528	5.49	0.000	9.060666 19.18933
_Isex_1	-3.9	2.570528	-1.52	0.131	-8.964334 1.164334
_IageXse~2_1	-6.525	3.635275	-1.79	0.074	-13.68705 .6370503
_IageXse~3_1	-2.7	3.635275	-0.74	0.458	-9.86205 4.46205
_cons	119.95	1.817638	65.99	0.000	116.369 123.531

A similar syntax can also be used to interact a categorical variable, such as `agegrp`, with a continuous variable. If you had a measurement of blood pressure from the previous year's examination (`bp0`), you could use the syntax

```
xi: regress bp i.agegrp*bp0
```

to fit a model where each `agegrp` has its own constant term and slope. The coefficient on `bp0` is the effect of that variable for `agegrp` = 1, and the coefficients on `_IageXbp0_2` and `_IageXbp0_3` are the contrasts between the slopes of `agegrp` = 2, 3 and `agegrp` = 1, respectively.

```
. xi: regress bp i.agegrp*bp0
i.agegrp          _Iagegrp_1-3         (naturally coded; _Iagegrp_1 omitted)
i.agegrp*bp0      _IageXbp0_#          (coded as above)
```

Source	SS	df	MS		
Model	39135.1347	5	7827.02693		
Residual	1777.66117	234	7.59684259		
Total	40912.7958	239	171.183246		

Number of obs = 240
F(5, 234) = 1030.30
Prob > F = 0.0000
R-squared = 0.9565
Adj R-squared = 0.9556
Root MSE = 2.7562

bp	Coef.	Std. Err.	t	P>|t|	[95% Conf. Interval]
_Iagegrp_2	-6.916522	4.092998	-1.69	0.092	-14.98036 1.147314
_Iagegrp_3	-.4387195	4.423308	-0.10	0.921	-9.153316 8.275876
bp0	.9269775	.0262564	35.30	0.000	.8752485 .9787066
_IageXbp0_2	.0527261	.0352761	1.49	0.136	-.0167733 .1222255
_IageXbp0_3	.0079293	.0368904	0.21	0.830	-.0647505 .0806091
_cons	13.58902	2.97342	4.57	0.000	7.730927 19.44712

One other type of term can be handled by `xi` for the interaction of categorical and continuous variables.[2] The command

```
xi: regress bp i.agegrp|bp0
```

specifies that the interaction variables should be included in the model but that the main effects of the categorical variable should not. The constant term of this regression pertains to all observations, whereas the slope with respect to `bp0` is allowed to differ by `agegrp` categories.

7.2.3   The statsby prefix

As mentioned previously, a limitation of the `by` prefix is that only one *command* can be specified. When performing an estimation command such as `regress`, you may want to both view the regression output for each by-group and save some of the statistics produced in each regression. These might include some of or all of the estimated coefficients, their standard errors, and summary statistics from the regression. This functionality is provided by the `statsby` ([D] **statsby**) prefix. With the syntax

statsby *exp_list*, by(*varlist*) [*options*] : *command*

command will be executed for each element of the by-list, as with `by`. However, the *exp_list* specifies that for each repetition, the results of evaluating one or more expressions are to be saved. By default, the dataset in memory will be replaced by the saved expressions. To preserve the existing contents of memory, you can use the saving(*filename*) option of `statsby` (before the colon) to specify the name of a new dataset.

Using the example from section 7.2.1, you could use the command

```
. statsby mean=r(mean) sd=r(sd) n=r(N), by(agegrp sex) saving(bpstats, replace):
> summarize bp
```

to produce a new dataset named `bpstats` with one observation per by-group. The dataset will contain the three variables specified in *exp_list*: the mean, standard deviation, and number of observations in each by-group. It will also contain one variable for each element of the by-list to identify from which combination of variables the observation was calculated (in this example, `agegrp` and `sex`).

---

2. If you are using `anova` ([R] **anova**), note that it has its own specific syntax for defining interaction effects.

```
. statsby mean=r(mean) sd=r(sd) n=r(N), by(agegrp sex) saving(bpstats, replace):
> summarize bp
(running summarize on estimation sample)

      command:  summarize bp
         mean:  r(mean)
           sd:  r(sd)
            n:  r(N)
           by:  agegrp sex

Statsby groups
 ───────┼─── 1 ───┼─── 2 ───┼─── 3 ───┼─── 4 ───┼─── 5
 ......

. use bpstats, clear
(statsby: summarize)

. list, sepby(agegrp)
```

	agegrp	sex	mean	sd	n
1.	30-45	Male	119.95	12.53498	40
2.	30-45	Female	116.05	9.483994	40
3.	46-59	Male	128.15	13.90914	40
4.	46-59	Female	117.725	10.10201	40
5.	60+	Male	134.075	10.27665	40
6.	60+	Female	127.475	12.02985	40

The total option of statsby specifies that an additional observation should be generated for the entire dataset specified by *command*. The values of the by-variables are set to missing for the observation corresponding to the total.

7.2.4   The rolling prefix

The statsby prefix computes statistics for *nonoverlapping* subsamples. In the context of time-series data, you often want to compute statistics for *overlapping* subsamples: so-called *rolling window* estimation using the rolling ([TS] **rolling**) prefix. For example, say you want to forecast a variable for several time periods, using only the observations available at the time to make a one-period-ahead forecast. The window can be an ever-widening window, including more observations over time (referred to as *recursive* estimation), or the window width can be fixed so that a chosen number of past time periods are considered in each computation. In the latter case, the number of available observations in each window can vary because of holidays, weekends, etc., for daily data.[3]

Like the statsby prefix, the rolling prefix can create a new dataset of results from the specified statistical command. With the syntax

rolling *exp_list* [*if*] [*in*] [, *options*]: *command*

---

3. The rolling prefix also supports *reverse recursive* analysis, where the endpoint is held fixed and the starting period varies.

command will be executed as specified by the options. For instance, if `window(12)` is specified, 12 time periods will be included in each estimation. *exp_list* are the expressions to be saved for each repetition. By default, the dataset in memory will be replaced by the saved expressions. To preserve the existing contents of memory, you can use the `saving(`*filename*`)` option of `rolling` (before the colon) to specify the name of a new dataset.

To illustrate, let's estimate a quantile regression (`qreg`, [R] **qreg**) of the daily return on IBM stock on the S&P 500 index return, using a 90-day window. This regression represents the celebrated Capital Asset Pricing Model (CAPM) in which the slope coefficient measures the sensitivity of the stock's return to that of the market (the so-called CAPM beta). Finance theory predicts that stocks with higher betas are more risky and require a higher return.

```
. use ibm, clear
(Source: Yahoo! Finance)
. rolling _b _se, window(90) saving(capm, replace) nodots: qreg ibm spx
file capm.dta saved
```

We can now examine the resulting Stata data file created by `rolling`. The `start` and `end` variables indicate the endpoints of each moving window. The variable of particular interest is `_b_spx`, the coefficient of the S&P 500 index return.

```
. use capm, clear
(rolling: qreg)
. label var _b_spx "quantile reg IBM beta"
. label var end "date"
. list in 1/5
```

	start	end	_b_spx	_b_cons	_se_spx	_se_cons
1.	02jan2003	01apr2003	1.066401	.1961641	.0866684	.1246594
2.	03jan2003	02apr2003	1.123538	.2001785	.0668068	.0978385
3.	04jan2003	03apr2003	1.123538	.2001785	.0667553	.097853
4.	05jan2003	04apr2003	1.094441	.1730048	.0870072	.126642
5.	06jan2003	05apr2003	1.094441	.1730048	.0870072	.126642

You could graph the movements of the rolling CAPM beta coefficients:

```
. tsset end
        time variable:  end, 01apr2003 to 16dec2004
                delta:  1 day
. format end %tdMonYY
. tsline _b_spx, yline(1) saving(fig7_1, replace) scheme(s2mono)
(file fig7_1.gph saved)
```

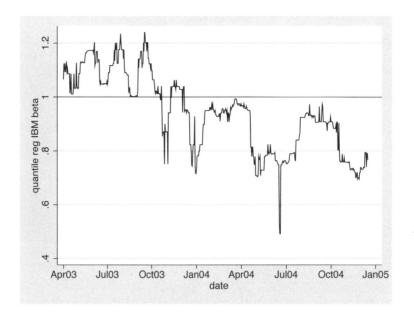

Figure 7.1. Rolling robust regression coefficients

In the CAPM, a beta of one implies that the volatility of the stock price is equal to that of the market index. Values higher than one imply that the stock price is more volatile than the market index, and values lower than one imply that the stock price is less volatile than the market index. From figure 7.1, it appears that IBM shares became less sensitive to stock market movements over the period.

7.2.5   The simulate and permute prefix

The simulate ([R] **simulate**) prefix provides Stata's facilities for Monte Carlo simulation. Like the other prefixes, simulate repeatedly executes a specified command. Here the reps() option specifies how many repetitions of the command are to be performed, each one producing a set of results in *exp_list*. With the formal syntax

simulate *exp_list*, reps(#) [*options*]: *command*

`simulate` works similarly to other prefix commands, with one important difference: *command* often will refer to a user-written program or ado-file.[4] You could, for example, simulate the behavior of several percentiles of a randomly distributed random variable with the following program. Here the program you must write is formulaic: it clears memory, sets the number of observations in the dataset, creates a random variable, and generates its descriptive statistics. Those statistics are passed back to `simulate` as scalars through the saved results (see section 5.3).

```
. type simpctile.ado
capture program drop simpctile
program simpctile, rclass
        version 10.1
        drop _all
        set obs 200
        generate z = rnormal()
        summarize z, detail
        return scalar p25 = r(p25)
        return scalar p50 = r(p50)
        return scalar p75 = r(p75)
end
```

Once we have written this program and stored it in `simpctile.ado`, we are ready to execute it, specifying that we want to perform 1,000 repetitions. Setting the `seed` of Stata's pseudorandom-number generator guarantees that the results will be identical for multiple runs of the experiment (preferable during debugging of our program).

```
. clear
. set seed 2007062926
. simulate p25=r(p25) p50=r(p50) p75=r(p75), reps(1000) nodots
> saving(pctiles, replace): simpctile
      command:  simpctile
          p25:  r(p25)
          p50:  r(p50)
          p75:  r(p75)
```

We can examine the distribution of these three percentiles of our $N(0, 1)$ random variables statistically and graph the distribution of the sample median in a histogram:

```
. use pctiles, clear
(simulate: simpctile)
. summarize
```

Variable	Obs	Mean	Std. Dev.	Min	Max
p25	1000	-.6727567	.0944072	-.9533168	-.3055624
p50	1000	-.001541	.0884508	-.2731787	.3048038
p75	1000	.6694791	.0912676	.3836729	.9612866

```
. label var p50 "Sample median"
. histogram p50, normal saving(fig7_2, replace) scheme(s2mono)
(bin=29, start=-.2731787, width=.01993043)
(file fig7_2.gph saved)
```

---

4. I defer a full discussion of writing an ado-file to chapter 11.

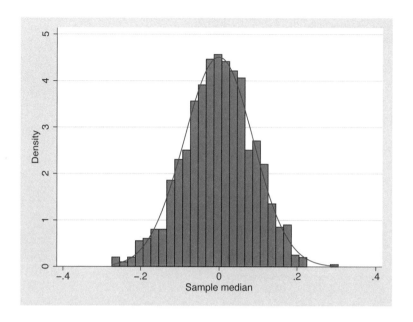

Figure 7.2. Distribution of the sample median via Monte Carlo simulation

Figure 7.2 illustrates that the distribution of the sample median is slightly skewed relative to that of a standard normal variate in this example. The overlay of the standard normal distribution is produced by the `normal` option. This is confirmed by the descriptive statistics for `p50`, in which nonzero skewness is apparent:

```
. summarize p50, detail

                        Sample median

          Percentiles      Smallest
 1%      -.2127989       -.2731787
 5%      -.1531975       -.2702311
10%      -.1147803        -.261639     Obs                1000
25%      -.0603764       -.2616217     Sum of Wgt.        1000

50%      -.0006763                     Mean            -.001541
                          Largest      Std. Dev.       .0884508
75%       .0589691         .217591
90%       .1119633        .2185251     Variance        .0078235
95%       .1406254        .2239791     Skewness       -.1278818
99%       .1871577        .3048038     Kurtosis        2.971572
```

Alternatively, we can examine the distribution of a sample median with a Q–Q plot (see figure 7.3), using `qnorm` ([R] **diagnostic plots**):

```
. use pctiles, clear
(simulate: simpctile)

. qnorm p50, saving(fig7_3, replace) scheme(s2mono)
(file fig7_3.gph saved)
```

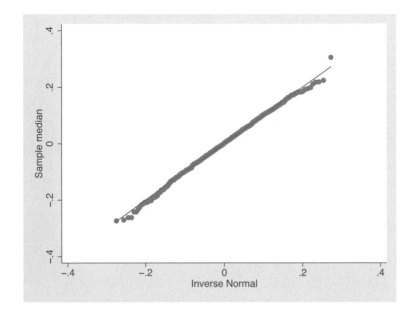

Figure 7.3. Q–Q plot of the distribution of the sample median

A related prefix, permute ([R] **permute**), is also based on Monte Carlo methods. It is used to estimate *p*-values for permutation tests from simulation results, where the randomness refers to reordering of the observations of a particular variable.

7.2.6   The bootstrap and jackknife prefixes

The bootstrap ([R] **bootstrap**) prefix is used to perform *bootstrap estimation*. Like simulate, it executes a command for a certain number of repetitions, gathering statistics in *exp_list* (by default, the estimated coefficient vector, _b, from a standard e-class command). The crucial difference is that while simulate draws (pseudo)random values from a specified distribution, bootstrap draws random samples *with replacement* from your dataset. For example, you can run a regression in which you doubt the classical assumption of normally distributed errors. The assumption of independent and identically distributed errors underlies the classical calculation of the covariance matrix of the regression coefficients. You could develop an alternative measure of the coefficients' precision by computing their *bootstrap standard errors*. These statistics are computed by randomly sampling (with replacement) from the regression residuals, taking their empirical distribution as a given. Alternatively, you can randomly sample (with replacement) paired observations on the dependent and independent variables. For estimation commands such as `regress`, you can use the vce(bootstrap) option to produce this estimator of the covariance matrix. However, if you are using a nonestimation (r-class) command, such as `summarize` or `correlate`, or if you are writing your own program as a command, the `bootstrap` prefix is essential. The formal syntax is

bootstrap $\big[$ *exp_list* $\big]$ $\big[$, *options* $\big]$: *command*

command can be a standard statistical command, or it can refer to a user-written program. By default, `bootstrap` conducts 50 replications. The bootstrap samples are, by default, size _N, the current sample size. They can be either smaller or larger than the sample size with the `size()` option.

Let's consider the case of computing a bootstrap standard error for a two-sample *t* test of sample means. We use the `bpress` dataset of fictitious blood pressure measurements and test whether mean blood pressure is affected by the patient's gender. For comparison, we present the standard `ttest` results as well as those produced by `bootstrap`. As shown, *exp_list* may contain expressions involving the return values of *command*.

```
. use bpress, clear
(fictional blood-pressure data)
. set seed 2007062926
. ttest bp, by(sex)
Two-sample t test with equal variances
```

Group	Obs	Mean	Std. Err.	Std. Dev.	[95% Conf. Interval]	
Male	120	127.3917	1.236031	13.54004	124.9442	129.8391
Female	120	120.4167	1.064357	11.65944	118.3091	122.5242
combined	240	123.9042	.8445493	13.0837	122.2405	125.5679
diff		6.975	1.631143		3.761679	10.18832

```
        diff = mean(Male) - mean(Female)                              t =    4.2761
Ho: diff = 0                                        degrees of freedom =      238

    Ha: diff < 0                  Ha: diff != 0                   Ha: diff > 0
 Pr(T < t) = 1.0000       Pr(|T| > |t|) = 0.0000         Pr(T > t) = 0.0000
```

```
. bootstrap diff=(r(mu_1) - r(mu_2)), nodots: ttest bp, by(sex)
```
Warning: Since ttest is not an estimation command or does not set e(sample),
           bootstrap has no way to determine which observations are used in
           calculating the statistics and so assumes that all observations are
           used. This means no observations will be excluded from the
           resampling because of missing values or other reasons.

           If the assumption is not true, press Break, save the data, and drop
           the observations that are to be excluded. Be sure that the dataset
           in memory contains only the relevant data.

```
Bootstrap results                                 Number of obs     =        240
                                                  Replications      =         50

        command:  ttest bp, by(sex)
           diff:  r(mu_1) - r(mu_2)
```

	Observed Coef.	Bootstrap Std. Err.	z	P>|z|	Normal-based [95% Conf. Interval]	
diff	6.975	1.271685	5.48	0.000	4.482544	9.467456

We see that the bootstrap standard error for the difference of means of 1.272 is considerably smaller than the classical standard error of 1.631, resulting in a narrower confidence interval.

You can also use the jackknife ([R] **jackknife**) prefix to calculate an alternative measure of the coefficients' precision. The concept of the jackknife is to execute *command* multiple times on the dataset, leaving out one observation each time. As with the bootstrap prefix, you usually will prefer to use the vce(jackknife) option on an estimation command that supports it. The usefulness of the jackknife prefix arises when you are using a nonestimation (r-class) command, such as summarize or correlate, or when you are writing your own program as a command.

7.2.7   Other prefix commands

As [U] **11.1.10 Prefix commands** indicates, there are three other prefix commands in official Stata: svy ([SVY] **svy**), nestreg ([R] **nestreg**), and stepwise ([R] **stepwise**). svy is the most widely used because it provides access to Stata's extensive list of commands for the handling of complex survey data. These datasets are not produced by simple random sampling from a population of interest. In contrast, their survey design uses an elaborate multistage sampling process, which is specified to Stata by the svyset ([SVY] **svyset**) command. Once the data have been svyset, many of Stata's standard estimation commands can be used following the svy prefix.

Four other commands are technically prefixes: capture, quietly, noisily, and version.[5] They are all useful for programming. For historical reasons, the first three do not require a trailing colon (:).

---

5. For more information on version, see http://www.stata.com/support/faqs/lang/version2.html, written by Nicholas J. Cox.

7.3   The forvalues and foreach commands

One of Stata's most powerful features is the ability to write a versatile Stata program without many repetitive statements. Many Stata commands contribute to this flexibility. As discussed in section 3.5, `egen` with a `by` prefix makes it possible to avoid typing many explicit statements, such as `summarize age if race==1`, `summarize age if race==2`, etc. Two of Stata's most useful commands are `forvalues` and `foreach`. These versatile tools have essentially supplanted other looping mechanisms in Stata. You could also use `while` ([P] **while**) to construct a loop, most commonly, when you are unsure how many times to repeat the loop contents. This is a common task when seeking the convergence of a numeric quantity to some target value. For example,

```
while reldif( newval, oldval ) > 0.001 {
    ...
}
```

would test for the *relative difference* between successive values of a criterion and would exit the loop when that difference was less than 0.1%. Conversely, if the computational method is not guaranteed to converge, this could become an infinite loop.

In contrast, when you have a defined set of values over which to iterate, `foreach` and `forvalues` are the tools of choice. These commands are followed by a left brace ({), one or more following command lines, and a terminating line containing only a right brace (}). You can place as many commands in the loop body as you wish. A simple numeric loop can be constructed as, for example,

```
. use gdp4cty, clear
. forvalues i = 1/4 {
  2.          generate double lngdp`i' = log(gdp`i')
  3.          summarize lngdp`i'
  4. }
```

Variable	Obs	Mean	Std. Dev.	Min	Max
lngdp1	400	7.931661	.59451	5.794211	8.768936

Variable	Obs	Mean	Std. Dev.	Min	Max
lngdp2	400	7.942132	.5828793	4.892062	8.760156

Variable	Obs	Mean	Std. Dev.	Min	Max
lngdp3	400	7.987095	.537941	6.327221	8.736859

Variable	Obs	Mean	Std. Dev.	Min	Max
lngdp4	400	7.886774	.5983831	5.665983	8.729272

In this example, we define the local macro `i` as the loop index. Following an equal sign, we give the range of values that `i` is to take on as a Stata numlist.[6] A range can be as simple as `1/4`, or it can be more complicated, such as `10(5)50`, indicating from 10 to 50 in steps of 5, or `100(-10)20`, indicating from 100 to 20 counting down by tens. Other syntaxes for the range are available; see [P] **forvalues** for details.

---

6. See section 3.2.2. Not all Stata numlists are compatible with `forvalues`.

This example provides one of the most important uses of `forvalues`: looping over variables where the variables have been given names with an integer component. This avoids the need for separate statements to transform each of the variables. The integer component need not be a suffix; we could loop over variables named ctyNgdp just as readily. Or, say that we have variable names with more than one integer component but we want to summarize the data only for countries 2 and 4:

```
. forvalues y = 1995(2)1999 {
  2.        forvalues i = 2(2)4 {
  3.                summarize gdp'i'_'y'
  4.        }
  5. }
```

Variable	Obs	Mean	Std. Dev.	Min	Max
gdp2_1995	400	3242.162	1525.788	133.2281	6375.105

Variable	Obs	Mean	Std. Dev.	Min	Max
gdp4_1995	400	3093.778	1490.646	288.8719	6181.229

Variable	Obs	Mean	Std. Dev.	Min	Max
gdp2_1997	400	3616.478	1677.353	153.0657	7053.826

Variable	Obs	Mean	Std. Dev.	Min	Max
gdp4_1997	400	3454.322	1639.356	348.2078	6825.981

Variable	Obs	Mean	Std. Dev.	Min	Max
gdp2_1999	400	3404.27	1602.077	139.8895	6693.86

Variable	Obs	Mean	Std. Dev.	Min	Max
gdp4_1999	400	3248.467	1565.178	303.3155	6490.291

As you can see, a nested loop is readily constructed with two `forvalues` statements.

As generally useful as `forvalues` can be, the `foreach` command is especially useful in constructing efficient do-files. This command interacts perfectly with some of Stata's most common constructs: the macro, the varlist, and the numlist. Like in `forvalues`, a local macro is defined as the loop index. Rather than cycling through a set of numeric values, `foreach` specifies that the loop index iterate through the elements of a local (or global) macro, or the variable names of a varlist, or the elements of a numlist. The list can also be an arbitrary list of elements on the command line or a new varlist of valid names for variables not present in the dataset.

This syntax allows `foreach` to be used in a flexible manner with any set of items, regardless of pattern. In several previous examples, we used `foreach` with the elements of a local macro defining the list. Let's illustrate its use here with a varlist from the `lifeexp` dataset, which is also used in some of the Stata reference manuals. We compute summary statistics, compute correlations with `popgrowth`, and generate scatterplots (not shown) for each element of a varlist versus `popgrowth`:

```
. sysuse lifeexp, clear
(Life expectancy, 1998)
```

```
. local mylist lexp gnppc safewater
. foreach v of varlist `mylist' {
  2.            summarize `v'
  3.            correlate popgrowth `v'
  4.            scatter popgrowth `v'
  5. }
```

Variable	Obs	Mean	Std. Dev.	Min	Max
lexp	68	72.27941	4.715315	54	79

(obs=68)

	popgro~h	lexp
popgrowth	1.0000	
lexp	-0.4360	1.0000

Variable	Obs	Mean	Std. Dev.	Min	Max
gnppc	63	8674.857	10634.68	370	39980

(obs=63)

	popgro~h	gnppc
popgrowth	1.0000	
gnppc	-0.3580	1.0000

Variable	Obs	Mean	Std. Dev.	Min	Max
safewater	40	76.1	17.89112	28	100

(obs=40)

	popgro~h	safewa~r
popgrowth	1.0000	
safewater	-0.4280	1.0000

Given the order of variables in this dataset, we also could have written the first line as

```
foreach v of varlist lexp-safewater {
```

Because you are free to reorder the variables in a dataset,[7] a do-file that depends on variable ordering can produce unpredictable results if variables have been reordered.

In the following example, we automate the construction of a `recode`[8] statement for a dataset containing values of gross domestic product (GDP) for several countries and several years. The country codes are labeled as 1, 2, 3, and 4 in the `cc` variable. We would like to embed their three-digit International Monetary Fund country codes into a copy of `cc`, `newcc`, so that we can use these values elsewhere, for example, in a `merge` operation. The resulting statement could just be typed out for four elements, but imagine its construction if we had 180 country codes! Note the use of `local ++i`, a shorthand way of incrementing the counter variable within the loop. The loop is used to build up the local macro `rc`, which defines the set of transformations that are to be applied to the `cc` variable.

---

7. See [D] **order**.
8. See section 3.3.3.

```
. use gdp4cty, clear
. local ctycode 111 112 136 134
. local i 0
. foreach c of local ctycode {
  2.          local ++i
  3.          local rc "`rc' (`i'=`c')"
  4. }
. display "`rc'"
 (1=111) (2=112) (3=136) (4=134)
. recode cc `rc', gen(newcc)
(400 differences between cc and newcc)
. tabulate newcc
```

RECODE of cc	Freq.	Percent	Cum.
111	100	25.00	25.00
112	100	25.00	50.00
134	100	25.00	75.00
136	100	25.00	100.00
Total	400	100.00	

The `foreach` statement also can be used to advantage with nested loops. You can combine `foreach` and `forvalues` in a nested loop structure, as illustrated here:

```
. use gdp4cty, clear
. local country US UK DE FR
. local yrlist 1995 1999
. forvalues i = 1/4 {
  2.          local cname: word `i' of `country'
  3.          foreach y of local yrlist {
  4.                  rename gdp`i'_`y' gdp`cname'_`y'
  5.          }
  6. }
. summarize gdpUS*
```

Variable	Obs	Mean	Std. Dev.	Min	Max
gdpUS_1995	400	3226.703	1532.497	328.393	6431.328
gdpUS_1999	400	3388.038	1609.122	344.8127	6752.894

It is a good idea to use indentation (either spaces or tabs) to align the loop body statements as shown here. Stata does not care as long as the braces appear as required, but it makes the do-file much more readable and easier to revise at a later date.

Alternatively, we could store the elements of the `country` list in numbered macros with `tokenize` ([P] **tokenize**), which places them into the macros 1, 2, 3, and 4:

```
. use gdp4cty, clear

. describe gdp*_*

              storage   display     value
variable name    type   format      label        variable label
---------------------------------------------------------------------------
gdp1_1995       float   %9.0g
gdp1_1999       float   %9.0g
gdp2_1995       float   %9.0g
gdp2_1999       float   %9.0g
gdp3_1995       float   %9.0g
gdp3_1999       float   %9.0g
gdp4_1995       float   %9.0g
gdp4_1999       float   %9.0g
gdp1_1997       float   %9.0g
gdp2_1997       float   %9.0g
gdp3_1997       float   %9.0g
gdp4_1997       float   %9.0g

. local country US UK DE FR

. local yrlist 1995 1999

. local ncty: word count country

. tokenize `country'

. forvalues i = 1/`ncty' {
  2.          foreach y of local yrlist {
  3.                  rename gdp`i'_`y' gdp``i''_`y'
  4.          }
  5. }

. summarize gdpUS*

    Variable |       Obs        Mean    Std. Dev.       Min        Max
-------------+--------------------------------------------------------
   gdpUS_1995 |       400    3226.703    1532.497    328.393   6431.328
   gdpUS_1999 |       400    3388.038    1609.122   344.8127   6752.894
```

Here the country names are stored as the values of the numbered macros. To extract those values on the right-hand side of the generate command, we must doubly dereference the macro i. The content of that macro the first time through the loop is the number 1. To access the first country code, we must dereference the macro `1'. Putting these together, within the loop we dereference i twice: ``i'' is the string US.

This latter technique, making use of tokenize, should be used with caution within an ado-file. Stata uses the numbered macros to reference positional arguments (the first, the second, etc., word on the command line) in an ado-file. Using tokenize within an ado-file will redefine those numbered macros, so care should be taken to move their contents into other macros. See chapter 11 for details of ado-file programming.

In summary, the foreach and forvalues commands are essential components of any do-file writer's toolkit. Whenever you see a set of repetitive statements in a Stata do-file, it is likely to mean that its author did not understand how one of these loop constructs could have made the program, its upkeep, and his or her life simpler. An excellent discussion of the loop commands can be found in Cox (2002a).

The `forvalues` and `foreach` commands often require the manipulation of local macros and lists. Stata's extended macro functions and macro list functions are useful in this regard. For more information on extended macro functions and macro list functions, see section 3.8.

8 Cookbook: Do-file programming III

This cookbook chapter presents for Stata do-file programmers several recipes using the programming features described in the previous chapter. Each recipe poses a problem and a worked solution. Although you may not encounter this precise problem, you should be able to recognize its similarities to a task that you would like to automate in a do-file.

8.1 Handling parallel lists

The problem. For each of a set of variables, you want to perform some steps that involve another group of variables, perhaps creating a third set of variables. These are *parallel lists*, but the variable names of the other lists cannot be deduced from those of the first list.[1] How can these steps be automated?

The solution. First, let's consider that we have two arbitrary sets of variable names and want to name the resulting variables based on the first set's variable names. For example, you might have some time series of population data for several counties and cities:

```
local county Suffolk Norfolk Middlesex Worcester Hampden
local cseat Boston Dedham Cambridge Worcester Springfield
local wc 0
foreach c of local county {
        local ++wc
        local sn: word `wc' of `cseat'
        generate seatshare`county' = `sn' / `c'
}
```

This `foreach` loop will operate on each pair of elements in the parallel lists, generating a set of new variables: `seatshareSuffolk`, `seatshareNorfolk`, ...

Another form of this logic would use a set of numbered variables in one of the loops. Then you could use a `forvalues` loop over the values (assuming they were consecutive or otherwise patterned) and the extended macro function `word # of` *string* to access the elements of the other loop. The `tokenize` command also could be used, as illustrated in section 7.3.

Alternatively, you could use a `forvalues` loop over both lists, using the `word count` *string* extended macro function:

---

1. This recipe borrows from Stata frequently asked question (FAQ) "How do I process parallel lists?" (http://www.stata.com/support/faqs/lang/parallel.html), written by Kevin Crow.

```
local n: word count 'county'
forvalues i = 1/'n' {
       local a: word 'i' of 'county'
       local b: word 'i' of 'cseat'
       generate seatshare'a' = 'b'/'a'
}
```

This yields the same results as the previous approach.

You may also find this logic useful in handling a set of constant values that align with variables. Let's say that you have a cross-sectional dataset of hospital budgets over various years, in the wide structure; that is, you have separate variables for each year (e.g., exp1994, exp1997, ...). You would like to apply a health care price deflator to each variable to place them in comparable terms. For example,

```
local yr 1994 1997 2001 2003 2005
local defl 87.6 97.4 103.5 110.1 117.4
local n: word count 'yr'
forvalues i = 1/'n' {
       local y: word 'i' of local yr
       local pd: word 'i' of local defl
       generate rexp'y' = exp'y' * 100 / 'pd'
}
```

This loop will generate a set of new variables measuring real expenditures (rexp1994, rexp1997, ...) by scaling each of the original (nominal-valued) variables by 100/defl for that year's value of the health care price deflator. This could also be achieved by using reshape to transform the data into the long structure, as described in section 5.5, but that is not really necessary in this context.

8.2   Calculating moving-window summary statistics

The problem. Suppose you would like to calculate moving-window statistics for a variable in your dataset. As discussed in section 7.2.4, these computations over overlapping subsamples cannot be handled with the statsby prefix. Each observation can appear in only one by-group. These statistics could be generated with the rolling prefix, but that approach creates a new dataset that must then be merged back into the original dataset.

The solution. A solution is provided by Cox and Baum's mvsumm routine, which calculates moving summary statistics.[2] Summary statistics include all statistics available from summarize, detail as well as several additional derived values, such as the interquartile range. The command works with either a single time series or a panel. Its only restriction is that there can be no internal gaps in the time series. Values can be missing, but an observation must be defined for each consecutive time period. If this is not the case in your data, you can use tsfill to rectify that problem.

---

2. The mvsumm routine is available from the Statistical Software Components (SSC) archive; see help ssc.

In a question raised on Statalist, a user wanted to calculate the average rate of growth over the past three years as a new variable. We can use the `grunfeld` dataset, containing 10 firms' time series, to illustrate how that might be done. The growth rate is approximated by the difference of the logarithm of the variable, here `invest`, the firms' capital investment expenditure. We want the average of the past three years' values, so we use the `LD.` time-series command to specify the lag of the first difference of `linvest` and the `stat(mean)` option to specify the statistic to be computed in the `mvsumm` command:

```
. use grunfeld, clear
. generate linvest = log(invest)
. mvsumm LD.linvest, generate(invrate) stat(mean) window(3) end
```

In the `mvsumm` command, the `window()` option specifies the window width, or how many time periods are to be included in the moving-window estimate. The `end` option specifies that the summary value should be aligned with the last period included in the window.

Having computed these moving-average growth rates for each firm, we can now display several firms' growth rate histories with `tsline` (see figure 8.1):

```
. tsline invrate if inrange(company, 1, 4) & year >= 1937, by(company, ti(" "))
> yline(0) ysc(range(-0.2 0.3)) ytick(-0.2(0.1)0.3) ylab(-0.2(0.1)0.3)
> scheme(s2mono)
```

(Continued on next page)

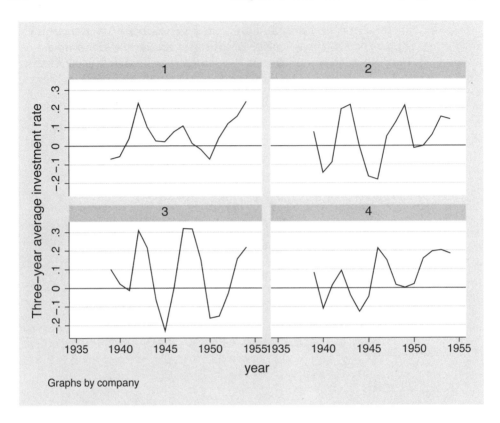

Figure 8.1. Growth rate histories of several firms

8.2.1   Producing summary statistics with rolling and merge

We could also use the `rolling` prefix to generate these summary statistics:

```
. qui rolling r(mean), window(3) saving(ldinvest, replace): summarize
> LD.linvest
```

Some additional work is needed to get the statistics produced by `rolling` back into our dataset. This prefix command creates the new variables `start` and `end`, indicating the first and last periods included in each window. Because we adopted the convention that a rolling-window statistic should be aligned with the last time period included in the window, we rename the `end` variable to `year` and give a more meaningful name to the summary variable `_stat_1`. Finally, we must take into account `rolling`'s convention of dealing with missing values. You could argue that a three-period window estimate of the mean of the lagged difference of a time-series variable can first be calculated from observations 3–5 and stored in observation 5, because observations 1 and 2 will be lost by the lag and differencing operations (and `mvsumm` makes that assumption). However, `rolling` will produce a value for observations 3 and 4, using just one or two values to compute the mean, respectively. We choose to omit those values (corresponding to years 1937–1938 in the Grunfeld dataset). We use `xtset` and save the revised dataset as `ldinvest`. We then restore our original dataset and use `merge` to include the summary statistics in the dataset:

```
. preserve
. use ldinvest, clear
(rolling: summarize)
. rename end year
. rename _stat_1 rolling_ldinvest
. keep if year >= 1939
(20 observations deleted)
. xtset company year
       panel variable:  company (strongly balanced)
        time variable:  year, 1939 to 1954
                delta:  1 unit
. save ldinvest, replace
file ldinvest.dta saved
. restore
. merge company year using ldinvest, unique
. drop _merge
. summarize invrate rolling_ldinvest
```

Variable	Obs	Mean	Std. Dev.	Min	Max
invrate	160	.0602769	.1303373	-.2292838	.475313
rolling_ld~t	160	.0602769	.1303373	-.2292838	.475313

We see that over the 160 observations (16 years per firm) in common, the `mvsumm` and `rolling` approaches produced the same summary statistics.

8.2.2 Calculating moving-window correlations

In several applications in finance, you may need to calculate a *moving correlation* between two time series. For example, the calculation of an optimal hedge ratio involves computing the correlation between two series in a moving-window context. A companion program to `mvsumm`, Cox and Baum's `mvcorr` uses a similar syntax to support the computation of moving-window correlations between two time series, x1 and x2. If the two time series are specified as x1 and Ln.x1, a moving nth-order autocorrelation is calculated.[3]

As shown in the previous section, we could also use the `rolling` prefix to produce moving-window correlations.

---

3. The `mvcorr` routine is available from the SSC archive; see `help ssc`.

8.3   Computing monthly statistics from daily data

The problem. Assume you have trading-day data from the stock market, with one observation for each day when the market is open. You want to calculate a measure of monthly volatility, defined by Merton (1980). His method uses the intramonth price movements to compute a measure of volatility for the month. This takes account of both the changes in the price series (Δp_t) and the number of days elapsed between the price observations (ϕ_t). Merton's daily volatility measure is computed from $100 \times$ the change in the price variable divided by the square root of the number of elapsed days. Algebraically, we want to calculate this quantity for each trading day:

$$\varsigma_t = \left(100 \frac{\Delta p_t}{\sqrt{\Delta \phi_t}} \right)^2$$

The solution. In our Stata dataset, each observation contains that day's closing price, p, and the date variable, `date`, a proper Stata date variable that represents each day as an integer value (see section 2.4.2). Because trading days are not consecutive calendar days (given weekends and holidays), we do not use the time-series lag command L. to calculate price changes since they would be missing at those times.[4] We calculate the elements of the formula for ς_t as the Stata variable `ds2`:

```
. use fictprice, clear
. qui generate dscorr = 100 * (p - p[_n-1])
. qui generate deld = date - date[_n-1]
. qui generate dscorrd = dscorr / sqrt(deld)
. qui generate ds2 = dscorrd^2
```

The series of daily values of `ds2` is then aggregated to monthly frequency to produce the monthly volatility measure:

$$\Phi_t = \sqrt{\sum_{t=1}^{T} \varsigma_t}$$

To compute monthly volatility, we generate a month variable, `mon`, and use the `bysort` prefix to cumulate the `ds2` variable over each month, recalling that the `sum()` function generates a running sum:

```
. generate mon = mofd(date)
. bysort mon: generate cumds2 = sum(ds2)
```

For each month, we take the square root of the last observation of the month and define it as `ssq`:

```
. qui bysort mon: generate ssq = sqrt(cumds2) if _n == _N
```

---

4. See section 2.4.3.

We now can drop all daily observations for days prior to the last day of each month, define the time-series calendar to be monthly, and save the new monthly dataset:

```
. drop if ssq == .
(2044 observations deleted)
. tsset mon, monthly
        time variable:  mon, 1998m5 to 2006m7
                delta:  1 month
. save myvolat, replace
file myvolat.dta saved
```

We can now graph our fictitious monthly volatility time series with the `tsline` command (see figure 8.2):

```
. tsline ssq, ti("Monthly volatility from Merton method")
```

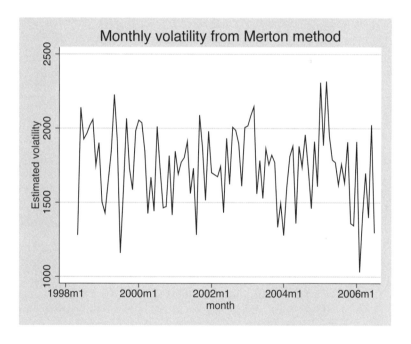

Figure 8.2. Estimated monthly volatility from daily data

8.4   Requiring at least n observations per panel unit

The problem. Per a common question on Statalist, if you have unbalanced panel data, how do you ensure that each unit has at least n observations available?

It is straightforward to calculate the number of available observations for each unit:

```
xtset patient date
by patient: generate nobs = _N
generate want = (nobs >=n)
```

These commands will produce an indicator variable, **want**, that selects those units which satisfy the condition of having at least *n* available observations.

This works well if all you care about is the number of observations available, but let's say you have a more subtle concern: you want to count *consecutive* observations. Perhaps you want to compute statistics based on changes in various measurements by using Stata's L. or D. time-series command.[5] Applying this command to series with gaps will create missing values.

The solution. A solution to this problem is provided by Nicholas J. Cox and Vince Wiggins in their response to the Stata FAQ "How do I identify runs of consecutive observations in panel data?"[6] The sequence of consecutive observations is often termed a *run* (a term commonly used in nonparametric statistics; see [R] **runtest**) or a *spell*.[7] Cox and Wiggins propose defining the runs in the time series for each panel unit:

```
generate run = .
by patient: replace run = cond(L.run == ., 1, L.run + 1)
by patient: egen maxrun = max(run)
generate wantn = (maxrun >=n)
```

The second command replaces the missing values of **run** with either 1 (denoting the start of a run) or the prior value + 1. For observations on consecutive dates, that will produce the integer series 1, . . . , *len*, where *len* is the last observation in the run. When a break in the series occurs, the prior (lagged) value of **run** will be missing, and **run** will be reset to 1. The variable **maxrun** then contains, for each patient, the highest value of **run** in that unit's sequence.

Although this identifies (with indicator **wantn**) those patients who do (or do not) have a spell of *n* consecutive observations, it does not allow you to immediately identify this spell. You may want to retain only this longest spell, or run of observations, and discard other observations from this patient. To carry out this sort of screening, you should become familiar with Nicholas J. Cox's **tsspell** program (available from the SSC archive; see **help ssc**), which provides comprehensive capabilities for spells in time series and panel data. A "canned" solution to the problem of retaining only the longest spell per patient is also available from my **onespell** routine (**findit onespell**), which makes use of **tsspell**.

---

5. See section 2.4.3.

6. See http://www.stata.com/support/faqs/data/panel.html.

7. See section 4.4.

8.5 Counting the number of distinct values per individual

The problem. Per a question on Statalist, if you have data on individuals that indicate their association with a particular entity, how do you count the number of entities associated with each individual? For instance, say we have a dataset of consumers who purchase items from various Internet vendors. Each observation identifies the consumer (pid) and the vendor (vid), where 1=amazon.com, 2=llbean.com, 3=overstock.com, and so on.

The solution. Several solutions were provided by Nicholas J. Cox in a Statalist posting.[8] For example,

```
bysort pid vid: generate count = (_n == 1)
by pid: replace count = sum(count)
by pid: replace count = count(_N)
```

Here we consider each combination of consumer and vendor, and we set `count = 1` for their first observation. We then replace `count` with its `sum()` for each consumer, keeping in mind that this is a running sum, so that it takes on 1 for the first vendor, 2 for the second, and so on. Finally, we replace `count` with its value for each consumer in observation _N, the maximum number of vendors with whom the consumer deals.

A second solution, somewhat less intuitive but shorter, is this one:

```
bysort pid (vid): generate count = sum(vid != vid[_n-1])
by pid: replace count = count(_N)
```

This solution takes advantage of the fact that when (vid) is used with the `bysort` prefix, the data are sorted in order of `vid` within each `pid`, even though the `pid` is the only variable defining the by-group. When the `vid` changes, another value of 1 is generated and summed. When subsequent transactions pertain to the same vendor, `vid != vid[_n-1]` evaluates to 0, and those zero values are added to the sum.

This problem is common enough that an official `egen` function has been developed to tag observations:

```
egen tag = tag(pid vid)
egen count = total(tag), by(pid)
```

The `tag()` function returns 1 for the first observation of a particular combination of `pid vid`, and zero otherwise. Thus its `total()` for each `pid` is the number of `vids` with whom the consumer deals.

As a third solution, Cox's `egenmore` package (see section 3.4) contains the `nvals()` function, which allows you to say

```
egen count = nvals(vid), by(pid)
```

---

8. See http://www.stata.com/statalist/archive/2007-05/msg00464.html.

For more information, see the response by Nicholas J. Cox and Gary Longton to the Stata FAQ "How do I compute the number of distinct observations?"[9]

---

9. See http://www.stata.com/support/faqs/data/distinct.html

9 Do-file programming: Other topics

9.1 Introduction

This chapter presents several do-file programming techniques that you can use to reduce your workload and improve the reliability and replicability of your work. The first section discusses the use of Stata matrices for the storage and presentation of computed quantities. The `post` ([P] **postfile**) and `postfile` commands' ability to create a new Stata dataset are discussed, followed by mention of several commands that can create external files for use in other software. The next two sections of the chapter discuss the automation of estimation and graphics.

9.1.1 What you should learn from this chapter

- How to use Stata matrices to organize and present computed results
- How to use `post` and `postfile` to generate Stata datasets
- How to use `outsheet` and `outfile` to export Stata variables' contents
- How to use `file` to create a completely flexible output file
- How to automate the production of standard-format tables
- How to automate the production of standard-format graphs
- How to use characteristics

9.2 Storing results in Stata matrices

Several Stata commands can be used to generate new Stata datasets containing the results of repetitive computations. For instance, the `statsby` prefix, discussed in section 7.2.3, can execute a statistical (r-class) or estimation (e-class) command for each by-group and save one or more results as observations in the new variables. The `rolling` prefix, described in section 7.2.4, performs the same function for rolling-window estimates in a time-series context.[1] But these prefix commands have a limitation: unless you have written your own statistical or estimation command, they can execute only

---

1. The `simulate` prefix, as discussed in section 7.2.5, does the same for the results specified in a user-written Monte Carlo simulation program. In that context, you can use any number of commands to generate the desired results, circumventing the limitations of the `by` prefix.

one Stata command, and they can only place results generated by that command in the resulting dataset.[2]

Often your goal may involve tabular output of statistical and estimation results, requiring the execution of several Stata commands to create each row of the table. It may be convenient to store these results in a Stata matrix rather than in a separate Stata dataset.

As an example, let's consider an enhanced version of the `grunfeld` dataset, which is used in the Stata reference manuals. This dataset is a balanced panel of 10 companies' time series of several variables. Our version of the `grunfeld` dataset, `grunfeldavg`, contains two additional variables: the average levels of firms' investment expenditures, `invavg`, and stock of fixed capital, `kapavg`. Let's say that we want to generate a table, one row per company, containing the 25th, 50th, and 75th percentiles of firm investment; the time-series correlation of the firm's investment with that of the average firm; the correlation of firm capital stock with that of the average firm; and a regression coefficient (and its standard error) from the regression of investment on the lagged values of the firm's capital stock and the average firm.

To produce this table from the `grunfeldavg` dataset, we first generate a list of the values of the `company` variable by using the `levelsof` command, storing that list in a local macro, `colist`. We find the number of companies with the extended macro function[3] `word count` *string*, which tells us how many rows we need in our matrix. We want to tabulate seven statistics per firm: three quantiles, two correlation coefficients, an estimated coefficient, and its standard error. We thus define the `table1` matrix with the `J(`*#rows*`, `*#cols*`, `*value*`)` function as a null matrix of `ncomp` rows and seven columns.

```
. use grunfeldavg, clear
. levelsof company, local(colist)
1 2 3 4 5 6 7 8 9 10
. local ncomp: word count 'colist'
. matrix table1 = J('ncomp', 7, 0)
```

We are now ready to set up a `foreach` loop over companies. Here the `company` variable merely takes on the values 1, ..., 10 but we want our program to work regardless of its values. Storing the distinct values of that variable in the local macro `colist` provides that feature. At the same time, we need a counter to address the successive rows of `table1`, which we provide with the local macro `i`, incremented within the loop.

The `summarize` command computes the desired percentiles, which we place in the first three columns of the company's row of `table1`. Likewise, we use `correlate` to generate the desired correlations and place them in columns 4 and 5, and we use `regress` to generate the desired statistics for the last two columns of the matrix. The entire set of commands is wrapped in a `quietly` block to suppress its output. Now we only need apply labels to the matrix with `matrix rownames` and `matrix colnames`.

---

2. For a solution circumventing this limitation, see section 12.1.
3. See section 3.8.

```
. local i 0

. levelsof company, local(colist)
1 2 3 4 5 6 7 8 9 10

. foreach c of local colist {
  2.          quietly {
  3.                  local ++i
  4.                  summarize invest if company == `c', detail
  5.                  matrix table1[`i', 1] = r(p25)
  6.                  matrix table1[`i', 2] = r(p50)
  7.                  matrix table1[`i', 3] = r(p75)
  8.                  correlate invest invavg if company == `c'
  9.                  matrix table1[`i', 4] = r(rho)
 10.                  correlate kstock kapavg if company == `c'
 11.                  matrix table1[`i', 5] = r(rho)
 12.                  regress invest L.kstock L.kapavg if company == `c'
 13.                  matrix table1[`i', 6] = _b["L.kstock"]
 14.                  matrix table1[`i', 7] = _se["L.kstock"]
 15.          }
 16. }

. matrix rownames table1 = `colist'

. matrix colnames table1 = p25 p50 p75 r_invest r_kap beta_k se_k
```

The matrix of results we have generated can be viewed with `matrix list`, which offers us limited functionality over its appearance.

```
. matrix list table1, format(%9.3f) ti("Grunfeld company statistics")

table1[10,7]:  Grunfeld company statistics
            p25       p50       p75  r_invest      r_kap    beta_k      se_k
  1     429.300   538.350   665.500     0.964      0.985     0.150     0.361
  2     321.750   419.550   471.350     0.808      0.898    -0.626     0.299
  3      59.050    93.550   146.750     0.899      0.991    -0.180     0.249
  4      55.990    71.085    95.010     0.928      0.957     0.298     0.160
  5      51.525    60.385    72.290     0.848      0.966    -0.019     0.059
  6      27.685    43.110    72.750     0.959      0.990    -0.281     0.265
  7      33.245    44.200    57.680     0.876      0.944     0.045     0.060
  8      30.305    38.540    53.920     0.928      0.963    -0.132     0.177
  9      29.715    38.110    55.405     0.723      0.975     0.350     0.118
 10       1.925     2.215     4.440     0.797      0.887     0.113     0.182
```

To produce a version of the table that we could include in our LaTeX research paper, we rely on Baum and Azevedo's `outtable` routine (`findit outtable`).[4] The resulting table is displayed as table 9.1.

```
. outtable using ch9.02t, mat(table1) replace format(%9.3f) center
> caption("Grunfeld company statistics")
```

---

4. If you wanted tab-delimited output, suitable for a word processor or spreadsheet, you could use Michael Blasnik's `mat2txt`, available from the Statistical Software Components (SSC) archive.

Table 9.1. Grunfeld company statistics

	p25	p50	p75	r invest	r kap	beta k	se k
1	429.300	538.350	665.500	0.964	0.985	0.150	0.361
2	321.750	419.550	471.350	0.808	0.898	-0.626	0.299
3	59.050	93.550	146.750	0.899	0.991	-0.180	0.249
4	55.990	71.085	95.010	0.928	0.957	0.298	0.160
5	51.525	60.385	72.290	0.848	0.966	-0.019	0.059
6	27.685	43.110	72.750	0.959	0.990	-0.281	0.265
7	33.245	44.200	57.680	0.876	0.944	0.045	0.060
8	30.305	38.540	53.920	0.928	0.963	-0.132	0.177
9	29.715	38.110	55.405	0.723	0.975	0.350	0.118
10	1.925	2.215	4.440	0.797	0.887	0.113	0.182

You might also use the flexibility of Stata matrices to juxtapose estimation results in a matrix with coefficients on the columns. This facility is not readily available from `estimates table` or Ben Jann's `estout` (see section 5.4.1). We again loop over companies with `forvalues` and run a regression of each company's investment expenditures on two lags of its capital stock and the lagged value of the average firm's capital stock. This will estimate four coefficients per firm, counting the constant term. Those coefficients will be made available in the saved results as Stata matrix `e(b)`, a row vector (see section 5.3). We use Stata's row join operator (\) to concatenate each firm's vector of coefficients into a matrix named `allbeta`.[5] Unlike in the previous example, we do not predefine this matrix. How, then, can this concatenation be done for the first firm? By using the `nullmat()` matrix function, as shown below, in which a null matrix of the appropriate size is prepended to the first firm's coefficient vector. We also build up the `local row` so that we can provide `matrix rownames` for the `allbeta` matrix.

```
. levelsof company, local(colist)
1 2 3 4 5 6 7 8 9 10

. local ncomp: word count 'colist'

. forvalues i = 1/'ncomp' {
  2.          local c: word 'i' of 'colist'
  3.          quietly regress invest L(1/2).kstock L.kapavg if company == 'c'
  4.          matrix beta'i' = e(b)
  5.          matrix allbeta = (nullmat(allbeta) \ beta'i')
  6.          local row "'row' 'c'"
  7. }

. matrix rownames allbeta = 'row'
```

We are now ready to invoke `outtable` to produce the desired LATEX table, which is displayed as table 9.2.

```
. outtable using ch9.03t, mat(allbeta) replace format(%9.4f) center
> caption("Grunfeld company estimates")
```

_____

5. A similar technique would make use of the column join operator (,) to concatenate matrices by column.

Table 9.2. Grunfeld company estimates

	L.kstock	L2.kstock	L.kapavg	cons
1	0.0070	0.0437	1.8377	118.8346
2	-0.8291	0.3487	0.9758	312.8891
3	-0.6049	0.4138	0.6065	37.7697
4	0.2517	0.0032	0.1267	27.2760
5	-0.2762	0.2523	0.1081	54.6390
6	-0.6712	0.4115	0.3612	-6.3536
7	0.4983	-0.4442	0.0695	5.0020
8	-0.5233	0.4040	0.1471	19.9655
9	0.2177	0.1351	-0.1390	-21.3947
10	0.0567	0.0110	0.0095	0.2617

A "canned" solution for the generation of Stata matrices of descriptive statistics is provided by Cox and Baum's `statsmat`, available from the SSC archive. That routine can provide a wide range of descriptive statistics as a matrix or in transposed form.[6] By default, `statsmat` will calculate the maximum, minimum, 25th, 50th, and 75th percentiles; the mean; and the standard deviation of each variable. This command can work with one variable to produce descriptive statistics for each value of a by-list and can use either the default casewise deletion or listwise deletion. Casewise deletion will cause any observation that is missing anywhere in the `statsmat` varlist to be omitted, while listwise deletion will use all available observations per variable.

9.3 The post and postfile commands

In most data-management and statistical tasks, you work with an existing Stata dataset and can create a new version of running the dataset in the process of running the do-file. But what if you want to create a completely new Stata dataset with different contents in the process, without disrupting the existing dataset? You could, of course, use the `preserve` command, perform whatever steps are needed to create the new dataset, save it, and then use the `restore` command to return to the preexisting structure of data in memory. That is the preferred way of using `collapse` or `contract`, which replaces the data in memory. Sometimes, however, it might be useful to be able to create a new Stata dataset while processing the existing data.

---

6. A similar functionality is available from `tabstat`, but that routine usually produces several matrices rather than one matrix.

This sort of task is handled by several of the prefix commands described in section 7.2. As discussed in that context, the `statsby`, `rolling`, and `simulate` prefixes each provide a `saving(`*filename*`)` option, which specifies the name of the new Stata dataset to be created. But what if the items you wish to place in the new dataset are not those readily computed by one of these prefix commands? Then you will want to learn how to use the `postfile` command package.

This command package is described as containing "utilities to assist Stata programmers in performing Monte Carlo–type experiments". But these commands are general and can be used to create new datasets with no relation to Monte Carlo experiments. They allow you to circumvent the restriction that Stata can work with only one dataset at a time, in that you can construct a new Stata dataset "on the fly" with these commands without disturbing the contents of the dataset in memory. To set this up, you first use `postfile` to specify a new varlist for the new dataset and a filename where it is to be stored. To reference this file, you must use a postname: an identifier or "handle" to be used in subsequent commands. It is best to obtain the postname from the `tempname` ([P] **macro**) command. Because each `postfile` command requires a postname, it is possible to post results to more than one postfile at a time.

After defining the structure and location of the new dataset, you use the `post` command to fill in each observation, referencing the postname. The individual variables' values are specified as *exp*s, which must appear in parentheses. After the dataset is complete, you use the `postclose` command to close the file, referenced by its postname.

Consider the example given in the previous section, where we created a Stata matrix containing several different firm-level statistics. Let's see how we might make that into a new Stata dataset with `postfile`.

We use the same logic as before to construct a list of the companies in local macro `colist`, and then we invoke the `postfile` command. The syntax of this command is

`postfile` *postname newvarlist* `using` *filename* [`, replace`]

The postname, used only within your do-file, is best created by using a temporary name with the `tempname` command. You then must define *newvarlist*: the new dataset's variables and, optionally, their data types. For instance, you could store one or more string variables in the dataset, declaring their data type before their name in the `postfile` variable list. The `using` clause specifies the external filename of the new Stata dataset.

```
. use grunfeldavg, clear
. levelsof company, local(colist)
1 2 3 4 5 6 7 8 9 10
. local ncomp: word count `colist'
. tempname p
. postfile `p' compnr p25 p50 p75 r_invest r_kap beta_k se_k using firmstats,
> replace
```

The command to write a single observation to the dataset is `post`, with the syntax

`post` *postname* (*exp*) (*exp*) ... (*exp*)

where each expression's value must be enclosed in parentheses. The value can be, as in this example, a local macro, or it can be a global macro, scalar, or constant. Rather than creating a matrix with one row per firm, you create a new dataset with one observation per firm. At the end of the loop over observations, you use `postclose` *postname* to close the file.

```
. forvalues i = 1/`ncomp' {
  2.          local c: word `i' of `colist'
  3.          quietly {
  4.                  summarize invest if company == `c', detail
  5.                  local p25 = r(p25)
  6.                  local p50 = r(p50)
  7.                  local p75 = r(p75)
  8.                  correlate invest invavg if company == `c'
  9.                  local r_invest = r(rho)
 10.                  correlate kstock kapavg if company == `c'
 11.                  local r_kap = r(rho)
 12.                  regress invest L.kstock L.kapavg if company == `c'
 13.          }
 14.          post `p' (`i') (`p25') (`p50') (`p75') (`r_invest') (`r_kap')
  >              (_b["L.kstock"]) (_se["L.kstock"])
 15. }
. postclose `p'
```

Because the `post` command accepts expressions as its arguments, we can reference the _b and _se elements from the `regress` command directly, without having to store them in local macros. You can verify that the new dataset contains the desired contents:

```
. use firmstats, clear

. summarize
```

Variable	Obs	Mean	Std. Dev.	Min	Max
compnr	10	5.5	3.02765	1	10
p25	10	104.049	146.2543	1.925	429.3
p50	10	134.9095	185.0054	2.215	538.35
p75	10	169.5095	218.1057	4.44	665.5
r_invest	10	.87306	.0785282	.7234149	.9635566
r_kap	10	.9555067	.0364744	.8865225	.9906394
beta_k	10	-.0282959	.2910943	-.6264817	.3501322
se_k	10	.1930768	.1002899	.0594801	.3614677

9.4 Output: The outsheet, outfile, and file commands

You may often encounter the need to export the contents of a Stata dataset to another application. Stata contains several commands that facilitate the export of a dataset's contents. Keep in mind that this sort of export generally will not be able to capture every aspect of Stata's variables: for instance, value labels will generally be lost on export, unless they are used in place of the numeric variable's values. For that reason, it is usually preferable to use a third-party application such as Stat/Transfer to move data between applications.[7] Nevertheless, you may need to export the data as an ASCII text file.

The `outsheet` ([D] **outsheet**) command writes data in "spreadsheet-style" format: either tab-delimited or comma-separated values. Tab-delimited format is usually a better choice because it avoids the issue of commas embedded in string variables.[8] The command syntax is

outsheet $\lceil$ *varlist* $\rceil$ using *filename* $\lceil$ *if* $\rceil$ $\lceil$ *in* $\rceil$ $\lceil$, *options* $\rceil$

If *varlist* is not provided, all the variables in memory are written to the output file in the order of their appearance (see [D] **order**). By default, the first line of the file will contain the variable names (which can be suppressed, but are useful if you plan to read this file into a spreadsheet). You can choose to write numeric values of labeled variables rather than their labels. If *filename* is specified without a suffix, the suffix `.out` is assumed.

Just as `insheet` reads tab-delimited (or comma-separated) values and `infile` reads space-delimited files, `outsheet` has a counterpart, `outfile` ([D] **outfile**), that writes an ASCII text–format file, which is, by default, space-delimited. Optionally, a comma-separated file can be written. The `outfile` command is also capable of writing a Stata *dictionary file*, or a separate `.dct` text file that documents the contents of the data file.[9] By default, `outfile` writes a file with the suffix `.raw`. The command syntax is identical to that of `outsheet`.

The most flexible form of output is provided by the `file` ([P] **file**) command package. Using the `file` commands, you can write and read ASCII text or binary files. Unlike `post`, `outsheet`, and `outfile`, which can write only a fixed number of fields per output record, `file` is capable of writing (or reading) a file of any arbitrary format. For example, some statistical packages use a format in which a set of variables is stored in a file with a header record containing two integer values: the number of observations (rows) and the number of variables (columns) in the subsequent data matrix. To write such a file from Stata, you must use the `file` command (or Mata functions with similar capability). You can also write arbitrary string content into an output file.

As an example, let's write the matrix created in section 9.2 into an external file, prefixed with two header lines: a time stamp and a line containing the matrix dimensions.

---

7. Those concerned with export to SAS should be aware of Stata's `fdasave` command, which is a general-purpose export routine that creates an SAS XPORT Transport file.

8. You do still have to be careful with string variables that could contain tab characters.

9. See [D] **infile (fixed format)** for the description of dictionary files.

To use the `file` commands, we must first set up a file handle,[10] best done as we did previously with `tempname`, and open that file handle:

`file open` *handle* `using` *filename,* {`read`|`write`|`read write`} [*options*]

The `file open` command must specify whether the file is to be used for input, output, or possibly both. We can also specify `text` or `binary` format (default is `text`, meaning ASCII).

```
. tempname fh
. file open 'fh' using matout.raw, write replace
```

Once the file is open, we can use `file write` to create each record. We must explicitly write line-end characters (denoted by _newline) where they should appear. The symbol _newline—which can be abbreviated to _n—in this context has nothing to do with the observation number (described in section 3.5.1). In the `file` command, _n is used to indicate that a new line symbol should be written to the file. If separate lines are to be written, the new line (or line-end) characters must be explicitly specified. Because Stata runs on systems with different new line conventions, it will write the appropriate new line characters for your operating system.

First, we determine how many rows and columns are in the matrix and assemble the two header records. We are then ready to write out each matrix element by using a double `forvalues` loop. We do not write the new line character (_newline) within the inner `forvalues` loop but instead at the completion of each row of the output file.

```
. local rows = rowsof(table1)
. local cols = colsof(table1)
. local timestamp "Matrix created 'c(current_date)' 'c(current_time)'"
. file write 'fh' "'timestamp'" _n
. file write 'fh' "'rows' 'cols'" _n
. forvalues i = 1/'rows' {
  2.          forvalues j = 1/'cols' {
  3.                  file write 'fh' (table1['i', 'j']) " "
  4.          }
  5.          file write 'fh' _n
  6. }
. file close 'fh'
```

---

10. *handle* plays the same role as *postname* does in the `postfile` package of commands.

This works well, but suppose we want to label the matrix rows and columns with their
rownames and colnames. Here the rownames are merely the integers 1, ..., 10, but
imagine that they are the firm code numbers or their stock market ticker symbols.
We enhance the previous do-file to extract the row and column names with macro
extended functions and add them to the output. We guard against quotation marks
appearing in the firm's name by using compound double quotes to display that item
(see section 3.2.5).

```
. tempname fh
. file open 'fh' using matout2.raw, write replace
. local rows = rowsof(table1)
. local cols = colsof(table1)
. local timestamp "Matrix created 'c(current_date)' 'c(current_time)'"
. file write 'fh' "'timestamp'" _n
. file write 'fh' "'rows' 'cols'" _n
. local coln: colnames table1
. local rown: rownames table1
. file write 'fh' "firmno 'coln'" _n
. forvalues i = 1/'rows' {
  2.         local rn: word 'i' of 'rown'
  3.         file write 'fh' '""'rn' "'
  4.         forvalues j = 1/'cols' {
  5.                 file write 'fh' (table1['i', 'j']) " "
  6.         }
  7.         file write 'fh' _n
  8. }
. file close 'fh'
```

You can now use type ([D] type) to examine the new file:

```
. type matout2.raw
Matrix created 26 Jul 2008 16:41:11
10 7
firmno p25 p50 p75 r_invest r_kap beta_k se_k
1 429.3 538.35001 665.5 .96355661 .98464086 .14971282 .36146776
2 321.75 419.54999 471.34999 .80775991 .89819793 -.62648164 .29878812
3 59.049999 93.549999 146.75 .89941256 .99063942 -.1804446 .24927241
4 55.99 71.085003 95.010002 .92792669 .9570222 .29802565 .16047287
5 51.525 60.385 72.289997 .8478581 .96597103 -.01904298 .05948012
6 27.685 43.110001 72.75 .9593111 .98986261 -.28145697 .26466183
7 33.244999 44.199999 57.679998 .87604314 .9442384 .04509232 .05995572
8 30.305 38.540001 53.92 .92841638 .96329808 -.13167999 .176705
9 29.715 38.109999 55.405001 .7234149 .97467447 .3501322 .11761613
10 1.925 2.215 4.4400001 .79690054 .88652247 .11318377 .1823479
```

This example only scratches the surface of the file package capabilities. You can use
file to produce any format of output file that you need, incorporating text strings,
local and global macros, scalar and matrix values, and the contents of Stata variables.
This same flexibility is afforded by file read, allowing you to read a file of arbitrary
format without disturbing the contents of Stata's memory.

9.5 Automating estimation output

One of the primary advantages to mastering do-file programming is that you gain the ability to automate the production of datasets, tables, and graphs. Prior sections of this chapter have dealt with various methods of producing new datasets in Stata or ASCII text format. In this section, we present an extended example illustrating how the production of a set of tables of statistical output can be fully automated. In this example, we create eight tables, each relating to underlying data for a specific year in a panel (longitudinal) dataset of individual workers' wages and hours. The same do-file could serve as a model for similar automation of statistical results for a set of hospitals, cancer trials, industries, or countries.

We illustrate with the `wagepan` dataset. These longitudinal data—4,360 observations on 48 variables—are presented in Rabe-Hesketh and Everitt (2006). We read in the data and generate one additional variable: `lhours`, the log of hours worked.

```
. use wagepan, clear
. xtset nr year
        panel variable:  nr (strongly balanced)
         time variable:  year, 1980 to 1987
                 delta:  1 unit
. generate lhours = log(hours)
```

As indicated by `xtset`, this dataset is a balanced panel of eight years' annual observations on 545 young males from the U.S. National Longitudinal Survey's Youth Sample. We want to estimate three wage equations for each year's wave of these data and eventually produce a separate table of regression results for each year. We can use Ben Jann's `estout` package's[11] (Jann 2007) `eststo` to generate stored estimates and `esttab` to tabulate them. Our target is a set of LaTeX tables, although we also could produce Stata Markup and Control Language, HTML, rich text format, or tab-delimited output.

We construct a `forvalues` loop over the years 1980–1987 with local macro y as the loop index. For each regression explaining `lwage`, the log of the worker's hourly wage, we fit the model with the qualifier `if year == 'y'`. For all three models, we want to add two statistics to the output: the semielasticity of the log wage with respect to the worker's years of education, `educ`, and its standard error, which we can calculate with the `mfx` command. For the third model, we also want to calculate an additional statistic. The model contains a measure of years of experience and its square. Thus the partial effect $\partial lwage/\partial exper$ must be computed from both of those coefficients for a given level of `exper`.

A digression on measures of marginal effects: the *elasticity* of y with respect to x is approximately the percentage change in y based on a one percent change in x, or $\partial \log y/\partial \log x$. If y or x is already in logarithmic terms, we need a *semielasticity* such as $\partial y/\partial \log x$ or $\partial \log y/\partial x$. The full elasticity measure is specified in the `mfx` command with the `eyex` option. The former semielasticity is specified as `dyex`—that is, we want

---

11. The `estout` package was presented in section 5.4.1.

the change in y (already in log terms) based on a one percent change in x—and the latter is `eydx`. The fourth option, `dydx`, is the regression coefficient itself. Although a regression coefficient is unchanged throughout the range of its regressor's values, any of the other measures will vary depending on the values of y and x. By default, `mfx` computes these values at the means of the estimation sample. A discussion of elasticities and semielasticities is provided in Wooldridge (2002, 16–18; 2009, 45–46).

Here we use `mfx, dyex` because the dependent variable is already in log terms. The `mfx` command stores its results in matrices, and from those matrices we can recover the semielasticity point estimates (from matrix `e(Xmfx_dyex)`, a row vector) and their standard errors (from matrix `e(Xmfx_se_dyex)`, also a row vector). We stack those vectors into matrix `eta` and extract the relevant estimates to use in the `ststo` command.

For the third model, we extract the mean level of `exper` from the `mfx` saved results' matrix `Xmfx_X` and use `lincom` to compute the expression $\widehat{\beta}_{\text{exper}} + 2\widehat{\mu}_{\text{exper}}\widehat{\beta}_{\text{exper}^2}$ in point and interval form. The `lincom` command leaves its scalars in the saved results, from which we extract `r(estimate)` and `r(se)`. Those values are passed to `ststo` for posting with the stored estimates.

Finally, we invoke `esttab` for the particular year's table, composing the output filename and table title to contain the year of analysis. The `scalar()` option allows us to include the statistics we have computed for each model in the table, and the `addnotes()` option is used to annotate the output, defining those statistics. We wrap the computation commands in `quietly` to suppress their output.

```
. forvalues y = 1980/1987 {
  2.           quietly {
  3.                    eststo clear
  4.                    regress lwage educ lhours  if year == 'y'
  5.                    mfx compute, dyex
  6.                    mat eta = e(Xmfx_dyex) \ e(Xmfx_se_dyex)
  7.                    eststo, addscalars(eta_educ eta[1, 1] etase eta[2, 1])
  8.                    regress lwage educ lhours black hisp if year == 'y'
  9.                    mfx compute, dyex
 10.                    mat eta = e(Xmfx_dyex) \ e(Xmfx_se_dyex)
 11.                    eststo, addscalars(eta_educ eta[1, 1] etase eta[2, 1])
 12.                    regress lwage educ exper expersq lhours black hisp if year
> == 'y'
 13.                    mfx compute, dyex
 14.                    mat eta = e(Xmfx_dyex) \ e(Xmfx_se_dyex)
 15.                    mat xbar = e(Xmfx_X)
 16.                    scalar mu2 = 2 * xbar[1, 2]
 17.                    lincom exper + mu2 * expersq
 18.                    eststo, addscalars(eta_educ eta[1, 1] etase eta[2, 1]
> exper r(estimate) se r(se))
 19.           }
 20.           esttab _all using lwage'y'.tex, replace ti("Wage equations for 'y'")
>                    nomtitles nodepvars not se noobs ar2 booktabs
>                    scalar(eta_educ etase exper se)
>                    addnotes("eta_educ: semielasticity of lwage with respect to
>                    educ" "etase: standard error of the semielasticity"
>                    "exper: effect of exper on lwage at mean exper")
>                    substitute("_cons" "Constant" "eta_educ" "$\eta_{\rm educ}$"
>                    "etase" "$\eta_{\rm se}$")
 21. }
(output written to lwage1980.tex)
(output written to lwage1981.tex)
(output written to lwage1982.tex)
(output written to lwage1983.tex)
(output written to lwage1984.tex)
(output written to lwage1985.tex)
(output written to lwage1986.tex)
(output written to lwage1987.tex)
```

We can now include the separate LaTeX tables produced by our do-file in our research paper with the LaTeX commands

```
\input{lwage1980}
\input{lwage1981}
\input{lwage1982}
...
\input{lwage1987}
```

This approach has the advantage that the tables themselves need not be included in the LaTeX document, so if we revise the tables (to include a different specification, for example), we need not copy and paste the tables. To illustrate, let's display one of the tables here as table 9.3.

Table 9.3. Wage equations for 1984

	(1)	(2)	(3)
educ	0.0759***	0.0758***	0.0923***
	(0.0124)	(0.0127)	(0.0158)
lhours	−0.162	−0.173	−0.173
	(0.0949)	(0.0950)	(0.0948)
black		−0.123	−0.126
		(0.0689)	(0.0691)
hisp		0.0160	0.00976
		(0.0619)	(0.0620)
exper			−0.0660
			(0.0765)
expersq			0.00551
			(0.00461)
Constant	2.046**	2.140**	2.122**
	(0.741)	(0.741)	(0.807)
adj. R^2	0.064	0.067	0.070
η_{educ}	0.893	0.892	1.086
η_{se}	0.146	0.150	0.185
exper			0.0113
se			0.0190

Standard errors in parentheses
* $p < 0.05$, ** $p < 0.01$, *** $p < 0.001$
η_{educ}: semielasticity of lwage with respect to educ
η_{se}: standard error of the semielasticity
exper: effect of exper on lwage at mean exper

LaTeX output is most convenient because LaTeX is itself a programming language, and Stata code can easily write programs in that language. The example above could be easily adapted to produce a set of tab-delimited files for use in Word or Excel or, alternatively, to produce a set of HTML web pages. For an excellent example of the automation of a web site containing many tables, see Gini and Pasquini (2006).

9.6   Automating graphics

Just as you can use do-files to automate the production of tabular output, you will find that Stata's graphical environment is particularly amenable to automation. In a research project, you may often need several essentially identical graphs: the same variables plotted for each unit of analysis or time period. Although Stata's graphics language is complex, it offers you the facility of complete customization of every aspect of the graph, including exporting the graph to a format usable by other applications. In this section, we will look at an extended example in the automation of graphs, building upon the example of the previous section. I present various examples of Stata graphics without explanation of the full syntax of Stata's graphics language. For an introduction to Stata graphics, see `help graph intro` and [G] **graph intro**. An in-depth presentation of Stata's graphics capabilities is provided by Michael Mitchell in his book, *A Visual Guide to Stata Graphics* (2008), and by Nicholas Cox's a number of his *Speaking Stata* columns in the *Stata Journal* (Cox 2004a,b,c,d, 2005a,b,c, 2006a).

Let's imagine that you want to produce a set of graphs for one of the wage equations estimated from the U.S. National Longitudinal Survey's Youth Sample. You would like two graphs for each year's regression. First, you want a scatterplot of the log wage (dependent) variable against the log hours variable, but you would like to distinguish minority workers (Blacks and Hispanics) from nonminority workers in the plot. Second, after estimating a wage equation,

```
regress lwage educ exper lhours if year == `y'
```

you would like to examine one of the *added-variable plots* described in [R] **regress postestimation**.

The added-variable plot, or `avplot`, is a graphical technique designed to identify the important variables in a relationship. The technique decomposes the multivariate relationship into a set of two-dimensional plots.[12] Taking each regressor (x^*) in turn, the added-variable plot is based on two residual series. The first series, e_1, contains the residuals from the regression of x^* on all other x, while the second series, e_2, contains the residuals from the regression of y on all x variables except x^*. That is, e_1 represents the part of x^* that is not linearly related to those other regressors, while e_2 represents the information in y that is left unexplained by a linear combination of the other regressors (excluding x^*). The added-variable plot for x^* is then the scatterplot of e_1 versus e_2. Two polar cases (as discussed by Cook and Weisberg [1994, 194]) are of interest. If most points are clustered around a horizontal line at the level of zero in the added-variable plot, x^* is irrelevant. On the other hand, if most points are clustered around a vertical line on the x axis at zero for that predictor, the plot would indicate that near-perfect collinearity is evident. Here, as well, the addition of x^* to the model would not be helpful.

---

12. An excellent discussion of the rationale for the added-variable plot is given by Cook and Weisberg (1994, 191–194).

The strength of a linear relationship between e_1 and e_2 (that is, the slope of a least-squares line through this scatter of points) represents the marginal value of x^* in the full model. If the slope is significantly different from zero, then x^* makes an important contribution to the model over and above that of the other regressors. The more closely the points are grouped around a straight line in the plot, the more important is the marginal contribution of x^*. As an added check, if the specification of the full model (including x^*) is correct, the plot of e_1 versus e_2 must exhibit linearity. Significant departures from linearity in the plot cast doubt on the appropriate specification of x^* in the model.

Suppose that you want to focus on the added-variable plot (`avplot`) of the log hours variable, `lhours`, in the wage equation. You would like the two graphs combined into one figure for each year of the analysis, 1980–1987.

To produce your automated graphics, you must first generate an indicator (dummy) variable denoting minority status and then use the **separate** command to split the log wage variable into minority and nonminority groups. The syntax of this command is

separate *varname* $\big[$ *if* $\big]$ $\big[$ *in* $\big]$, by(*byvar*| *exp*) $\big[$ *options* $\big]$

The `by` clause specifies the groups into which *varname* is to be separated.[13] You might, for example, want to separate the variable into a number of categories. Alternatively, *exp*, or an algebraic expression, can be given to create two categories: for example, `exper > 5`. The `shortlabel` option specifies that the variable name be omitted from the new variable labels. By default, *varname* will be used as the *stub* of the new variables to be created. With `minority` as an indicator variable taking on the values 0 and 1, two variables will be created: `lwage0` and `lwage1`. These variables can then be placed on the y axis of a scatterplot in place of the variable `lwage`.

Creating the set of eight combined figures then involves, as in our prior example, setting up a `forvalues` loop over `year` with y as the loop variable, running the regression, and producing the two figures for each year. You can use the `nodraw` option on each graph command to prevent their display and the `name` option to name them. Although the *Stata Reference Manual's* description of `avplot` ([R] **regress postestimation**) does not list all these options, the command allows any `graph twoway` ([G] **graph twoway**) option to be used. Finally, `graph combine` ([G] **graph combine**) puts the graphs together into one figure. You can choose to juxtapose them horizontally or vertically by using the `rows()` and `cols()` options. You might have four graphs or six graphs per year. Here we have only the two, so they are combined and the resulting graph is saved to a `.gph` file named `lwage‘y’` for each value of year.

---

13. The `separate` command was discussed in section 5.9.4.

```
. separate lwage, by(black | hisp) shortlabel

              storage  display    value
variable name  type    format     label      variable label
─────────────────────────────────────────────────────────────────
lwage0          float  %9.0g                  !(black | hisp)
lwage1          float  %9.0g                  black | hisp
. forvalues y = 1981/1987 {
  2.          quietly {
  3.              regress lwage educ exper lhours if year == `y'
  4.              scatter lwage0 lwage1 lhours if year==`y', msize(small)
  >                  scheme(s2mono) nodraw name(scat`y',replace) ytitle(lwage)
  >                  legend(pos(5) ring(0) col(1))
  5.              avplot lhours, msize(small) nodraw scheme(s2mono)
  >                  name(avplot`y', replace)
  6.              graph combine scat`y' avplot`y', nodraw col(2)
  >                  saving(lwage`y', replace) ti("Log wage vs. log hours for `y'")
  7.          }
  8. }
```

If you are satisfied with the graphs and want to include them in a document, you can translate the native-format Stata .gph file into another format. The highest-quality format for many purposes is Encapsulated PostScript, or .eps. You can translate into this format by typing

```
graph export lwage`y'.eps, replace
```

within the `forvalues` loop. Alternative high-quality export formats available from `graph export` ([G] **graph export**) include .png for HTML pages, .tif for page-layout programs, and .pdf (only available in Stata for Macintosh).

You can also change the aspect of the graph before exporting it with a command such as

```
graph display, xsize(5) ysize(3)
```

where the arguments of these two options are in inches. An example of the graphs produced is given in figure 9.1.

(Continued on next page)

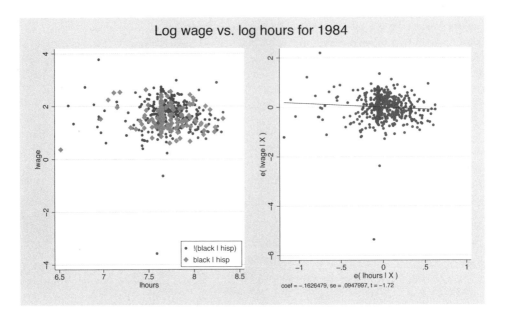

Figure 9.1. Automated graphics

In summary, the complete flexibility of Stata's graphics language lends itself to generating graphs automatically, without resorting to manual modification of each graph.[14] In this example, we demonstrated how to produce a set of graphs for subsets of the data on which statistical analyses have been performed. Alternatively, you may need to produce the same set of graphs in a repetitive process, such as a monthly update of a web site based on the latest statistics. For an excellent example of a graph-rich web site automatically updated when new data become available, see Gini and Pasquini (2006).

Many user-written graphic commands are conveniences that help you produce the graphics you need without having to master more-complex details of Stata's graphics language. A useful guide to Nicholas Cox's many contributions in this area is available in the SSC archive as njc_stuff (Cox 2007a).

9.7 Characteristics

Characteristics are described in [U] **12.8 Characteristics** as "an arcane feature of Stata but ... of great use to Stata programmers". Each saved dataset and each variable within the dataset can have any number of characteristics associated with them. The characteristics are saved with the data so that they are persistent. Dataset characteristics include indicators for data saved as panel (longitudinal) data with xtset

---

14. Stata's Graph Editor permits complete manual customization of the graph. The Graph Recorder allows the changes made to be written to a script. See **help graph editor**.

or `tsset`, saved as survival-time data with `stset`, or saved as complex survey data with `svyset`. Storing these characteristics with the data allows the datasets to retain this useful information so that they can be exchanged with other users who need not reestablish the panel (or survival-time, or complex survey) qualifiers of the data.

Characteristics can be examined, defined, or renamed with the `char` ([P] **char**) command. `char list` will show the characteristics of the current dataset as `_dta[`*charname*`]` and will also show charteristics defined for any variables in the dataset. `char list` *varname* will display characteristics associated with *varname*. Characteristics can be defined with

`char` *evarname*[*charname*] *"text"*

where *evarname* can be either `_dta` or the name of a variable. This command will define characteristic *charname* for *evarname* as the string *text*. If you are defining your own characteristics, you should include at least one capital letter in *charname* because lowercase characters are reserved for official Stata. Characteristics' values can contain 8,681 characters in Small Stata and 67,784 characters in all other flavors of Stata. The `char rename` command allows you to move a set of characteristics from one variable to another.

One interesting use for characteristics involves defining the `varname` characteristic for one or more variables. This characteristic interacts with the `subvarname` option of the `list` command. With that option invoked, the name of the variable is replaced in the output with the value of the `varname` characteristic. As an example, let's define characteristics for several variables in the Stata sample dataset `auto.dta`:

```
. sysuse auto, clear
(1978 Automobile Data)
. char make[varname] "modle"
. char price[varname] "prix"
. char weight[varname] "poids"
. char length[varname] "longueur"
```

(Continued on next page)

We can now display the data with either English or French labels:

```
. list make price weight length in 1/10, sep(0)
```

	make	price	weight	length
1.	AMC Concord	4,099	2,930	186
2.	AMC Pacer	4,749	3,350	173
3.	AMC Spirit	3,799	2,640	168
4.	Buick Century	4,816	3,250	196
5.	Buick Electra	7,827	4,080	222
6.	Buick LeSabre	5,788	3,670	218
7.	Buick Opel	4,453	2,230	170
8.	Buick Regal	5,189	3,280	200
9.	Buick Riviera	10,372	3,880	207
10.	Buick Skylark	4,082	3,400	200

```
. list make price weight length in 1/10, sep(0) subvarname
```

	modle	prix	poids	longueur
1.	AMC Concord	4,099	2,930	186
2.	AMC Pacer	4,749	3,350	173
3.	AMC Spirit	3,799	2,640	168
4.	Buick Century	4,816	3,250	196
5.	Buick Electra	7,827	4,080	222
6.	Buick LeSabre	5,788	3,670	218
7.	Buick Opel	4,453	2,230	170
8.	Buick Regal	5,189	3,280	200
9.	Buick Riviera	10,372	3,880	207
10.	Buick Skylark	4,082	3,400	200

For a large-scale example of the use of characteristics in data validation, consider Bill Rising's `ckvar` command, described in Rising (2007).

10 Cookbook: Do-file programming IV

This cookbook chapter presents for Stata do-file programmers several recipes using the programming features described in the previous chapter. Each recipe poses a problem and a worked solution. Although you may not encounter this precise problem, you may be able to recognize its similarities to a task that you would like to automate in a do-file.

10.1 Computing firm-level correlations with multiple indices

The problem. A user on Statalist posed a question involving a very sizable dataset of firm-level stock returns and a set of index fund returns. He wanted to calculate, for each firm, the average returns and the set of correlations with the index funds, and determine with which fund they were most highly correlated.

The solution. Let's illustrate this problem with some actual daily stock returns data from 1992–2006 for 291 firms, from the Center for Research in Security Prices; the data contain 311,737 daily observations in total. We have constructed nine simulated index funds' returns. The hypothetical funds, managed by a group of Greek investment specialists, are labeled the Kappa, Lambda, Nu, Xi, Tau, Upsilon, Phi, Chi, and Psi funds.

To solve the problem, we define a loop over firms. For each firm of the `nf` firms, we want to calculate the correlations between firm returns and the set of `nind` index returns, and find the maximum value among those correlations. The `hiord` variable takes on the values 1–9, while `permno` is an integer code assigned to each firm by the Center for Research in Security Prices. We set up a Stata matrix, `retcorr`, to hold the correlations, with `nf` rows and `nind` columns. The number of firms and number of indices are computed by the **word count** *string* extended macro function (see section 3.8) applied to the local macro produced by `levelsof`.

```
. use crspsubseta, clear
. label def ind 1 Kappa 2 Lambda 3 Nu 4 Xi 5 Tau 6 Upsilon 7 Phi 8 Chi 9 Psi
. qui levelsof hiord, local(indices)
. local nind: word count `indices'
. qui levelsof permno, local(firms)
. local nf: word count `firms'
. matrix retcorr = J(`nf', `nind', .)
```

We calculate the average return for each firm with `summarize, meanonly`. In looping over firms, we use `correlate` to compute the correlation matrix of each firm's returns, `ret`, with the set of index returns. For firm n, we move the elements of the last row of the matrix corresponding to the correlations with the index returns into the nth row of the `retcorr` matrix. We also place the mean for the nth firm into that observation of the `meanret` variable.

```
. local n 0
. qui gen meanret = .
. qui gen ndays = .
. local row = 'nind' + 1
. foreach f of local firms {
  2.          qui correlate index1-index'nind' ret if permno == 'f'
  3.          matrix sigma = r(C)
  4.          local ++n
  5.          forvalues i = 1/'nind' {
  6.                  matrix retcorr['n', 'i'] = sigma['row', 'i']
  7.          }
  8.          summarize ret if permno == 'f', meanonly
  9.          qui replace meanret = r(mean) in 'n'
 10.          qui replace ndays = r(N) in 'n'
 11. }
```

We can now use the `svmat` command to convert the `retcorr` matrix into a set of variables, `retcorr1`–`retcorr9`. The `egen` function `rowmax()` computes the maximum value for each firm. We then must determine which of the nine elements is matched by that maximum value. This maximum value is stored in `highcorr`.

```
. svmat double retcorr
. qui egen double maxretcorr = rowmax(retcorr*)
. qui generate highcorr = .
. forvalues i = 1/'nind' {
  2.          qui replace highcorr = 'i' if maxretcorr == retcorr'i'
>                 & !missing(maxretcorr)
  3. }
```

We now can sort the firm-level data in descending order of `meanret` by using `gsort`, and list firms and their associated index fund numbers. These values show, for each firm, which index fund their returns most closely resemble. For brevity, we only list the 50 best-performing firms.

```
. gsort -meanret highcorr
. label values highcorr ind
. list permno meanret ndays highcorr in 1/50, noobs sep(0)
```

permno	meanret	ndays	highcorr
24969	.0080105	8	Nu
53575	.0037981	465	Tau
64186	.0033149	459	Upsilon
91804	.0028613	1001	Psi
86324	.0027118	1259	Chi
60090	.0026724	1259	Upsilon
88601	.0025065	1250	Chi
73940	.002376	531	Nu
84788	.0023348	945	Chi
22859	.0023073	1259	Lambda
85753	.0022981	489	Chi
39538	.0021567	1259	Nu
15667	.0019581	1259	Kappa
83674	.0019196	941	Chi
68347	.0019122	85	Kappa
81712	.0018903	1259	Chi
82686	.0017555	987	Chi
23887	.0017191	1259	Lambda
75625	.0017182	1259	Phi
24360	.0016474	1259	Lambda
68340	.0016361	1259	Upsilon
34841	.001558	1259	Nu
81055	.0015497	1259	Lambda
85631	.0015028	1259	Chi
89181	.0015013	1259	Chi
76845	.0014899	1006	Phi
48653	.0014851	1259	Xi
90879	.0014393	1259	Psi
85522	.0014366	454	Chi
80439	.0014339	1186	Chi
85073	.0014084	1259	Phi
86976	.0014042	1259	Chi
51596	.0014028	1259	Tau
77971	.0013873	1259	Xi
25487	.0013792	1259	Chi
14593	.0013747	1072	Kappa
79950	.0013615	1259	Nu
79879	.0013607	127	Phi
12236	.0012653	858	Kappa
77103	.0012513	648	Lambda
81282	.0012314	1259	Chi
75034	.0012159	1259	Phi
46922	.0012045	1259	Xi
82488	.0011911	359	Chi
75912	.0011858	1173	Phi
82307	.0011574	1259	Kappa
83985	.0011543	1259	Kappa
79328	.0011498	1259	Phi
11042	.0011436	1259	Lambda
92284	.0011411	1259	Psi

An alternative approach to the computations, taking advantage of Mata, is presented in section 14.3.

10.2   Computing marginal effects for graphical presentation

The problem. Suppose you would like to produce a graph showing how a regressor's marginal effect and its associated confidence interval vary across a range of values for the regressor. As discussed in section 9.5, in a linear regression, the coefficients' point estimates are fixed, but derived estimates such as the elasticities (mfx, eyex) vary throughout the range of the regressor and response variable.

The solution. To illustrate, let's fit a model of median housing prices as a function of several explanatory factors: nox, the concentration of pollutants; dist, the distance from an urban center; rooms, the number of rooms in the house; stratio, the student–teacher ratio in that community; and proptax, the level of local property taxes. We compute elasticities (by default, at the point of means) with the eyex option for each explanatory variable.

```
. use hprice2a, clear

. regress price nox dist rooms stratio proptax
```

Source	SS	df	MS
Model	2.6717e+10	5	5.3434e+09
Residual	1.6109e+10	500	32217368.7
Total	4.2826e+10	505	84803032

Number of obs =	506
F(5, 500) =	165.85
Prob > F =	0.0000
R-squared =	0.6239
Adj R-squared =	0.6201
Root MSE =	5676

price	Coef.	Std. Err.	t	P>|t|	[95% Conf. Interval]
nox	-2570.162	407.371	-6.31	0.000	-3370.532 -1769.793
dist	-955.7175	190.7124	-5.01	0.000	-1330.414 -581.021
rooms	6828.264	399.7034	17.08	0.000	6042.959 7613.569
stratio	-1127.534	140.7653	-8.01	0.000	-1404.099 -850.9699
proptax	-52.24272	22.53714	-2.32	0.021	-96.52188 -7.963555
_cons	20440.08	5290.616	3.86	0.000	10045.5 30834.66

```
. mfx, eyex

Elasticities after regress
     y  = Fitted values (predict)
        =    22511.51
```

variable	ey/ex	Std. Err.	z	P>|z|	[95% C.I.]	X
nox	-.6336244	.10068	-6.29	0.000	-.830954 -.436295	5.54978
dist	-.1611472	.03221	-5.00	0.000	-.224273 -.098022	3.79575
rooms	1.906099	.1136	16.78	0.000	1.68344 2.12876	6.28405
stratio	-.9245706	.11589	-7.98	0.000	-1.15171 -.697429	18.4593
proptax	-.0947401	.04088	-2.32	0.020	-.174871 -.014609	40.8237

The significance levels of the elasticities are similar to those of the original coefficients. The regressor `rooms` is elastic, with an increase in `rooms` having almost twice as large an effect on `price` in percentage terms.[1] The other three regressors are inelastic, with estimated elasticities within the unit interval. The 95% confidence interval for `stratio` includes values less than −1.0, so we cannot conclude that housing prices are inelastic with respect to the student–teacher ratio. They might drop by more than one percent with a one percent increase in `stratio`.

The `at()` option of `mfx` is capable of computing point and interval estimates of the marginal effects or elasticities at any point. For ease of use, you can specify that one variable takes on a specific value while all others are held at their sample means or medians to trace out the effects of that regressor. For example, we can calculate a house price elasticity over the range of values of `nox` in the sample. The command also handles the discrete changes appropriate for indicator variables.

In the example below, we evaluate the variation in the elasticity of median house prices with respect to the community's student–teacher ratio in both point and interval form. We first run the regression and compute selected percentiles of `stratio` by using the `detail` option of `summarize`, saving the percentiles in a temporary variable, x, obtained from the `tempvar` command.

```
. use hprice2a, clear
. // run regression
. quietly regress price nox dist rooms stratio
. // compute appropriate t statistic  for 95% confidence interval
. scalar tmfx = invttail(e(df_r), 0.975)
. tempvar y x eyex seyex1 seyex2
. foreach v in `y' `x' `eyex' `seyex1' `seyex2'  { // generate variables needed
  2.           qui generate `v' = .
  3. }
. // summarize, detail computes percentiles of stratio
. quietly summarize stratio if e(sample), detail
. local pct  1 10 25 50 75 90 99
. local i = 0
. foreach p of local pct {
  2.           local pc`p' = r(p`p')
  3.           local ++i
  4. // set those percentiles into tempvar x
  .           quietly replace `x' = `pc`p'' in `i'
  5. }
```

To produce the graph, we must compute elasticities at the selected percentiles and store the `mfx` results in the temporary variable, y. The `mfx` command, like all estimation commands, leaves results behind that are described in the list of saved results. The saved quantities include scalars such as e(`Xmfx_y`), which is the predicted value of y generated from the regressors, and several matrices containing the marginal effects or elasticities. In the example above, we computed the elasticities with the `eyex` op-

---

1. In an elastic relationship, the elasticity is greater than one in absolute value, so a one percent change in x causes more than a one percent change in y.

tion. They are returned in the `e(Xmfx_eyex)` matrix with standard errors returned in
the `e(Xmfx_se_eyex)` matrix. The do-file extracts the appropriate values from those
matrices and uses them to create variables containing the percentiles of `stratio`, the
corresponding predicted values of `price`, the elasticity estimates, and their confidence
interval bounds.

```
 . local i = 0
 . foreach p of local pct {
 2. // compute elasticities at those points
 .     quietly mfx compute, eyex at(mean stratio = `pc`p'')
 3.     local ++i
 4. // save predictions at these points in tempvar y
 .     quietly replace `y' = e(Xmfx_y) in `i'
 5. // retrieve elasticities
 .     matrix Meyex = e(Xmfx_eyex)
 6.     matrix eta = Meyex[1, "stratio"]            // for the stratio column
 7.     quietly replace `eyex' = eta[1, 1] in `i'   // and save in tempvar eyex
 8. // retrieve standard errors of the elasticities
 .     matrix Seyex = e(Xmfx_se_eyex)
 9.     matrix se = Seyex[1,"stratio"]              // for the stratio column
10. // compute upper and lower bounds of confidence interval
 .     quietly replace `seyex1' = `eyex' + tmfx*se[1, 1] in `i'
11.     quietly replace `seyex2' = `eyex' - tmfx*se[1, 1] in `i'
12. }
```

These series are then graphed (as shown in figure 10.1), combining three `twoway` graph
types: `scatter` for the elasticities, `rline` for their standard errors, and `connected` for
the predicted values, with a second axis labeled with their magnitudes.[2]

```
 . label variable `x' "Student/teacher ratio (percentiles `pct')"
 . label variable `y' "Predicted median house price, dollars"
 . label variable `eyex' "Elasticity"
 . label variable `seyex1' "95% c.i."
 . label variable `seyex2' "95% c.i."
 . // graph the scatter of elasticities vs. percentiles of stratio
 . // as well as the predictions with rline
 . // and the 95% confidence bands with connected
 . twoway (scatter `eyex' `x', ms(Oh) yscale(range(-0.5 -2.0)) ylabel(,nogrid))
 > (rline `seyex1' `seyex2' `x')
 > (connected `y' `x', yaxis(2) yscale(axis(2) range(18000 35000))),
 > ytitle(Elasticity of price v. student/teacher ratio)
```

---

2. For a presentation of Stata's more sophisticated graphics capabilities, including overlaying several
   plot types, see *A Visual Guide to Stata Graphics, Second Edition* (Mitchell 2008).

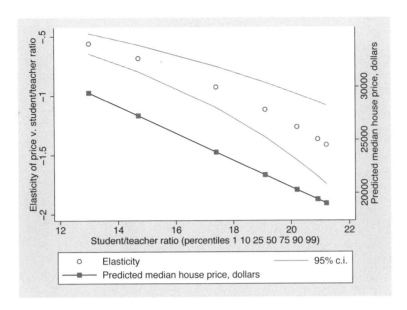

Figure 10.1. Point and interval elasticities computed with `mfx`

The predictions of the model for various levels of the student–teacher ratio demonstrate that more crowded schools are associated with lower housing prices, with all other things being equal. The elasticities vary considerably over the range of `stratio` values.

10.3   Automating the production of LATEX tables

The problem. Suppose a group of researchers is working with a frequently updated dataset containing information on U.S. corporate directors. One of the deliverables for the research project is a set of tables of variables' descriptive statistics and tests on their subsample means to be included in a LATEX document. The researchers want to automate the production of these tables so that when the data are updated, the tables can be immediately regenerated without any manual effort.

(Continued on next page)

The solution. The researchers can use Stata matrices as housekeeping devices in conjunction with the `file` command (discussed in section 9.4) to achieve these goals. A particularly useful tool is Nicholas J. Cox's `makematrix` command (Cox 2003b). This command makes a Stata matrix from the output of other Stata commands that do not store their results in a directly usable matrix form. For example, consider the `tabstat` command, used here to tabulate the number of observations that fall into two categories defined by the indicator variable `insider` (signaling whether the director is an "inside" director, employed by the firm, or an "outside" director, independent of the firm). The command will produce the subsample and total counts needed, but with the `save` option, it places them in three separate matrices, as `return list` shows.[3]

```
. use litgov_estsample.dta, clear

. tabstat insider, by(insider) stat(N) save

Summary for variables: insider
     by categories of: insider

  insider |           N
----------+----------
        0 |        1736
        1 |         984
----------+----------
    Total |        2720
----------+----------

. return list

macros:
             r(name2) : "1"
             r(name1) : "0"

matrices:
            r(Stat2) :  1 x 1
            r(Stat1) :  1 x 1
        r(StatTotal) :  1 x 1
```

We use `makematrix` to place the tabulated counts into a row vector, matrix `t2a`, and use the column-join operator (`,`) to add the total count:

```
. makematrix t2a, from(r(Stat1) r(Stat2)): tabstat insider, by(insider) stat(N)
> save

t2a[1,2]
         Stat1   Stat2
insider   1736     984

. summarize insider, meanonly

. matrix temp = r(N)

. matrix t2a = t2a, temp
```

We use a similar command to produce the fraction of directors who are their company's CEO; the variable `ceo` is an indicator variable for that characteristic. We assemble that information in matrix `t2aa`, and we use the row-join operator (`\`) to combine that matrix with matrix `t2a`.

---

3. If there were more than two categories of the variable being tabulated, additional matrices would be created by the `save` option. All the subsample matrices can be combined by using this technique.

```
. makematrix t2aa, from(r(Stat1) r(Stat2)): tabstat ceo, by(insider) stat(mean)
> save
t2aa[1,2]
            Stat1        Stat2
ceo             0   .27845528
. mat t2aa[1,1] = .
. summarize ceo if insider, meanonly
. mat temp = r(N)
. mat t2aa = t2a \ (t2aa, temp)
```

We now want to produce descriptive statistics for a set of variables and conduct
two statistical tests on their subsample means: a standard t test for the difference of
means, **ttest**, and a Mann–Whitney two-sample test, **ranksum** ([R] **ranksum**). For
each test, we want to tabulate the p-value. This value is available in **r(p)** for the t test
and can be calculated after **ranksum** from its computed z statistic, stored in **r(z)**. We
use one other useful function in this code fragment: the **nullmat()** matrix function,
which allows you to include a matrix[4] in an expression even if the matrix does not yet
exist.

```
. foreach v of varlist audit defendant_ANY ins_trade_ANY departed age tenure
> stkholding {
  2.        qui makematrix t2b, from(r(Stat1) r(Stat2)): tabstat 'v', by(insider)
> stat(mean) save
  3.        summarize 'v', meanonly
  4.        matrix 'v'1 = r(N)
  5.        qui ttest 'v', by(insider)
  6.        matrix 'v'2 = r(p)
  7.        qui ranksum 'v', by(insider)
  8.        matrix 'v'3 = 1 - normal(abs(r(z)))
  9.        matrix 'v' = t2b, 'v'1, 'v'2, 'v'3
 10.        matrix t2bb = (nullmat(t2bb) \ 'v')
 11. }
```

Before working on the output routine, we should check to see whether the two
matrices we have constructed look sensible:

```
. matrix colnames t2aa = Outsider Insider Total
. matrix rownames t2aa = Observations CEO
. matrix list t2aa
t2aa[2,3]
                Outsider     Insider       Total
Observations        1736         984        2720
        CEO            .   .27845528         984
. matrix colnames t2bb = Outsider Insider N t_pval MW_pval
```

---

4. See section 9.2.

```
. matrix list t2bb

t2bb[7,5]
                 Outsider    Insider          N      t_pval     MW_pval
        audit   .49884793   .14126016       2720    4.498e-82          0
defendant_ANY    .1013986   .50102249       2694    8.83e-132          0
ins_trade_ANY   .03484062   .21733333       2099    2.260e-42          0
     departed   .38652074    .4949187       2720    3.656e-08   1.977e-08
          age   59.703341   54.963415       2720    1.608e-41          0
       tenure    7.734764   9.3958844       2720    9.011e-10    .00005246
    stkholding   .17106612   2.6251817       2720    3.311e-27          0
```

The very small p-values result from the considerable differences between the means for outside and inside directors.

Having validated the matrices' contents, we are now ready to use the `file` command to produce the output file `table2.tex`. We set up a `tempname` for the file handle as local macro `hh`. This file handle is referenced in each subsequent invocation of `file`.[5] When using this command to produce an ASCII text file, we must explicitly write line-end markers (`_newline`) at the end of each line.[6] Using standard LaTeX table syntax, we separate column entries with ampersands (`&`) and mark the table line endings with a double backslash (`\\`).[7] Where a dollar sign (`$`) is needed in the output, we must "quote" it with a preceding backslash so that the dollar sign is not considered the beginning of a global macro.[8] Within each row of the table, we reference the appropriate cells of the `t2aa` and `t2bb` matrices.

```
. local inv Audit Defendant Ins\_Trading Departed Age Tenure Voting\_Share
. local inv1 "Member audit committee (0/1)"
. local inv2 "Defendant (0/1)"
. local inv5 "Age (years)"
. local inv6 "Board tenure (years)"
. local inv7 "Voting share (\%)"
. tempname hh
. file open 'hh' using table2.tex, write replace
. file write 'hh' "\begin{table}[htbp]\caption{Director-level
variables}\medskip"
> _newline
. file write 'hh' "\begin{tabular}{lrrrrr}" _newline
"\hline \noalign\smallskip" _newline
. file write 'hh' " Variable & Outside & Inside & &
> \multicolumn{2}{c}{\$p\$-value of difference} \\" _newline
. file write 'hh' "& directors & directors & & \multicolumn{2}{c}{in location}
> \\" _newline
```

---

5. You can open more than one file handle and write to different handles in turn. See section 9.3 for a discussion of file handles.

6. You can use the abbreviation `_n` for the line-end marker. I avoid that notation to prevent confusion with the use of `_n` in Stata to refer to the current observation.

7. Where a percent sign (`%`) is needed in LaTeX, it must be "escaped" with a preceding backslash.

8. See section 4.6.

```
. file write 'hh' "\noalign\smallskip  \hline \noalign\smallskip & Mean & Mean &
> \$N\$ & \$t\$ test & Mann--Whitney \\" _newline
. file write 'hh' "\noalign\smallskip \hline \noalign\smallskip" _newline
. file write 'hh' "Observations (\$N\$) & " (t2aa[1, 1]) " & " (t2aa[1, 2]) " & "
> (t2aa[1, 3]) " \\" _newline
. file write 'hh' "\\{\sl Involvement} \\" _newline
. file write 'hh' "CEO (0/1) & N/A & " %7.3f (t2aa[2, 2]) " & " (t2aa[2, 3]) "
> \\" _newline
. forvalues i = 1/2 {
  2.        file write 'hh' "'inv'i'' & "  %7.3f (t2bb['i', 1]) " & "  %7.3f
> (t2bb['i', 2]) " & " (t2bb['i', 3])  " & " %7.3f (t2bb['i', 4])  " & " %7.3f
> (t2bb['i', 5]) " \\" _newline
  3. }
. file write 'hh' "\\{\sl Outcome} \\" _newline
. local i 4
. file write 'hh' "Departed (0/1) & "  %7.3f (t2bb['i', 1]) " & "  %7.3f
> (t2bb['i', 2]) " & " (t2bb['i', 3])  " & " %7.3f (t2bb['i', 4])  " & "
> %7.3f (t2bb['i', 5]) " \\" _newline
. file write 'hh' "\\{\sl Demographics} \\" _newline
. forvalues i = 5/7 {
  2.        file write 'hh' "'inv'i'' & "  %7.3f (t2bb['i', 1]) " & "  %7.3f
> (t2bb['i', 2]) " & " (t2bb['i', 3])  " & " %7.3f (t2bb['i', 4])  " & " %7.3f
> (t2bb['i', 5]) " \\" _newline
  3. }
. file write 'hh' "\noalign\smallskip \hline" _n "\end{tabular}" "\medskip"
> _newline
. file write 'hh'  "\end{table}" _newline
. file close 'hh'
```

We are ready to view the finished product; see table 10.1. We could readily add information to the table in the form of headings or notes stored in separate LATEX files referenced in the table environment.

(Continued on next page)

Table 10.1. Director-level variables

Variable	Outside directors	Inside directors		*p*-value of difference in location	
	Mean	Mean	*N*	*t* test	Mann–Whitney
Observations (*N*)	1736	984	2720		
Involvement					
CEO (0/1)	N/A	0.278	984		
Member audit committee (0/1)	0.499	0.141	2720	0.000	0.000
Defendant (0/1)	0.101	0.501	2694	0.000	0.000
Outcome					
Departed (0/1)	0.387	0.495	2720	0.000	0.000
Demographics					
Age (years)	59.703	54.963	2720	0.000	0.000
Board tenure (years)	7.735	9.396	2720	0.000	0.000
Voting share (%)	0.171	2.625	2720	0.000	0.000

10.4   Tabulating downloads from the Statistical Software Components archive

The problem. The web server log of Statistical Software Components archive activity counts the number of downloads of each ado-file. I want to present the number of *package downloads*, rather than individual ado-file downloads, because many packages contain multiple ado-files. Packages have one or more authors, so I would also like to present the download statistics for each author, adjusting for multiple authorship.

The solution. The raw data are of the following form:

```
1464:  1.38%:  51.428:  3.36%: Nov/ 1/07  5:11 AM: /repec/bocode/o/outreg.ado
1244:  1.17%:  96.083:  6.28%: Nov/ 1/07  4:34 AM: /repec/bocode/o/outreg2.ado
1173:  1.11%: 143.543:  9.38%: Nov/ 1/07  5:37 AM: /repec/bocode/e/estout.ado
1168:  1.10%:  47.396:  3.10%: Nov/ 1/07  5:37 AM: /repec/bocode/e/esttab.ado
1164:  1.10%:   0.915:  0.06%: Nov/ 1/07  5:37 AM: /repec/bocode/_/_eststo.ado
```

For each observation, the first variable gives the number of ado-file downloads, while the last variable contains the name of the ado-file. I read those fields with `infix` (see section 2.7.1) and save them as a Stata dataset, `hits.dta`:

```
. local mmyy oct2007
. infix nhit 1-5 str url 53-92 using "`mmyy'.ssc.raw", clear
. format url %40s
. sort url
. drop if nhit == .
. save hits, replace
```

Separately, I have prepared another ASCII file, `extrAU.raw`, which contains five variables: `package`, `url`, `nmods`, `author`, and a sequence number. This file contains one observation for each package/author combination, with `nmods` listing the number of ado-files in the package. These data are of the following form:

```
_PEERS   /repec/bocode/_/_gpeers.ado   1   Amine Ouazad   1
XTMIS    /repec/bocode/x/xtmis.ado     1   Minh Nguyen    2
FTEST    /repec/bocode/f/ftest.ado     1   Maarten L. Buis   3
POWERQ   /repec/bocode/p/powerq.ado    1   Nikolaos A Patsopoulos   4
POWERQ   /repec/bocode/p/powerq.ado    1   Tiago V Pereira   4
A2REG    /repec/bocode/a/a2reg.ado     2   Amine Ouazad   5
A2REG    /repec/bocode/a/a2group.ado   2   Amine Ouazad   5
```

In this example, each package except `A2REG` contains one ado-file; `A2REG` contains two. I read this file with `insheet` (see section 2.7.1):

```
. insheet using  "extrAU.raw", clear
. gen module = reverse(substr(reverse(url), 1, strpos(reverse(url), "/") -1))
. sort url
. save authors, replace
```

This code fragment defines `module` as the last segment of the ado-file's URL; for example, from `/repec/bocode/a/acplot.ado`, we need just `acplot.ado`. The string manipulation functions `reverse()`, `substr()`, and `strpos()` are used to extract the desired substring. This dataset is then saved as `authors.dta`.

I am now ready to combine the `hits` and `authors` datasets by using `merge` (see section 5.8) with the `uniqmaster` option:

```
. use hits
. merge url using authors, uniqmaster
. drop if _merge < 3
. drop _merge
```

Any observations found in either file that are not matched in the `merge` can be discarded based on the value of the `_merge` variable. Those with `_merge=2` are ado-files that were not downloaded. If there are any with `_merge=1`, problems with the `extrAU.raw` file would be indicated.

The number of package hits can now be generated as the ratio of hits (downloads) to modules, or ado-files, in the package. The number of package hits can be noninteger, because a particular use of the `ssc` command can download only some of the ado-files in the package (those which have been updated on the archive). Given the number of package hits, the `collapse` command can generate a single record giving the sum of downloads for each author/package combination. Unlike some tallies of multiply authored materials such as journal articles, the total number of package hits are assigned to each author for a package with multiple authors.

```
. generate npkghit = nhit / nmods
. collapse (sum) npkghit, by(author package)
. gsort -npkghit
```

I can now list package downloads in descending order by typing

```
. list npkghit author package, noobs
```

producing a list like this:

```
npkghit                         author             package
2187.00                       Roy Wada             OUTREG2
1149.00              John Luke Gallup              OUTREG
1125.00                      Ben Jann              ESTOUT
 914.33            Christopher F Baum               IVREG2
 914.33              Mark E Schaffer                IVREG2
 914.33              Steven Stillman               IVREG2
 900.75               David Roodman              XTABOND2
 835.67             Barbara Sianesi              PSMATCH2
 835.67                 Edwin Leuven             PSMATCH2
 741.00          Sophia Rabe-Hesketh               GLLAMM
 481.50              Mark E Schaffer             XTIVREG2
```

To produce a summary listing by authors, I can merely carry out another `collapse`
by `author`,

```
. collapse (sum) npkghit, by(author)
. gsort -npkghit
. list if author != "", noobs
```

producing a list like this:

```
author      npkghit
         Christopher F Baum     6620.70
            Nicholas J. Cox     6610.82
                  Ben Jann     3044.17
           Mark E Schaffer     2824.83
                  Roy Wada     2187.00
           Steven Stillman     1886.33
             David Roodman     1406.45
         Stephen P. Jenkins     1268.32
           Thomas Steichen     1201.66
          John Luke Gallup     1168.00
             Adrian Mander     1073.00
             Vince Wiggins     1072.00
              Roger Newson     1056.20
```

These statistics are saved in a Stata dataset for combination with two prior months'
values, underlying the values displayed by the `ssc hot` command (`help ssc`).

10.5   Extracting data from graph files' sersets

The problem. To revise a paper submitted to a professional journal some time ago, you
must produce new graphs. You have the do-file that produced the graphs, its log file,
and the graph (`.gph`) files themselves, but the Stata dataset was lost in a hard disk
crash (and was not backed up elsewhere—not a recommended practice).

The solution. Fortunately, Stata graph (.gph) files are not merely bitmaps or lists of vector graphics instructions: a graph file is actually a program to reproduce the graph.[9] This program also contains the data series that appear on the graph, stored in one or more sersets ([P] **serset**). Sersets are like datasets in that they contain information about one or more variables. Each serset associated with a graph is assigned a sequential number starting with zero. Your references to sersets use those numbers.

To illustrate the use of sersets to retrieve the data behind a graph file, let's first consider a do-file that creates the graph files:

```
. use airquality, clear
. drop if town == "Alburq"
(1 observation deleted)
. generate tabrv = upper(substr(town, 1, 3))
. summarize temp, meanonly
. generate hightemp = (temp > r(mean) & !missing(temp))
. label def tlab 0 "below mean temp" 1 "above mean temp"
. label values hightemp tlab
. scatter so2 temp, msize(tiny) mlabel(tabrv) mlabsize(vsmall)
> saving(fig10_4_1, replace) scheme(s2mono)
(file fig10_4_1.gph saved)
. scatter so2 precip, msize(tiny) mlabel(tabrv) mlabsize(vsmall)
> by(hightemp) saving(fig10_4_2, replace) scheme(s2mono)
(file fig10_4_2.gph saved)
. scatter so2 wind, msize(tiny) mlabel(tabrv) mlabsize(vsmall)
> by(hightemp) saving(fig10_4_3, replace) scheme(s2mono)
(file fig10_4_3.gph saved)
```

The `airquality` dataset was also used in section 5.4. For pedagogical purposes, we drop one city (Albuquerque) so that all the cities' names can be distinctly abbreviated to three letters. The do-file produces three figures: The first includes variables so2 and temp, with points labeled by tabrv. The second includes so2 and precip, with hightemp used as a by-variable. The third includes so2 and wind, also with hightemp as a by-variable. Thus the three graphs contain the city names (tabrv) and four measures: so2, temp, precip, and wind. For illustration, here is the second graph as figure 10.2.

---

9. Strictly speaking, this pertains only to live graph files. If the `asis` option is used in **graph save** or as a suboption to **saving()**, the file is "frozen" in its current state and can no longer be edited by the Graph Editor.

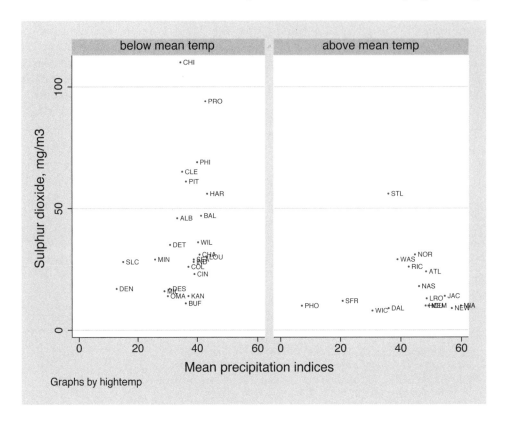

Figure 10.2. Air quality in U.S. cities

We now imagine that the original `airquality.dta` dataset is no longer accessible, but you have the three `.gph` files produced above. In the following do-file, `graph use` accesses one of the graph files. `serset dir` illustrates the available sersets from that graph file, here a single serset. `serset use` adds those variables to the dataset, which we then save as `f10_4_1.dta`.

```
. clear
. serset clear
. graph use fig10_4_1
. serset dir
  0.  40 observations on 3 variables
      so2 temp tabrv
. serset use
. sort tabrv
. save f10_4_1, replace
file f10_4_1.dta saved
```

We then go through the same steps for the second and third graph files. In these files, because there are two graph panels (see figure 10.2), there are two sersets, numbered 0 and 1. Because the by() option has been used to separate the dataset, each serset contains a subset of the original dataset. We access each serset and store its contents in a separate Stata dataset:

```
. serset clear
. graph use fig10_4_2
. serset dir
  0.   24 observations on 3 variables
         so2 precip tabrv
  1.   16 observations on 3 variables
         so2 precip tabrv
. serset set 0
. serset use
. sort tabrv
. save f10_4_2, replace
file f10_4_2.dta saved
. serset set 1
. serset use
. sort tabrv
. save f10_4_2a, replace
file f10_4_2a.dta saved
. serset clear
. graph use fig10_4_3
. serset dir
  0.   24 observations on 3 variables
         so2 wind tabrv
  1.   16 observations on 3 variables
         so2 wind tabrv
. serset set 0
. serset use
. sort tabrv
. save f10_4_3, replace
file f10_4_3.dta saved
. serset set 1
. serset use
. sort tabrv
. save f10_4_3a, replace
file f10_4_3a.dta saved
```

We are now ready to combine all the series and reconstruct a significant chunk of the missing Stata dataset. We use the first dataset, which contains observations for all the cities, and merge datasets f10_4_2 and f10_4_3, which contain observations for a subset of observations of the precip and wind variables (those related to hightemp = 0). Because two using datasets are being merged, three status variables are created: _merge, _merge1, and _merge2. We use the wildcard (*) to drop them all.

At this point, the dataset in memory contains complete information for all observations on so2, temp, and tabrv, but contains only a subset of observations on precip and wind. To incorporate the observations from the other subset (for which hightemp = 1), we use merge with the update option. This causes the additional observations to replace missing values in the precip and wind variables. At the end of this process, we have complete information on all five variables.

```
. use f10_4_1, clear
. merge tabrv using f10_4_2 f10_4_3, unique
. drop _merge
. sort tabrv
. merge tabrv using f10_4_2a, update unique
. drop _merge
. sort tabrv
. merge tabrv using f10_4_3a, update unique
. drop _merge
. save f10_4_all, replace
file f10_4_all.dta saved
. describe
Contains data from f10_4_all.dta
  obs:             40
  vars:             7                              6 Sep 2008 13:24
  size:         1,000 (99.9% of memory free)
```

variable name	storage type	display format	value label	variable label
so2	float	%9.0g		
temp	float	%9.0g		
tabrv	str3	%9s		
precip	float	%9.0g		
_merge1	byte	%8.0g		_merge representing f10_4_2
wind	float	%9.0g		
_merge2	byte	%8.0g		_merge representing f10_4_3

```
Sorted by:
. summarize
```

Variable	Obs	Mean	Std. Dev.	Min	Max
so2	40	30.525	23.56985	8	110
temp	40	55.7375	7.317864	43.5	75.5
tabrv	0				
precip	40	37.494	10.9554	7.05	59.8
_merge1	40	.6	.4961389	0	1
wind	40	9.4575	1.444155	6	12.7
_merge2	40	.6	.4961389	0	1

We are now in a position to produce tables, statistics, or graphs for these with the reconstructed dataset, f10_4_all.dta.

10.6   Constructing continuous price and returns series

The problem. Many important commodities and financial instruments trade on organized futures markets, where buyer and seller agree to exchange the commodity at a future date by means of a futures contract. Futures contract prices are quoted just as stock prices are quoted, but for futures contracts, the price quotes refer to an instrument with a specific expiration date. Contracts expire on a fixed schedule: for the most heavily traded commodities, monthly. Many market participants never actually take delivery of the underlying commodity—such as crude oil—but rather maintain a continuous position in crude oil futures, "rolling over" that position when the nearest-to-maturity contract approaches its expiration date. This complicates analysis of data derived from futures prices, such as rates of return from maintaining a futures position. Let's say would like to produce a continuous price series and returns series from the set of futures contract price quotations available, spanning the various contract maturity dates.

The solution. Industry analysts have noted that to avoid market disruptions, the large participants in the crude oil futures market roll over their positions from the near contract to the next-near contract over several days before the near contract's expiration date. We can thus define a method of producing a continuous price series from a set of price quotations on multiple months' contracts. Consider the last five trading days before expiration, and consider the midpoint of that interval to be the rollover date. Prior to the rollover date, the price is taken from the near contract, that which will imminently expire. On and after the rollover date, the price is taken from the next-near contract, the contract expiring next month.

For all but the rollover date, the return is the *log price relative*, that is, the logarithm of the ratio of price today to price yesterday, or $\log(P_t/P_{t-1})$. For the rollover date, we assume that the near-contract position is liquidated at the previous trading day's settlement price and instantaneously rolled into the next-near contract. Thus the return on the rollover date is the log price relative of the next-near contract for the rollover day and the previous trading day.

To illustrate, let's display an excerpt of the crude oil futures contract quotations:

```
. use crudeoil, clear
. list in 1/51, sepby(qmdy)
```

	qmdy	cmdy	settle	contract	qm	cm
1.	02jan1985	01feb1985	25.92	CL1985G	300	301
2.	02jan1985	01mar1985	25.81	CL1985H	300	302
3.	02jan1985	01apr1985	25.69	CL1985J	300	303
4.	02jan1985	01may1985	25.63	CL1985K	300	304
5.	02jan1985	01jun1985	25.60	CL1985M	300	305
6.	02jan1985	01jul1985	25.59	CL1985N	300	306
7.	02jan1985	01aug1985	25.57	CL1985Q	300	307
8.	02jan1985	01sep1985	25.57	CL1985U	300	308
9.	02jan1985	01oct1985	25.57	CL1985V	300	309
10.	02jan1985	01nov1985	25.57	CL1985X	300	310
11.	02jan1985	01dec1985	25.57	CL1985Z	300	311
12.	02jan1985	01jan1986	25.57	CL1986F	300	312
13.	02jan1985	01feb1986	25.57	CL1986G	300	313
14.	02jan1985	01mar1986	25.57	CL1986H	300	314
15.	02jan1985	01apr1986	25.57	CL1986J	300	315
16.	02jan1985	01may1986	25.57	CL1986K	300	316
17.	02jan1985	01jun1986	25.57	CL1986M	300	317
18.	03jan1985	01feb1985	25.84	CL1985G	300	301
19.	03jan1985	01mar1985	25.79	CL1985H	300	302
20.	03jan1985	01apr1985	25.68	CL1985J	300	303
21.	03jan1985	01may1985	25.65	CL1985K	300	304
22.	03jan1985	01jun1985	25.59	CL1985M	300	305
23.	03jan1985	01jul1985	25.58	CL1985N	300	306
24.	03jan1985	01aug1985	25.56	CL1985Q	300	307
25.	03jan1985	01sep1985	25.56	CL1985U	300	308
26.	03jan1985	01oct1985	25.56	CL1985V	300	309
27.	03jan1985	01nov1985	25.56	CL1985X	300	310
28.	03jan1985	01dec1985	25.56	CL1985Z	300	311
29.	03jan1985	01jan1986	25.56	CL1986F	300	312
30.	03jan1985	01feb1986	25.56	CL1986G	300	313
31.	03jan1985	01mar1986	25.56	CL1986H	300	314
32.	03jan1985	01apr1986	25.56	CL1986J	300	315
33.	03jan1985	01may1986	25.56	CL1986K	300	316
34.	03jan1985	01jun1986	25.56	CL1986M	300	317
35.	04jan1985	01feb1985	25.18	CL1985G	300	301
36.	04jan1985	01mar1985	25.19	CL1985H	300	302
37.	04jan1985	01apr1985	25.16	CL1985J	300	303
38.	04jan1985	01may1985	25.13	CL1985K	300	304
39.	04jan1985	01jun1985	25.10	CL1985M	300	305
40.	04jan1985	01jul1985	24.90	CL1985N	300	306
41.	04jan1985	01aug1985	25.06	CL1985Q	300	307
42.	04jan1985	01sep1985	25.06	CL1985U	300	308
43.	04jan1985	01oct1985	25.06	CL1985V	300	309
44.	04jan1985	01nov1985	25.06	CL1985X	300	310
45.	04jan1985	01dec1985	25.06	CL1985Z	300	311
46.	04jan1985	01jan1986	25.06	CL1986F	300	312
47.	04jan1985	01feb1986	25.06	CL1986G	300	313
48.	04jan1985	01mar1986	25.06	CL1986H	300	314
49.	04jan1985	01apr1986	25.06	CL1986J	300	315
50.	04jan1985	01may1986	25.06	CL1986K	300	316
51.	04jan1985	01jun1986	25.06	CL1986M	300	317

In this listing of three days' quotations, `qm` is the month number of the quote date, and `cm` is the month number in which the contract expires. Stata's dates start from 1 January 1960, so month 300 is January 1985 and month 301 is February 1985. `qmdy` is the quote date and `cmdy` is the maturity month of the contract; that is, contract `CL1985G` is the "Feb 85" contract that last traded on 18 January 1985, because contracts expire in roughly the third week of the previous calendar month. That contract was quoted at $25.92 (per barrel of crude oil) on 2 January 1985 and $25.84 on 3 January 1985.

We first must define the near contract (that closest to expiration) for each month:

```
. // identify last day of trading for near contract in each month
. bysort qmdy (cmdy): generate near = contract if _n == 1
(155907 missing values generated)
```

We now can identify the rollover date and the prior trading date:[10]

```
. // qmdy is the first date when the near contract is no longer quoted
. // that minus 3 trading days is the target rollover date
. bysort cmdy (qmdy): generate rolldate = qmdy[_n-3] if near[_n] != near[_n-1]
(161354 missing values generated)

. bysort cmdy (qmdy): generate roll1date = qmdy[_n-4] if near[_n] != near[_n-1]
(161354 missing values generated)

. bysort cmdy (qmdy): generate nnear = contract if near[_n] != near[_n-1]
(161353 missing values generated)

. // fixup for first obs
. replace nnear = . in 1
(1 real change made, 1 to missing)
```

We use the `egen mean()` function to set the `rolldate`, `roll1date`, and `nnear` values into each trading day of the quote month:

```
. bysort qm: egen rollover = mean(rolldate)

. bysort qm: egen rollover1 = mean(roll1date)

. bysort qm: egen nextnear = mean(nnear)
```

With these variables defined, we are ready to calculate the continuous price and returns series:

```
. // calculate price series as settle(near) for prerollover dates
. bysort qm: generate futprice = settle if contract == near & qmdy < rollover
(158572 missing values generated)

. // calculate price series as settle(nnear) for rollover date et seq.
. bysort qm: replace futprice = settle if contract==nextnear & qmdy >= rollover
(2665 real changes made)

. format futprice %9.2f

. // calculate return series for the rollover date
. bysort qm: generate settleprev = settle if contract == nextnear & qmdy ==
> rollover1
(161356 missing values generated)
```

---

10. In the code, note that references to `near[_n]` can be replaced with **near**. I use the explicit subscript to clarify the meaning of the code.

```
. bysort qm: egen sprev = mean(settleprev)

. bysort qm: generate double futret = log(settle) - log(sprev) if qmdy ==
> rollover
(153860 missing values generated)

. // drop obs no longer needed, flagged by missing settle
. drop if futprice == .
(155907 observations deleted)

. // calc returns for all non-settle dates
. sort qmdy

. replace futret = log(settle) - log(settle[_n-1]) if missing(futret)
(5449 real changes made)
```

We now verify that the proper series have been constructed:

```
. sort qmdy cmdy

. list qmdy contract futprice futret if !missing(futprice) in 1/62,  noobs
> sepby(qm)
```

qmdy	contract	futprice	futret
02jan1985	CL1985G	25.92	.
03jan1985	CL1985G	25.84	-.00309119
04jan1985	CL1985G	25.18	-.02587364
07jan1985	CL1985G	25.56	.01497857
08jan1985	CL1985G	25.48	-.0031348
09jan1985	CL1985G	25.43	-.00196422
10jan1985	CL1985G	25.76	.01289332
11jan1985	CL1985G	25.77	.00038813
14jan1985	CL1985G	26.12	.01349029
15jan1985	CL1985G	25.91	-.00807235
16jan1985	CL1985H	25.57	-.01243699
17jan1985	CL1985H	25.69	.00468205
18jan1985	CL1985H	25.75	.0023328
21jan1985	CL1985H	25.97	.00850737
22jan1985	CL1985H	25.55	-.01630471
23jan1985	CL1985H	25.40	-.00588813
24jan1985	CL1985H	25.28	-.00473556
25jan1985	CL1985H	25.25	-.00118744
28jan1985	CL1985H	25.23	-.00079241
29jan1985	CL1985H	25.38	.00592768
30jan1985	CL1985H	25.67	.01136157
31jan1985	CL1985H	26.41	.02841972

01feb1985	CL1985H	26.74	.01241784
04feb1985	CL1985H	26.52	-.00826138
05feb1985	CL1985H	26.78	.00975618
06feb1985	CL1985H	27.07	.01077073
07feb1985	CL1985H	27.21	.00515843
08feb1985	CL1985H	27.59	.01386887
11feb1985	CL1985H	28.04	.0161787
12feb1985	CL1985H	27.36	-.02454998
13feb1985	CL1985J	27.06	.0186503
14feb1985	CL1985J	27.04	-.00073932
15feb1985	CL1985J	27.38	.0124955
19feb1985	CL1985J	27.29	-.00329242
20feb1985	CL1985J	27.18	-.00403895
21feb1985	CL1985J	27.14	-.00147279
22feb1985	CL1985J	26.76	-.01410039
25feb1985	CL1985J	26.44	-.01203021
26feb1985	CL1985J	26.79	.01315068
27feb1985	CL1985J	26.69	-.00373973
28feb1985	CL1985J	26.73	.00149753
01mar1985	CL1985J	27.20	.01743049
04mar1985	CL1985J	27.74	.01965841
05mar1985	CL1985J	27.55	-.0068729
06mar1985	CL1985J	27.77	.00795381
07mar1985	CL1985J	28.08	.01110126
08mar1985	CL1985J	27.74	-.01218217
11mar1985	CL1985J	27.57	-.00614719
12mar1985	CL1985J	27.92	.01261507
13mar1985	CL1985J	28.06	.00500178
14mar1985	CL1985J	28.19	.00462227
15mar1985	CL1985J	28.32	.00460093
18mar1985	CL1985K	28.25	.02182948
19mar1985	CL1985K	28.19	-.00212613
20mar1985	CL1985K	27.99	-.00712003
21mar1985	CL1985K	28.32	.01172096
22mar1985	CL1985K	28.24	-.00282885
25mar1985	CL1985K	28.09	-.00532576
26mar1985	CL1985K	28.45	.01273454
27mar1985	CL1985K	28.16	-.01024566
28mar1985	CL1985K	28.25	.00319093
29mar1985	CL1985K	28.29	.00141496

You should note in this recipe how business-daily data have been handled: we did not use any time-series operators (such as L.), because the trading-day data are not evenly spaced in calendar time. Also the multiple price quotations per trading day have been transformed into simple time series of prices and returns in the process.

11 Ado-file programming

11.1 Introduction

In this chapter, I discuss *ado-file programming*: creating your own Stata commands to extend Stata's functionality. In section 7.2, I discussed some special-purpose ado-file programs in presenting the `simulate` and `bootstrap` prefixes. I now turn to a more general use of Stata's `program` statement and discuss the advantages of developing an ado-file rather than a set of do-files.

Stata's design makes it easy for you to develop a new Stata command and equip that command with all the standard elements of Stata syntax, such as if *exp* and in *range* qualifiers, weights, and command options. Why should you consider writing your own command? Many users find that they want to automate a frequently executed procedure in their workflow and that procedure may depend on items that change. You could write a do-file that relies on global macros,[1] but you would still have to insert new values in those macros every time you ran the do-file. Writing a command that implements this procedure is a better approach.

Alternatively, you may want to implement a statistical, data-management, or graphical technique not available in official Stata or from user-contributed routines in the *Stata Journal*, the Statistical Software Components (SSC) archive, or elsewhere. New graphical procedures, statistical estimation techniques, and tests are continually appearing in the literature. You may decide that it is easy enough to implement one of those techniques or tests as a Stata command and perhaps even share it with other Stata users.[2]

This chapter provides an overview of how to develop your own ado-file, or Stata command, by using the `program` statement. I discuss both r-class and e-class programs, as well as programs for use with `egen`, `ml` ([R] **ml**), `nl` ([R] **nl**), `nlsur` ([R] **nlsur**), and the `simulate`, `bootstrap`, and `jackknife` prefixes. If you want to further develop your understanding of the mechanics of developing Stata commands, you might consider enrolling in one of StataCorp's NetCourses on programming techniques.[3]

---

1. See section 3.7.
2. See `ssc describe sscsubmit`.
3. Visit http://www.stata.com/netcourse/ for more information.

11.1.1   What you should learn from this chapter

- How to develop a Stata command as an ado-file program
- How to use the `syntax` and `return` statements
- How to provide program options and sample qualifiers
- How to write a Stata Markup and Control Language (SMCL) help file
- How to make programs byable
- How to write an `egen` function
- How to write an e-class program
- How to certify your program's reliability
- How to write programs for `ml`, `nl`, `nlsur`, `simulate`, `bootstrap`, and `jackknife`
- How to follow good ado-file programming style

11.2   The structure of a Stata program

A Stata program follows a simple structure. You must declare the name of the program, optionally specify its characteristics, and handle the program's arguments. When you invoke the program in Stata by typing its name, the arguments are the items, separated by spaces, following the program name on the command line.[4] You then perform the steps that this program is to take (compute one or more quantities, display some output, read or write a file, transform the data, or produce a graph) and, optionally, return various items to the user (or to the program that calls this one). As an example, I write the `parrot` command, which merely echoes whatever it is given as an argument:

```
. type parrot.ado
program parrot
        version 10.1
        display "`1'"
end
. parrot "This is Stata..."
This is Stata...
```

This program takes one argument. When you call a program, the arguments are assigned to local macros 1, 2, ..., 9, and so forth, in order, while the local macro 0 contains everything typed after the command name. Thus, within the program, local macro 1 refers to the string typed on the command line as the first (and only) argument. We have placed the `parrot` command in an ado-file, `parrot.ado`, in a directory on the ado-path (see section 2.2.2). If the name of the ado-file does not match the name of the program defined, Stata will refuse to execute the program. Placing the ado-file in the appropriate `PERSONAL` directory will make the program permanently available in your copy of Stata.

---

4. If an argument is a string containing spaces, it should be quoted on the command line.

The `parrot` program carries out one rudimentary step: it takes an argument and operates on it (albeit trivially). It does not access Stata's variables, nor does it return any computed quantities. Nevertheless, it illustrates the notion: once you have defined a Stata command with `program`, it becomes indistinguishable from any other Stata command. You should, of course, ensure that your command's name is not the same as any existing command, either in official Stata or in the user-written programs (including yours) on your system. Before choosing a name, use the `which` command to ensure that the program does not already exist on your system. You also might want to use the `findit` command to ensure that the name has not been used by a user-written routine (in the *Stata Journal*, the SSC archive, or user sites) in case you decide to install that routine at some later date.

In the next sections of this chapter, I discuss the `program` statement, followed by a discussion of handling the program's arguments with the `syntax` statement and managing the sample to be used in the program. Finally, we will consider how to return computed quantities to the calling program. We will develop a program, step by step, to generate several percentile-based statistics of interest.

11.3   The program statement

Let's say that you want to compute several statistics from the percentiles p# of a continuous variable. Researchers often use the interquartile range, (p75 − p25), as an alternative to the standard deviation as a measure of a variable's spread. Those concerned with income distributions often use (p90 − p10) as a measure of inequality. If we are concerned about outliers, we might compute (p95 − p5) or (p99 − p1) to compare the variable's range $(x_{\max} - x_{\min})$ with these percentile ranges.

Computing these percentile ranges in a do-file is easy enough. You merely need to use `summarize, detail` and access the appropriate percentiles in its saved results. But you might like to have a program that would calculate the ranges from these percentiles and make them available for later use. What must you do to write one? The first step is to choose a name. As suggested above, you should first use `which` and `findit` to ensure that the name is not already in use. Those commands indicate that `pctrange` is free for your use, so open the Do-file Editor (or your favorite text editor[5]) and create the file `pctrange.ado`. Here is our first crack at the code:

---

5. See `ssc describe texteditors`.

```
. type pctrange.ado
*! pctrange v1.0.0  CFBaum 11aug2008
program pctrange, rclass
        version 10.1
        syntax varlist(max=1 numeric)
        quietly summarize `varlist', detail
        scalar range = r(max) - r(min)
        scalar p7525 = r(p75) - r(p25)
        scalar p9010 = r(p90) - r(p10)
        scalar p9505 = r(p95) - r(p5)
        scalar p9901 = r(p99) - r(p1)
        display as result _n "Percentile ranges for `varlist'"
        display as txt "75-25: " p7525
        display as txt "90-10: " p9010
        display as txt "95-05: " p9505
        display as txt "99-01: " p9901
        display as txt "Range: " range
end
```

Once a program has been loaded from an ado-file into Stata's memory, it is retained for the duration of the session. Because we will be repeatedly changing the program's ado-file during its development, we must tell Stata to reload the program to ensure that we are working with the latest version. To do that, we use the discard ([P] **discard**) command before we try out changes to our program. The comment line starting with *! (termed *star-bang* in geekish) is a special comment that will show up if you issue the findfile ([P] **findfile**) or which command. It is always a good idea to document an ado-file with a sequence number, author name, and date.

The program statement identifies the program name as pctrange. We define the program as rclass. Unless a program is defined as rclass or eclass, it cannot return values with the return or ereturn statements, respectively (see section 5.3). Although we do not have any return statements in the current version of the program, we will want to add them in later versions. The version ([P] **version**) line states that the ado-file requires Stata 10.1 and ensures that the program will obey Stata 10.1 syntax when executed by newer releases of Stata.

11.4   The syntax and return statements

The next line of the program, containing the syntax ([P] **syntax**) statement, provides the ability for a Stata program to parse its command line and extract the program's arguments for use within the program. In this simple example, we use only one element of syntax, specifying that the program has a mandatory varlist with a maximum of one numeric variable. Stata will enforce the constraint that only one name appears on the command line and that the name refers to an existing numeric variable. Here we also could have used syntax varname(numeric) to allow one variable, but we would still refer to `varlist' to access its name.

To calculate percentiles of the variable given as `pctrange`'s argument, we must use the `detail` option of `summarize`. We use the `quietly` prefix to suppress output. The following five lines define scalars[6] containing the four percentile ranges and the standard range. We display a header, list each range, and end the program.

Let's try out the first version of `pctrange` on the Stock–Watson `mcas` data.[7] Variable `tot_day` measures total spending per elementary-school pupil in each of the 220 Massachusetts school districts.

```
. use mcas, clear
. pctrange tot_day
Percentile ranges for tot_day
75-25: 1070
90-10: 2337.5
95-05: 3226
99-01: 4755
Range: 6403
. display p7525
1070
```

The scalars defined in the program have global scope (unlike local macros), so they are accessible after the program has run. We can `display` them, as shown above, or use them in later computations.

There is one flaw in that logic, however: what if you already are using one of these names for a scalar or for a variable in your dataset? Must you remember all the names of scalars returned by various programs and avoid reusing them? That is not convenient, so let's name those scalars within the program with `tempname`s. A `tempname` is actually a local macro, assigned automatically by Stata, that allows you to create the equivalent of a local scalar or matrix. Because the `tempname`s are local macros, we refer to their contents just as we would a local macro, with left and right single quotes.[8]

---

6. See section 3.9.
7. `mcas` refers to the Massachusetts Comprehensive Assessment System, a system of standardized tests administered in that state to all primary and secondary school students. See http://www.doe.mass.edu/mcas/.
8. See section 3.6.

```
. type pctrange.ado
*! pctrange v1.0.1  CFBaum 11aug2008
program pctrange
        version 10.1
        syntax varlist(max=1 numeric)
        local res range p7525 p9010 p9505 p9901
        tempname `res'
        display as result _n "Percentile ranges for `varlist'"
        quietly summarize `varlist', detail
        scalar `range' = r(max) - r(min)
        scalar `p7525' = r(p75) - r(p25)
        scalar `p9010' = r(p90) - r(p10)
        scalar `p9505' = r(p95) - r(p5)
        scalar `p9901' = r(p99) - r(p1)
        display as txt "75-25: " `p7525'
        display as txt "90-10: " `p9010'
        display as txt "95-05: " `p9505'
        display as txt "99-01: " `p9901'
        display as txt "Range: " `range'
end
```

One of the important features of Stata commands is their ability to return results for use in further computations. We now add that feature to the pctrange command by using the return command to place the computed scalars in the saved results. The scalars will be accessible after the program runs but will not collide with any other objects. One bit of trickery is necessary with these tempnames: while the left-hand side of the return scalar refers to the macro's name, the right-hand side must dereference the macro one further time to extract the value stored in that name. Here is the modified code:

```
. type pctrange.ado
*! pctrange v1.0.2  CFBaum 11aug2008
program pctrange, rclass
        version 10.1
        syntax varlist(max=1 numeric)
        local res range p7525 p9010 p9505 p9901
        tempname `res'
        display as result _n "Percentile ranges for `varlist'"
        quietly summarize `varlist', detail
        scalar `range' = r(max) - r(min)
        scalar `p7525' = r(p75) - r(p25)
        scalar `p9010' = r(p90) - r(p10)
        scalar `p9505' = r(p95) - r(p5)
        scalar `p9901' = r(p99) - r(p1)
        display as txt "75-25: " `p7525'
        display as txt "90-10: " `p9010'
        display as txt "95-05: " `p9505'
        display as txt "99-01: " `p9901'
        display as txt "Range: " `range'
        foreach r of local res {
                return scalar `r' = ``r''
        }
end
```

To reduce a bit of typing (and to make the program easier to maintain), we have listed the items to be created and used a `foreach` loop to issue the `return` statements. When we run the program and view the saved results, we see

```
. discard
. quietly pctrange tot_day
. return list
scalars:
              r(p9901) =   4755
              r(p9505) =   3226
              r(p9010) =   2337.5
              r(p7525) =   1070
              r(range) =   6403
```

11.5   Implementing program options

In the last output, we used the `quietly` prefix to suppress output. But if you are going to use the `pctrange` command to compute (but not display) these ranges, perhaps you should be able to produce the output as an option. Square brackets, [ ], in `syntax` signify an optional component of the command. Our third version of the program adds [, PRINT], indicating that the command has a print option and that it is truly optional (you can define nonoptional, or required, options for a Stata command).

```
. type pctrange.ado

*! pctrange v1.0.3  CFBaum 11aug2008
program pctrange, rclass
        version 10.1
        syntax varlist(max=1 numeric) [, PRINT]
        local res range p7525 p9010 p9505 p9901
        tempname 'res'
        quietly summarize 'varlist', detail
        scalar 'range' = r(max) - r(min)
        scalar 'p7525' = r(p75) - r(p25)
        scalar 'p9010' = r(p90) - r(p10)
        scalar 'p9505' = r(p95) - r(p5)
        scalar 'p9901' = r(p99) - r(p1)
        if "'print'" == "print" {
                display as result _n "Percentile ranges for 'varlist'"
                display as txt "75-25: " 'p7525'
                display as txt "90-10: " 'p9010'
                display as txt "95-05: " 'p9505'
                display as txt "99-01: " 'p9901'
                display as txt "Range: " 'range'
        }
        foreach r of local res {
                return scalar 'r' = ''r''
        }
end
```

If you now execute the program with `pctrange tot_day, print`, its output is printed.

We might also want to make the command print by default and allow an option to suppress printing. We do this with a `noprint` option:

```
. type pctrange.ado

*! pctrange v1.0.4  CFBaum 11aug2008
program pctrange, rclass
        version 10.1
        syntax varlist(max=1 numeric) [, noPRINT]
        local res range p7525 p9010 p9505 p9901
        tempname 'res'
        quietly summarize 'varlist', detail
        scalar 'range' = r(max) - r(min)
        scalar 'p7525' = r(p75) - r(p25)
        scalar 'p9010' = r(p90) - r(p10)
        scalar 'p9505' = r(p95) - r(p5)
        scalar 'p9901' = r(p99) - r(p1)
        if "'print'" != "noprint" {
                display as result _n "Percentile ranges for 'varlist'"
                display as txt "75-25: " 'p7525'
                display as txt "90-10: " 'p9010'
                display as txt "95-05: " 'p9505'
                display as txt "99-01: " 'p9901'
                display as txt "Range: " 'range'
        }
        foreach r of local res {
                return scalar 'r' = ''r''
        }
end
```

To test the option, we examine whether the local macro `print` contains the word `noprint`, that is, whether the name of the option is `print`. You can add many other types of options, some of which we will present below, to a Stata program.

11.6   Including a subset of observations

A statistical command should accept `if` *exp* and `in` *range* qualifiers if it is to be useful. Very little work is needed to add these features to our program. The definition of `if` *exp* and `in` *range* qualifiers and program options is all handled by the `syntax` statement. In the improved program, `[if]` and `[in]` denote that each of these qualifiers can be used.

With an `if` *exp* or `in` *range* qualifier, something less than the full sample will be analyzed. Before doing any computations on the subsample, we must ensure that the subsample is not empty. Accordingly, we calculate `r(N)` to indicate the sample size used in the computations, check to see that it is not zero, and add it to the display header.

The `marksample touse` command uses the information provided in an `if` *exp* or `in` *range* qualifier if one or both were given on the command line. The `marksample` command marks those observations that should enter the computations in an indicator variable, '`touse`', equal to 1 for the desired observations, and 0 otherwise. The '`touse`' variable is a temporary variable, or `tempvar`, which will disappear when the ado-file ends, like a local macro.

After defining this temporary variable, we use `count if` '`touse`' to calculate the number of observations after applying the qualifiers and display an error if there are no observations (see Cox [2007c]). We must add `if` '`touse`' to each statement in the program that works with the input varlist. Here we need modify only the `summarize` statement to include `if` '`touse`'. The new version of the program is

```
. type pctrange.ado
*! pctrange v1.0.5  CFBaum 11aug2008
program pctrange, rclass
        version 10.1
        syntax varlist(max=1 numeric) [if] [in] [, noPRINT]
        marksample touse
        quietly count if 'touse'
        if 'r(N)' == 0 {
                error 2000
        }
        local res range p7525 p9010 p9505 p9901
        tempname 'res'
        quietly summarize 'varlist' if 'touse', detail
        scalar 'range' = r(max) - r(min)
        scalar 'p7525' = r(p75) - r(p25)
        scalar 'p9010' = r(p90) - r(p10)
        scalar 'p9505' = r(p95) - r(p5)
        scalar 'p9901' = r(p99) - r(p1)
        if "'print'" != "noprint" {
                display as result _n ///
                    "Percentile ranges for 'varlist', N = 'r(N)'"
                display as txt "75-25: " 'p7525'
                display as txt "90-10: " 'p9010'
                display as txt "95-05: " 'p9505'
                display as txt "99-01: " 'p9901'
                display as txt "Range: " 'range'
        }
        foreach r of local res {
                return scalar 'r' = ''r''
        }
        return scalar N = r(N)
        return local varname 'varlist'
end
```

We might want to compare the percentile ranges in communities with above-average per capita incomes with those from the overall sample. The variable `percap` measures per capita income in each school district. We compute its statistics from a subsample of communities with above-average `percap`:

```
. discard

. summarize percap, meanonly

. pctrange tot_day if percap > r(mean) & !missing(percap)
Percentile ranges for tot_day, N = 78
75-25: 1271
90-10: 2572
95-05: 3457
99-01: 5826
Range: 5826
```

11.7 Generalizing the command to handle multiple variables

Perhaps by this time you are pleased with the `pctrange` command, but realize that it would be really handy to run it for several variables with just one command. You could always loop over those variables with a `foreach` loop, but assembling the output afterward might be a bit of work. Because the program produces five statistics for each variable, perhaps a nicely formatted table would be useful—and that will require some rethinking about how the command's results are to be displayed and returned.

First, we must tell the `syntax` statement that more than one numeric variable is allowed. The program will perform as it does now for one variable or produce a table if given several variables. Because we are constructing a table, a Stata matrix[9] is a useful device to store the results we generate from `summarize`. Rather than placing the elements in scalars, we declare a matrix with the J() function, calculating the number of rows needed with the extended macro function `word count` *string*.[10] The `foreach` loop then cycles through the varlist, placing the percentile ranges for each variable into one row of the matrix. The local macro `rown` is used to build up the list of row names, applied with `matrix rownames`.

We have added two additional options in the `syntax` statement for this version: a `format()` option, which allows you to specify the Stata format[11] used to display the matrix elements, and the `mat` option, which is discussed below.

```
. type pctrange.ado
*! pctrange v1.0.6  CFBaum 11aug2008
program pctrange, rclass byable(recall)
        version 10.1
        syntax varlist(min=1 numeric ts) [if] [in] [, noPRINT ///
                FORmat(passthru) MATrix(string)]
        marksample touse
        quietly count if 'touse'
        if 'r(N)' == 0 {
                error 2000
        }
        local nvar : word count 'varlist'
        if 'nvar' == 1 {
                local res range p7525 p9010 p9505 p9901
```

---

9. See sections 3.10 and 9.2.
10. See section 3.8.
11. See [D] **format**.

```
                    tempname 'res'
                    quietly summarize 'varlist' if 'touse', detail
                    scalar 'range' = r(max) - r(min)
                    scalar 'p7525' = r(p75) - r(p25)
                    scalar 'p9010' = r(p90) - r(p10)
                    scalar 'p9505' = r(p95) - r(p5)
                    scalar 'p9901' = r(p99) - r(p1)
                    if "'print'" != "noprint" {
                            display as result _n ///
                                    "Percentile ranges for 'varlist', N = 'r(N)'"
                            display as txt "75-25: " 'p7525'
                            display as txt "90-10: " 'p9010'
                            display as txt "95-05: " 'p9505'
                            display as txt "99-01: " 'p9901'
                            display as txt "Range: " 'range'
                    }
                    foreach r of local res {
                            return scalar 'r' = ''r''
                    }
                    return scalar N = r(N)
            }
            else {
                    tempname rmat
                    matrix 'rmat' = J('nvar',5,.)
                    local i 0
                    foreach v of varlist 'varlist' {
                            local ++i
                            quietly summarize 'v' if 'touse', detail
                            matrix 'rmat'['i',1] = r(max) - r(min)
                            matrix 'rmat'['i',2] = r(p75) - r(p25)
                            matrix 'rmat'['i',3] = r(p90) - r(p10)
                            matrix 'rmat'['i',4] = r(p95) - r(p5)
                            matrix 'rmat'['i',5] = r(p99) - r(p1)
                            local rown "'rown' 'v'"
                    }
                    matrix colnames 'rmat' = Range P75-P25 P90-P10 P95-P05 P99-P01
                    matrix rownames 'rmat' = 'rown'
                    if "'print'" != "noprint" {
                            local form ", noheader"
                            if "'format'" != "" {
                                    local form "'form' 'format'"
                            }
                            matrix list 'rmat' 'form'
                    }
                    if "'matrix'" != "" {
                            matrix 'matrix' = 'rmat'
                    }
                    return matrix rmat = 'rmat'
            }
            return local varname 'varlist'
    end
```

You can now invoke the program on a set of variables and, optionally, specify a format for the output of matrix elements:

```
. discard

. pctrange regday specneed bilingua occupday tot_day tchratio, form(%9.2f)
                 Range      P75-P25     P90-P10     P95-P05      P99-P01
     regday    5854.00       918.50     2037.00     2871.00      4740.00
    specneed  49737.01      2282.78     4336.76     5710.46     10265.45
    bilingua 295140.00         0.00     6541.00     8817.00     27508.00
    occupday  15088.00         0.00     5291.50     8096.00     11519.00
     tot_day   6403.00      1070.00     2337.50     3226.00      4755.00
    tchratio     15.60         3.25        5.55        7.55        10.60
```

The `mat` option allows the matrix to be automatically saved as a Stata matrix with that name. This is useful if you are running `pctrange` several times (perhaps in a loop) and want to avoid having to rename the result matrix, `r(rmat)`, each time. If we use this feature, we can use Baum and Azevedo's `outtable` routine (available from the SSC archive) to convert the matrix into a LaTeX table (shown in table 11.1):

```
. pctrange regday specneed bilingua occupday tot_day tchratio, mat(MCAS) noprint

. outtable using MCAS, mat(MCAS) caption("{\smrm MCAS} percentile ranges")
> format(%9.2f) nobox replace
```

Table 11.1. MCAS percentile ranges

	Range	P75-P25	P90-P10	P95-P05	P99-P01
regday	5854.00	918.50	2037.00	2871.00	4740.00
specneed	49737.01	2282.78	4336.76	5710.46	10265.45
bilingua	295140.00	0.00	6541.00	8817.00	27508.00
occupday	15088.00	0.00	5291.50	8096.00	11519.00
tot day	6403.00	1070.00	2337.50	3226.00	4755.00
tchratio	15.60	3.25	5.55	7.55	10.60

Other SSC routines can be used to produce a table in tab-delimited, rich text, or HTML formats.

11.8   Making commands byable

As a final touch, you might want the `pctrange` command to be byable: to permit its use with a by prefix. Because we are not creating any new variables with this version of the program, this can be done by simply adding `byable(recall)`[12] to the `program` statement. The new `program` statement becomes:

```
program pctrange, rclass byable(recall)
```

---

12. See [P] **byable** for details. You can also use `byable(onecall)`, but that option requires more work on your part. `byable(recall)` is usually suitable.

The other enhancement you might consider is allowing the varlist to contain variables with time-series operators, such as `L.gdp` or `D.income`. We can easily incorporate that feature by changing the `syntax` statement to add the `ts` suboption:

```
syntax varlist(min=1 numeric ts) [if] [in] [, noPRINT FORmat(passthru)
MATrix(string)]
```

With these modifications, we can apply `pctrange` with the `by` prefix or use time-series operators in the varlist. To illustrate the byable nature of the program, let's generate an indicator for teachers' average salaries above and below the mean, and calculate the `pctrange` statistics for those categories:

```
. discard

. summarize avgsalry, meanonly

. generate byte highsal = avgsalry > r(mean) & !missing(avgsalry)

. label define sal 0 low 1 high

. label val highsal sal

. tabstat avgsalry, by(highsal) stat(mean N)
```

Summary for variables: avgsalry
        by categories of: highsal

highsal	mean	N
low	33.5616	101
high	38.60484	94
Total	35.9927	195

```
. bysort highsal: pctrange regday specneed bilingua occupday tot_day tchratio
```

-> highsal = low

	Range	P75-P25	P90-P10	P95-P05	P99-P01
regday	4858	703	1740	2526	3716
specneed	49737.008	2030.8198	3997.9497	5711.2104	11073.81
bilingua	295140	0	6235	8500	13376
occupday	11519	0	5490	7095	11286
tot_day	5214	780	1770	2652	4597
tchratio	11.6	3.1999989	6.2999992	7.8000002	9.3999996

-> highsal = high

	Range	P75-P25	P90-P10	P95-P05	P99-P01
regday	5433	1052	2189	2807	5433
specneed	8570.4004	2486.3604	4263.9702	5620.54	8570.4004
bilingua	33968	0	8466	11899	33968
occupday	15088	0	5068	8100	15088
tot_day	5928	1179	2572	3119	5928
tchratio	15.6	2.4000006	4.7000008	6.2999992	15.6

We see that average salaries in low-salary school districts are over $5,000 less than those in high-salary school districts. These differences carry over into the percentile ranges, where the ranges of `tot_day`, total spending per pupil, are much larger in the high-salary districts than in the low-salary districts.

Program properties

User-written programs can also have *properties* associated with them. Some of Stata's prefix commands use these properties for command validation. If you are interested in writing programs for use with `nestreg`, `svy`, or `stepwise`, you should read [P] **program properties**.

Separately, several prefix commands (namely, `bootstrap`, `jackknife`, and `svy`) can report exponentiated coefficients, such as hazard ratios or odds ratios, when the `eform()` option is used. To make this feature available in your own program, it must have the associated properties defined (for example, `hr` for the hazard ratio or `or` for the odds ratio). Again see [P] **program properties** for details. The extended macro function `properties` *command* will report on the properties of any ado-file. For example,

```
local logitprop: properties logit
```

will provide a list of properties associated with the `logit` command.

11.9   Documenting your program

Stata's development tools make it easy for you to both write a useful program and document it in the way Stata users expect: as readily accessible online help. Even if the program is solely for your use, producing a help file is straightforward and useful. You need not resort to scrutinizing the code to remember how the program is called if you document it now and keep that documentation up to date as the program is modified or enhanced. Just create a text file—for example, `pctrange.sthlp`[13]—with your favorite text editor.[14]

The best way to document a Stata program is to learn a bit of SMCL. Writing the help file in SMCL allows you to produce online help indistinguishable from that of official commands and other user-written programs. It also wraps properly when you resize the Viewer window or use a different size font in that window. A bare-bones help file for our `pctrange` program can be constructed with the following SMCL code:

```
{smcl}
{* *! version 1.0.0 31jul2007}{...}
{cmd:help pctrange}
{hline}

{title:Title}

{p2colset 5 18 20 2}{...}
{p2col :{hi:pctrange} {hline 2}}Calculate percentile ranges{p_end}
{p2colreset}{...}
```

---

13. Before version 10, Stata's help files were suffixed `.hlp`. Unless you are writing a program that must be compatible with earlier versions of Stata, you should use the newer `.sthlp` suffix.

14. See `ssc describe texteditors`.

```
{title:Syntax}

{p 8 17 2}
{cmd:pctrange} {varlist} {ifin} [{cmd:,} {cmd:noprint}
{cmdab:for:mat(}{it:string}{cmd:)}
{cmdab:mat:rix(}{it:string}{cmd:)}]

{p 4 6 2}
{cmd:by} is allowed; see {manhelp by D}.{p_end}
{p 4 6 2}
{it:varlist} may contain time-series operators; see {help tsvarlist}.{p_end}


{title:Description}

{pstd}{cmd:pctrange} computes four percentile ranges of the specified
variable(s): the 75-25 (interquartile) range, the 90-10, 95-05, and 99-01
ranges as well as the conventional range. These ranges are returned as
scalars.  If multiple variables are included in the {it:varlist}, the results
are returned as a matrix.


{title:Options}

{phang}{opt noprint} specifies that the results are to be returned
but not printed.

{phang}{opt format(string)} specifies the format to be used
in displaying the matrix of percentile ranges for multiple variables.

{phang}{opt matrix(string)} specifies the name of the Stata matrix
to be created for multiple variables.


{title:Examples}

{phang}{stata "sysuse auto" : . sysuse auto}{p_end}
{phang}{stata "pctrange mpg" : . pctrange mpg}{p_end}
{phang}{stata "pctrange price mpg turn, format(%9.2f)" :. pctrange price mpg turn,
format(%9.2f)}{p_end}


{title:Author}

{phang}Christopher F. Baum, Boston College{break}
 baum@bc.edu{p_end}


{title:Also see}

{psee}
Online:  {manhelp summarize R}, {manhelp pctile D}
{p_end}
```

As you can see, the left and right braces ({ }) are the key markup characters in SMCL, playing the role of the angle brackets (< >) in HTML web page markup. There are many similarities between SMCL and HTML (and LaTeX, for that matter).[15] As in LaTeX, blank lines in SMCL influence vertical spacing, but in SMCL, multiple blank lines are taken literally. The formatted help for `pctrange` can now be viewed:

```
help pctrange
```

Title

    pctrange — Calculate percentile ranges

Syntax

        pctrange varlist [if] [in] [, noprint format(string) matrix(string)]

    by is allowed; see [D] by.
    varlist may contain time-series operators; see tsvarlist.

Description

    pctrange computes four percentile ranges of the specified variable(s):
    the 75-25 (interquartile) range, the 90-10, 95-05, and 99-01 ranges as
    well as the conventional range. These ranges are returned as scalars. If
    multiple variables are included in the varlist, the results are returned
    as a matrix.

Options

    noprint specifies that the results are to be returned but not printed.

    format(string) specifies the format to be used in displaying the matrix
        of percentile ranges for multiple variables.

    matrix(string) specifies the name of the Stata matrix to be created for
        multiple variables.

Examples

    . sysuse auto
    . pctrange mpg
    . pctrange price mpg turn, format(%9.2f)

Author
    Christopher F Baum, Boston College
        baum@bc.edu

Also see

    Online: [R] summarize, [D] pctile

---

15. Mechanical translation of SMCL log files into HTML is reliably performed by the `log2html` package of Baum, Cox, and Rising, available from the SSC archive.

For full details on the construction of SMCL help files (and SMCL output in general), see [P] **smcl** and [U] **18.11.6 Writing online help**.

11.10   egen function programs

As I discussed in section 3.4, the **egen** (extended generate) command is open-ended, in that any Stata user can define an additional **egen** function by writing a specialized ado-file program. The name of the program (and of the file in which it resides) must start with _g; that is, _gcrunch.ado will define the crunch() function for **egen**.

To illustrate **egen** functions, let's create a function to generate one of the pctrange values, the (p90 – p10) percentile range. The program follows the same pattern as our stand-alone pctrange command with one important difference: in an **egen** function, you must deal with the new variable to be created. The syntax for **egen** is

egen [*type*] *newvarname* = *fcn*(*arguments*) [*if*] [*in*] [, *options*]

The **egen** command, like **generate**, can specify a data type. The syntax indicates that a new variable name must be provided, followed by an equal sign and *fcn*, or a function, with arguments. **egen** functions can also have if *exp* and in *range* qualifiers, and options.

The computation for our **egen** function is the same as that for pctrange, using **summarize** with the **detail** option. On the last line of the function, we **generate** the new variable (of the appropriate type, if specified[16]) under the control of the 'touse' temporary indicator variable, limiting the sample as specified.

```
. type _gpct9010.ado
*! _gpct9010 v1.0.0  CFBaum 11aug2008
program _gpct9010
        version 10.1
        syntax newvarname =/exp [if] [in]
        tempvar touse
        mark 'touse' 'if' 'in'
        quietly summarize 'exp' if 'touse', detail
        quietly generate 'typlist' 'varlist' = r(p90) - r(p10) if 'touse'
end
```

This function works perfectly well, but it creates a new variable containing one scalar value. As noted earlier, that is a profligate use of Stata's memory (especially for large _N) and often can be avoided by retrieving the single scalar, which is conveniently stored by our pctrange command. To be useful, we would like the **egen** function to be byable so that it can compute the appropriate percentile-range statistics for several groups defined in the data.

---

16. When *newvarname* or *newvarlist* is specified on **syntax**, the macro *typlist* is filled with the data type(s) to be used for the new variable(s).

The changes to the code are relatively minor. We add an options descriptor to the
syntax statement, because egen will pass the by prefix variables as a by option to our
program. Rather than using summarize, we use egen's own pctile() function, which
is documented as allowing the by prefix, and pass the options to this function. The
revised function reads

```
. type _gpct9010.ado
*! _gpct9010 v1.0.1  CFBaum 11aug2008
program _gpct9010
        version 10.1
        syntax newvarname =/exp [if] [in] [, *]
        tempvar touse p90 p10
        mark 'touse' 'if' 'in'
        quietly {
                egen double 'p90' = pctile('exp') if 'touse', 'options' p(90)
                egen double 'p10' = pctile('exp') if 'touse', 'options' p(10)
                generate 'typlist' 'varlist' = 'p90' - 'p10' if 'touse'
        }
end
```

These changes permit the function to produce a separate percentile range for each group
of observations defined by the by-list. To illustrate, we use auto.dta:

```
. discard
. sysuse auto, clear
(1978 Automobile Data)
. bysort rep78 foreign: egen pctrange = pct9010(price)
```

Now if we want to compute a summary statistic (such as the percentile range) for each
observation classified in a particular subset of the sample, we can use the pct9010()
function to do so.

11.11 Writing an e-class program

The ado-file programs I have discussed in earlier sections are all r-class programs; that
is, they provide results in the return list.[17] Many statistical procedures involve
fitting a model (rather than computing one or more statistics) and are thus termed
estimation commands, or e-class commands. One of Stata's great strengths derives
from the common nature of its estimation commands, which follow a common syntax,
leave behind the same objects, and generally support the same postestimation tools
(such as test, lincom, and mfx to compute marginal effects and predict to compute
predicted values, residuals, and similar quantities). Although e-class commands are
somewhat more complicated than r-class commands, it is reasonably simple for you to
implement an estimation command as an ado-file. Many of the programming concepts
discussed in earlier sections are equally useful when dealing with e-class commands. The
additional features needed generally relate to postestimation capabilities.

---

17. There is one exception: an egen program, such as _gpct9010, cannot return results in the saved
    results.

As detailed in [U] **18.9 Accessing results calculated by estimation commands**, there are a number of conventions that an e-class command must follow:

- The command must save its results in e(), accessed by `ereturn list`, rather than in r().

- It should save its name in e(cmd).

- It should save the contents of the command line in e(cmdline).

- It should save the number of observations in e(N) and identify the estimation sample by setting the indicator variable (or "function") e(sample).

- It must save the entire coefficient vector as Stata matrix e(b) and the variance–covariance matrix of the estimated parameters as Stata matrix e(V).

Correct capitalization of these result names is important. The coefficient vector is saved as a $1 \times k$ row vector for single-equation estimation commands, with additional rows added for multiple-equation estimators. The variance–covariance matrix is saved as a $k \times k$ symmetric matrix. The presence of e(b) and e(V) in standardized locations enables Stata's postestimation commands (including those you write) to work properly. Estimation commands can set other e() scalars, macros, or matrices.

Whereas an r-class program, such as `pctrange`, uses the **return** ([P] **return**) command to return its results in r(), an e-class program uses the **ereturn** ([P] **ereturn**) command. The command `ereturn` *name* = *exp* returns a scalar value, while `ereturn local` *name value* and `ereturn matrix` *name matname* return a macro and a Stata matrix, respectively. You do not use `ereturn` for the coefficient vector or estimated variance–covariance matrix, as I now discuss.

The `ereturn post` command posts the estimates of b and V to their official locations. To return the coefficient vector and its variance–covariance matrix, you need to create the coefficient vector, say 'beta', and its variance–covariance matrix, say 'vce', and pass them back in the following fashion. We also can define the estimation sample flagged by the sample indicator temporary variable, 'touse':

```
ereturn post 'beta' 'vce', esample('touse')
```

You can now save anything else in e(), using the `ereturn scalar`, `ereturn local`, or `ereturn matrix` commands, as described above. It is best to use the commonly used names for various quantities. For instance, e(df_m) and e(df_r) are commonly used to denote the numerator (model) and denominator (residual) degrees of freedom. e(F) commonly refers to the test against the null (constant-only) model for nonasymptotic results, while e(chi2) is used for an asymptotic estimator. e(r2) or e(r2_p) refer to the R^2 or pseudo-R^2, respectively. Although you are free to choose other names for your `ereturn` values, it is most helpful if they match those used in common Stata commands. See [U] **18.10.2 Saving results in e()** for more details.

11.11.1   Defining subprograms

If a user-written Stata command is to be executed, the file defining that command must be on the ado-path.[18] However, one ado-file can contain more than one program. The subsequent programs are *local* in the sense that they cannot be called independently. So, for instance, `one.ado` could contain

```
program one
...
end
program two
...
end
program three
...
end
```

Presumably, the `one` command calls the `two` and `three` programs in the course of its execution. Those subprogram names are not visible to other commands and can be used only by `one`. Using subprograms allows you to isolate sections of your program that are used repeatedly or separate sections that perform distinct tasks within your command. We use this facility in section 11.13.

11.12   Certifying your program

All computer programs have the potential to misbehave. Some will crash when presented with certain input values; some will never stop running; some will generate output that turns out to be incorrect. If you are writing your own Stata programs, how can you guard against these types of misbehavior and be reasonably satisfied with your program's reliability? You can use the same tools that StataCorp developers use to certify official Stata's reliability by constructing and running a *certification script*.

In the first issue of the *Stata Journal*, lead software developer William Gould formally discussed Stata's certification tools (Gould 2001) and illustrated how they could be used to certify the performance of a user-written Stata program. Stata's `assert` command, discussed in section 5.2 in the context of data validation, is the key component of a certification script. The result of a computation is compared with a known value or with the result of another routine known to be correct.

Gould describes the certification test script as a collection of do-files, executed in turn by a master do-file. If no errors are detected by any of the do-files, the script will run to completion. However, at the first error, the do-file will abort. To illustrate, he presents a fragment of a typical test script (Gould 2001, 37):

---

18. Or the program must be loaded into memory from a do-file or entered interactively. I do not consider those possibilities in this book.

```
cscript summarize
which summarize
use auto
summarize mpg
assert r(N) == 74 & r(sum_w) == 74 & r(min) == 12 & r(max) == 41 & r(sum) == 1576
assert reldif(r(mean), 21.29729729729730) < 1e-14
assert reldif(r(Var), 33.47204738985561) < 1e-14
assert r(sd) == sqrt(r(Var))
summarize mpg if foreign
assert r(N) == 22 & r(sum_w) == 22 & r(min) == 14 & r(max) == 41 & r(sum) == 545
...
```

The `assert` commands in this fragment of the script all ensure that results left behind in the saved results from `summarize` match known values and are internally consistent (e.g., the computed standard deviation, `r(sd)`, is the square root of the computed variance, `r(Var)`). Note the use of the `reldif()` function to compare a constant with the computed value. As mentioned in section 2.4 in the context of finite-precision arithmetic, we cannot perform exact comparisons against noninteger values (see Gould [2006b] and Cox [2006b]).

The certification script should not only test against known values but also test that error messages are correctly produced when they should be. For instance, in the `auto.dta` dataset, applying `summarize` outside the defined observation range should produce an `error 198` code:

```
summarize price in 75/99
```

How can we test this statement, given that its execution will cause an error and stop the do-file? With the `rcof` command (see `help rcof`). Coupled with the `noisily` prefix, we can use

```
rcof "noisily summarize price in 75/99" == 198
```

to ensure that an error (with the proper error code) is produced. If no error is produced, or if an error with another error code is produced the certification script will fail. You can also test for appropriate syntax.

```
rcof "noisily summarize mpg, detail meanonly" != 0
```

tests whether the `summarize` command properly rejects the combination of the `detail` and `meanonly` options, which should be mutually exclusive.

(Continued on next page)

As another example of certification practices, Baum, Schaffer, and Stillman's `ivreg2` routine for instrumental-variables estimation is an extension of Stata's earlier `ivreg` command and parallels the current `ivregress` command. We trust that Stata's developers have fully certified the behavior of official `ivregress` and its predecessor, `ivreg`, as well as that of `regress` and `newey` (`ivreg2` can estimate ordinary least squares and Newey–West regressions). Therefore, we have constructed a certification script for `ivreg2`, available as an ancillary do-file of the `ivreg2` package. The certification script conducts several tests to ensure that the computed values are correct. Many of those tests rely on comparing the detailed results generated by `ivreg2` with those from `ivregress`, `ivreg`, `regress`, and `newey`. See `ssc describe ivreg2`; chapter 8 of Baum (2006a); and Baum, Schaffer, and Stillman (2003, 2007).

In summary, best practices in Stata programming involve setting up a certification script for your program. The script should evolve because every flaw in the program that is corrected should translate into an additional check against that condition. Every time the program is enhanced by adding features, additional certification tests should be added to ensure that those features work properly under all conditions, including conditions that may seem implausible. Just as the documentation of a program is as important as the code, validation of the program's reliability should be automated and redone whenever changes are made to the code.

11.13   Programs for ml, nl, nlsur, simulate, bootstrap, and jackknife

The ado-file programming techniques discussed in earlier sections carry over to those Stata commands and prefixes that involve writing a program. For example, maximum likelihood estimation, performed with the `ml` command, requires that you write a likelihood function evaluator.[19] This is a formulaic program that calculates either one term of the log-likelihood function (LLF) or its total over the estimation sample. In this context, you need only follow the template established for the particular form of `ml` you choose to specify the parameters to be estimated and the way in which they enter the LLF. Stata's `ml` routines support four methods of coding the LLF: the *linear form* (`lf`) method and methods `d0`, `d1`, and `d2`. The linear form is the easiest to work with but requires that the statistical model meets the linear-form restrictions (see [R] **ml**). Methods `d0`, `d1`, and `d2` require coding the LLF, the LLF and its first derivatives, or the LLF and its first and second derivatives, respectively.

For example, here is an LLF linear-form evaluator for a linear regression model with normally distributed errors:

---

19. An essential reference for maximum-likelihood programming in Stata is Gould, Pitblado, and Sribney (2006).

```
. type mynormal_lf.ado
*! mynormal_lf v1.0.0  CFBaum 11aug2008
program mynormal_lf
  version 10.1
  args lnf mu sigma
  quietly replace 'lnf' = ln(normalden($ML_y1,  'mu', 'sigma'))
end
```

In this program, we use the **args** command[20] to retrieve the three items that are passed to the LLF evaluator: a variable, **lnf**, whose values are to be computed in the routine, and the two parameters to be estimated (the conditional mean of the dependent variable [referred to as **$ML_y1**], mu, and its variance, **sigma**). The linear-form restrictions imply that we need not work explicitly with the elements of the conditional mean $X\widehat{\beta}$. To invoke the LLF evaluator, using the **auto.dta** dataset, we can use the command

```
ml model lf mynormal_lf (mpg = weight displacement) /sigma
```

which estimates the regression of **mpg** on **weight** and **displacement** with a constant term under the assumption of homoskedasticity (a constant error variance, estimated as **sigma**). The flexibility of this approach is evident if we consider a heteroskedastic regression model in which $\sigma_i = \gamma_0 + \gamma_1 \text{ price}_i$.[21] We can estimate that model with the command

```
ml model lf mynormal_lf (mpg = weight displacement turn) (price)
```

where the second "equation" refers to the specification of the σ_i term. The ado-file need not be modified even though we changed the list of regressors and made a different assumption on the distribution of the errors.

Writing an ml-based command

The ado-file above can be used interactively with the **ml** command. What if you wanted to create a new Stata estimation command that implemented this particular maximum likelihood estimator? You merely need to write an additional ado-file, **mynormal.ado**:

```
. type mynormal.ado
*! mynormal v1.0.0  CFBaum 11aug2008
program mynormal
        version 10.1
        if replay()  {
                if ("'e(cmd)'" != "mynormal") error 301
                Replay '0'
        }
        else Estimate '0'
end
```

---

20. In the **ml** context, we need only a subset of the capabilities of the **syntax** command. The **args** command is better suited for use in a likelihood-evaluator program. Its syntax when used with **ml** is defined by that command; see [R] **ml**.

21. For a discussion of homoskedasticity and heteroskedasticity, see Baum (2006a, chap. 6).

```
program Replay
        syntax [, Level(cilevel) ]
        ml display, level('level')
end

program Estimate, eclass sortpreserve
        syntax varlist [if] [in]  [, vce(passthru) Level(cilevel) * ]
        mlopts mlopts, 'options'
        gettoken lhs rhs: varlist
        marksample touse
        ml model lf  mynormal_lf (mu: 'lhs' = 'rhs') /sigma  ///
            if 'touse', 'vce' 'mlopts' maximize
        ereturn local cmd "mynormal"
        Replay, level('level')
end
```

The `mynormal` program is a wrapper for two subprograms as discussed in section 11.11.1: `Replay` and `Estimate`.

The `Replay` program permits our `mynormal` command to emulate all Stata estimation commands in supporting the replay feature. After you use a standard estimation command, you can always replay the estimation results if you have not issued another estimation command in the interim by merely giving the estimation command's name (for example, `regress`). The `mynormal` program checks to see that the previous estimation command was indeed `mynormal` before executing `Replay`.

The `Estimate` command does the work of setting the stage to call `ml` in its noninteractive mode, as signaled by the `maximize` option. We use the `sortpreserve` option to specify that the sort order of the dataset should be preserved and restored after our program has ended. The `syntax` command parses the variable list given to `mynormal` into the left-hand side (dependent) and right-hand side (covariates). In this example, only the homoskedastic case of a fixed parameter, `/sigma`, is supported. The `mlopts` command allows you to specify one or more of the maximum-likelihood options (see `help mlopts` and Gould, Pitblado, and Sribney [2006, 180–183]).

To illustrate `mynormal`, let's use `auto.dta`:

```
. mynormal price mpg weight turn
initial:       log likelihood =      -<inf>  (could not be evaluated)
feasible:      log likelihood = -814.40522
rescale:       log likelihood = -731.80124
rescale eq:    log likelihood = -701.88231
Iteration 0:   log likelihood = -701.88231
Iteration 1:   log likelihood = -693.55438
Iteration 2:   log likelihood = -678.01692
Iteration 3:   log likelihood = -677.74653
Iteration 4:   log likelihood = -677.74638
Iteration 5:   log likelihood = -677.74638

                                              Number of obs   =        74
                                              Wald chi2(3)    =     46.26
Log likelihood = -677.74638                   Prob > chi2     =    0.0000
```

price	Coef.	Std. Err.	z	P>|z|	[95% Conf. Interval]	
mu						
mpg	-72.86501	79.0677	-0.92	0.357	-227.8348	82.10482
weight	3.524339	.7947479	4.43	0.000	1.966661	5.082016
turn	-395.1902	119.2837	-3.31	0.001	-628.9819	-161.3985
_cons	12744.24	4629.664	2.75	0.006	3670.269	21818.22
sigma						
_cons	2298.005	188.8948	12.17	0.000	1927.778	2668.232

We can use the replay feature, change the level of significance reported for confidence intervals, and invoke `test`, just as we could with any Stata estimation command:

```
. mynormal, level(90)
```

```
                                    Number of obs   =        74
                                    Wald chi2(3)    =     46.26
Log likelihood = -677.74638         Prob > chi2     =    0.0000
```

price	Coef.	Std. Err.	z	P>|z|	[90% Conf. Interval]	
mu						
mpg	-72.86501	79.06771	-0.92	0.357	-202.9198	57.18979
weight	3.524339	.7947481	4.43	0.000	2.217094	4.831583
turn	-395.1902	119.2837	-3.31	0.001	-591.3944	-198.986
_cons	12744.24	4629.665	2.75	0.006	5129.123	20359.37
sigma						
_cons	2298.005	188.8949	12.17	0.000	1987.301	2608.709

```
. test weight = 5
 ( 1)  [mu]weight = 5
           chi2(  1) =    3.45
         Prob > chi2 =   0.0633
. predict double pricehat, xb
. summarize price pricehat
```

Variable	Obs	Mean	Std. Dev.	Min	Max
price	74	6165.257	2949.496	3291	15906
pricehat	74	6165.257	1829.306	1988.606	10097.71

We can also use `predict` to compute the fitted values from our maximum likelihood estimation and `summarize` to compare them with the original series.

11.13.1   Programs for the nl and nlsur commands

Similar issues arise when performing nonlinear least-squares estimation for either a single equation (nl) or a set of equations (nlsur). Although these commands can be used interactively or in terms of "programmed substitutable expressions", most serious use is likely to involve your writing a *function-evaluator program*. That program will compute the dependent variable(s) as a function of the parameters and variables specified.

The techniques used for a maximum-likelihood function evaluator, as described above, are similar to those used by nl and nlsur function-evaluator programs. For instance, we might want to estimate a constant elasticity of substitution (CES) production function,

$$\ln Q_i = \beta_0 - \frac{1}{\rho} \ln \left\{ \delta K_i^{-\rho} + (1 - \delta) L_i^{-\rho} \right\} + \epsilon_i$$

which relates a firm's output, Q_i, to its use of capital, or machinery, K_i, and labor, L_i (see Greene [2008, 119]). The parameters in this highly nonlinear relationship are β_0, ρ, and δ.

We store the function-evaluator program as nlces.ado, because nl requires a program name that starts with the letters nl. As described in [R] nl, the syntax statement must specify a varlist, allow for an if *exp* qualifier, and allow for an at(*name*) option. The parameters to be estimated are passed to your program in the row vector at. In our CES example, the varlist must contain exactly three variables, which are extracted from the varlist by the args command. This command assigns its three arguments to the three variable names provided in the varlist. For ease of reference, we assign tempnames to the three parameters to be estimated. The generate and replace statements make use of the if *exp* clause. The function-evaluator program must replace the observations of the dependent variable, here the first variable passed to the program, referenced within as logoutput.

```
. type nlces.ado
*! nlces v1.0.0   CFBaum 11aug2008
program nlces
    version 10.1
    syntax varlist(numeric min=3 max=3) if, at(name)
    args logoutput K L
    tempname b0 rho delta
    tempvar kterm lterm
    scalar `b0' = `at'[1, 1]
    scalar `rho' = `at'[1, 2]
    scalar `delta' = `at'[1, 3]
    gen double `kterm' = `delta' * `K'^(-(`rho')) `if'
    gen double `lterm' = (1 - `delta') *`L'^(-(`rho')) `if'
    replace `logoutput' = `b0' - 1 / `rho' * ln(`kterm' + `lterm') `if'
end
```

We invoke the estimation process with the nl command by using Stata's production dataset.[22] You specify the name of your likelihood function evaluator by including only

---

22. This dataset can be accessed with webuse production.

the unique part of its name (here `ces`, not `nlces`), followed by @. The order in which the parameters appear in the `parameters()` and `initial()` options defines their order in the `at` vector.[23] The `initial()` option is not required but is recommended.

```
. use production, clear
. nl ces @ lnoutput capital labor, parameters(b0 rho delta)
> initial(b0 0 rho 1 delta 0.5)
(obs = 100)
Iteration 0:   residual SS =   29.38631
Iteration 1:   residual SS =   29.36637
Iteration 2:   residual SS =   29.36583
Iteration 3:   residual SS =   29.36581
Iteration 4:   residual SS =   29.36581
Iteration 5:   residual SS =   29.36581
Iteration 6:   residual SS =   29.36581
Iteration 7:   residual SS =   29.36581
```

Source	SS	df	MS
Model	91.1449924	2	45.5724962
Residual	29.3658055	97	.302740263
Total	120.510798	99	1.21728079

```
                                  Number of obs =        100
                                  R-squared     =     0.7563
                                  Adj R-squared =     0.7513
                                  Root MSE      =   .5502184
                                  Res. dev.     =   161.2538
```

lnoutput	Coef.	Std. Err.	t	P>|t|	[95% Conf. Interval]	
/b0	3.792158	.099682	38.04	0.000	3.594316	3.989999
/rho	1.386993	.472584	2.93	0.004	.4490443	2.324941
/delta	.4823616	.0519791	9.28	0.000	.3791975	.5855258

```
Parameter b0 taken as constant term in model & ANOVA table
```

After execution, you have access to all of Stata's postestimation commands. For example, the elasticity of substitution $\sigma = 1/(1 + \rho)$ of the CES function is not directly estimated, but is rather a nonlinear function of the estimated parameters. We can use Stata's `nlcom` ([R] **nlcom**) command to generate point and interval estimates of σ by using the *delta method*:[24]

```
. nlcom (sigma: 1 / (1 + [rho]_b[_cons]))
        sigma:  1 / (1 + [rho]_b[_cons])
```

lnoutput	Coef.	Std. Err.	t	P>|t|	[95% Conf. Interval]	
sigma	.4189372	.0829424	5.05	0.000	.2543194	.583555

This value, falling below unity in point and interval form, indicates that in the firms studied, the two factors of production (capital and labor) are not perfectly substitutable for one another.

---

23. Alternatively, you can use the `nparameters(#)` option and merely specify the number of parameters. They are then named `b1`, `b2`, etc.

24. See [R] **predictnl**.

The programming techniques illustrated here for `nl` carry over to the `nlsur` command (new in Stata version 10), which allows you to apply nonlinear least squares to a system of nonsimultaneous (or *seemingly unrelated*) equations. Likewise, you could write a wrapper for `nlces`, just as we illustrated above for maximum likelihood, to create a new Stata command.

11.13.2   Programs for the simulate, bootstrap, and jackknife prefixes

The `simulate`, `bootstrap`, and `jackknife` prefixes can be used with many Stata commands, but often entail writing your own command to specify the quantities to be simulated or the statistic for which bootstrap or jackknife standard errors are required.[25]

`simulate` is used to perform Monte Carlo simulations of a random process. The command or program to be simulated is executed many times, as specified by the `reps()` option. Each time it is executed, a new random draw is performed and one or more quantities are calculated. Monte Carlo simulation is often used to analyze the properties of a particular estimator or test procedure. In this context, the command or program called by `simulate` will generate a new value of the estimator or test statistic in each replication, and return one or more quantities to the `simulate` command. `simulate` then constructs a new dataset in which each observation corresponds to one replication. That artificial dataset can then be used to compute various statistics of interest relating to the estimator or test statistic being studied.

As an example, let's consider a simulation of the power of a standard t test for sample mean under homoskedastic and groupwise heteroskedastic errors.[26] The test statistic is computed under the former assumption that the error distribution has a common variance, σ^2. To set up a simulation experiment, we use Stata's `census2` dataset,[27] which contains 50 U.S. states' data on a number of demographic variables, as well as a region indicator that takes on the integer values 1, 2, 3, and 4 for four U.S. regions. We create a groupwise heteroskedastic error from the region indicator.

The simulation program, `mcsimul.ado`, takes a variable name of an existing variable that is to be used in the simulation experiment. Optionally, a `mu(#)` value can be provided to be used in the null hypothesis of the t test. If no value is provided, a default value of 75 is used.[28]

The simulation program expects to find two variables, `zmu` and `zfactor`, defined for each observation. It creates two temporary random variables, `y1` and `y2`, as normally distributed with means of $\overline{varname}+0.20$ `zmu` and $\overline{varname}+0.20$ `zfactor`, respectively. The simulation program performs t tests with the null hypothesis that the mean of the

---

25. This should not be confused with the calculation of bootstrap or jackknife standard errors in many Stata estimation commands via specification of the `vce()` option. Here we consider computation of bootstrap (jackknife) standard errors for a statistic that cannot be computed by executing just one Stata command.

26. See the `robvar` command ([R] **sdtest**) and Baum (2006c).

27. This dataset can be accessed with `webuse census2`.

28. Hard-coding this value as a default value for a numeric option is done for pedagogical purposes.

variable is mu. We record the *p*-value with which that null can be rejected; that is, we record the significance level of the test. The two `return scalar` statements indicate that the program produces two return values: the *p*-values for the two `ttest`s.

```
. type mcsimul.ado
*! mcsimul v1.0.0  CFBaum 11aug2008
program mcsimul, rclass
      version 10.1
      syntax varname(numeric) [, mu(real 75)]
      tempvar y1 y2

      generate 'y1' = 'varlist' + rnormal() * 0.20 * zmu
      generate 'y2' = 'varlist' + rnormal() * 0.20 * z_factor
      ttest 'y1' = 'mu'
      return scalar p1 = r(p)
      ttest 'y2' = 'mu'
      return scalar p2 = r(p)
end
```

We now must set up the simulation experiment and execute the `simulate` command. The `set seed` command will cause the pseudorandom-number generator to generate the same sequence of values each time this program is executed, which is useful for debugging purposes. We generate the `z_factor` variable as $10 \times$ `region` and compute the scalar `zmu` to have the same value as the mean of `z_factor`. We now can invoke `simulate`, storing the two return values in new variables `p1` and `p2`. The `saving()` option indicates that a new dataset, `mcsimul.dta`, should be produced, and `reps(1000)` defines the experiment's number of replications. The argument of the `simulate` prefix is merely the name of the simulation program, `mcsimul`, followed by the required variable name. We execute the simulation for the variable `drate`, the death rate for each state, expressed as the number of deaths per 10,000 population. It has a sample mean of 84.3 and a range from 40 to 107.[29] We use a hypothesized value of 80 in the `mu()` option.

```
. set seed 20070731
. use census2, clear
(1980 Census data by state)
. generate z_factor = 10 * region
. summarize z_factor, meanonly
. scalar zmu = r(mean)
. quietly simulate  p1=r(p1) p2=r(p2), saving(mcsimul,replace) nodots
> reps(1000): mcsimul drate, mu(80)
```

Following the simulation, we can examine the resulting dataset:

```
. use mcsimul, clear
(simulate: mcsimul)
. generate R5pc_1 = (p1 < 0.05)
. generate R5pc_2 = (p2 < 0.05)
```

---

29. Analysis of these state-level descriptive statistics should be performed using weights. For pedagogical purposes, we treat them as raw data.

```
. summarize
      Variable |       Obs        Mean    Std. Dev.        Min         Max
```
```
            p1 |      1000    .0484347     .044073    .0009723    .3428818
            p2 |      1000    .0572194    .0607124    .0007912    .4788797
        R5pc_1 |      1000        .642     .479652           0           1
        R5pc_2 |      1000        .597    .4907462           0           1
```

As expected, the power of the test (which is based on the assumption of normally distributed, homoskedastic errors) varies when heteroskedastic errors are encountered, with the average p-value almost one percentage point larger in the latter case. The R5pc_1 and R5_pc2 series, which flag the p-values below 0.05, indicate that the heteroskedastic case demonstrates slightly lower power (597 out of 1,000 rejections at five percent, versus 642 for the homoskedastic case).

The same programming techniques apply when writing programs for the bootstrap or jackknife prefixes. In contrast to Monte Carlo simulation, which is used to analyze an artificial dataset, bootstrap techniques (and the closely related jackknife techniques) use the empirical distribution of the errors in an estimated model. This is performed by resampling with replacement from the existing estimation results for a specified number of replications. Most commonly, bootstrap standard errors are computed by specifying vce(bootstrap) on an estimation command. However, if you want to compute bootstrap standard errors for a quantity that cannot be directly estimated by an existing Stata command, you must write a program to implement the estimation procedure you wish to study. Like simulate, bootstrap and jackknife create new datasets with one observation for each replication.

11.14   Guidelines for Stata ado-file programming style

To highlight the importance of good programming style practices in writing ado-files, I present here an edited excerpt from Nicholas J. Cox's excellent essay "Suggestions on Stata programming style" (Cox 2005f). The rest of this section is quoted from that essay.

11.14.1   Presentation

In this section, I give a list of basic guidelines for formal Stata programs.

- Always include a comment containing the version number of your program, your name or initials, and the date the program was last modified above the program line, for example,

      ```
      *! 1.0.0 Tom Swift 21jan2006
      program myprog
      ```

(As said, this line is indeed just a comment line; it bears no relation to the Stata `version` command. However, `which myprog` will echo this comment line back to you whenever this `myprog` is visible to Stata along your ado-path. Both this comment and a `version` command should be used.)

- Use sensible, intelligible names where possible for programs, variables, and macros.

- Choose a name for your program that does not conflict with anything already existing. Suppose that you are contemplating *newname*. If typing either `which` *newname* or `which` *newname*`.class` gives you a result, StataCorp is already using the name. Similarly, if `ssc type` *newname*`.ado` gives you a result, a program with your name is already on SSC. No result from either does not guarantee that the program is not in use elsewhere: `findit` *newname* may find such a program, although often it will also find much that is irrelevant to this point.

- Brevity of names is also a virtue. However, no platform on which Stata is currently supported requires an 8-character limit. Tastes are in consequence slowly shifting: an intelligible long name for something used only occasionally would usually be considered preferable to something more cryptic.

- Actual English words for program names are supposedly reserved for StataCorp.

- Use the same names and abbreviations for command options that are in common use in official Stata's commands. Try to adopt the same conventions for options' syntax; for example, allow a *numlist* where similar commands use a *numlist*. Implement sensible defaults wherever possible.

- Group `tempname`, `tempvar`, and `tempfile` declarations.

- Use appropriate `display` styles for messages and other output. All error messages (and no others) should be displayed as `err`; that is, type `di as err`. In addition, attach a return code to each error message; `198` (syntax error) will often be fine.

11.14.2   Helpful Stata features

Stata has several features that make programming easier. Examples of ways a programmer can use these features are as follows:

- Stata is tolerant through version control of out-of-date features, but that does not mean that you should be. To maximize effectiveness and impact and to minimize problems, write programs using the latest version of Stata and exploit its features.

- Make yourself familiar with all the details of `syntax`. It can stop you from reinventing little wheels. Use wildcards for options to pass to other commands when appropriate.

- Support `if` *exp* and `in` *range* where applicable. This is best done using `marksample touse` (or occasionally `mark` and `markout`). Have `touse` as a temporary variable if and only if `marksample` or a related command is used. See `help marksample`.

- Make effective use of information available in `e()` and `r()`. If your program is to run in a context that implies that results or estimates are available (say, after `regress`), make use of the stored information from the prior command.

- Where appropriate, ensure that your command returns the information that it computes and displays so that another user may use it `quietly` and retrieve that information.

- Ensure that programs that focus on time series or panel data work with time-series operators if at all possible. In short, exploit `tsset`.

- Familiarize yourself with the built-in material revealed by `creturn list`. Scrolling right to the end will show several features that may be useful to you.

- SMCL is the standard way to format Stata output.

11.14.3 Respect for datasets

In general, make no change to the data unless that is the direct purpose of your program or that is explicitly requested by the user.

- Your program should not destroy the data in memory unless that is essential for what it does.

- You should not create new permanent variables on the side unless notified or requested.

- Do not use variables, matrices, scalars, or global macros whose names might already be in use. There is absolutely no need to guess at names that are unlikely to occur, because temporary names can always be used (type `help macro` for details on `tempvar`, `tempname`, and `tempfile`).

- Do not change the variable type unless requested.

- Do not change the sort order of data; use `sortpreserve`.

11.14.4 Speed and efficiency

Here is a list of basic ways to increase speed and efficiency:

- `foreach` and `forvalues` are cleaner and faster than most `while` loops and much faster than the old `for` that still satisfies some devotees. Within programs, avoid `for` like the plague. (Note to new Mata users: this does not refer to Mata's `for`.)

- Avoid `egen` within programs; it is usually slower than a direct attack.

- Try to avoid looping over observations, which is slow. Fortunately, it can usually be avoided.

- Avoid `preserve` if possible. `preserve` is attractive to the programmer but can be expensive in time for the user with large data files. Programmers should learn to master `marksample`.

- Specify the type of temporary variables to minimize memory overhead. If a `byte` variable can be used, specify `generate byte 'myvar'` rather than letting the default type be used, which would waste storage space.

- Temporary variables will be automatically dropped at the end of a program, but also consider dropping them when they are no longer needed to minimize memory overhead and to reduce the chances of your program stopping because there is no room to add more variables.

11.14.5   Reminders

In this section, I describe a few general procedures that will improve one's code:

- Remember to think about string variables as well as numeric variables. Does the task carried out by your program make sense for string variables? If so, will it work properly? If not, do you need to trap input of a string variable as an error, say, through `syntax`?

- Remember to think about making your program support `by` *varlist*: when this is natural. See `help byable`.

- Remember to think about weights and implement them when appropriate.

- The job is not finished until the `.sthlp` is done. Use SMCL to set up your help files. Old-style help files, while supported, are not documented, while help files not written in SMCL cannot take advantage of its paragraph mode, which allows lines to autowrap to fit the desired screen width. For an introduction to the SMCL required to write a basic help file, see [U] **18.11.6 Writing online help** or `help examplehelpfile`.

11.14.6   Style in the large

Style in the large is difficult to prescribe, but here are some generalities:

- Before writing a program, check that it has not been written already! `findit` is the broadest search tool.

- The best programs do just one thing well. There are exceptions, but what to a programmer is a Swiss army knife with a multitude of useful tools may look to many users like a confusingly complicated command.

- Get a simple version working first before you start coding the all-singing, all-dancing version that you most desire.

- Very large programs become increasingly difficult to understand, build, and maintain, roughly as some power of their length. Consider breaking such programs into subroutines or using a structure of command and subcommands.

- More general code is often both shorter and more robust. Sometimes programmers write to solve the most awkward case, say, to automate a series of commands that would be too tedious or error-prone to enter interactively. Stepping back from the most awkward case to the more general one is often then easier than might be thought.

- Do not be afraid to realize that at some point you may be best advised to throw it all away and start again from scratch.

11.14.7 Use the best tools

Find and use a text editor that you like and that supports programming directly. A good editor, for example, will be smart about indenting and will allow you to search for matching braces. Some editors even show syntax highlighting. For much more detailed comments on various text editors for Stata users, see `ssc describe texteditors`.

12 Cookbook: Ado-file programming

This cookbook chapter presents for Stata ado-file programmers several recipes using the programming features described in the previous chapter. Each recipe poses a problem and a worked solution. Although you may not encounter this precise problem, you should be able to recognize its similarities to a task that you would like to automate in an ado-file.

12.1 Retrieving results from rolling:

The problem. The `rolling` prefix (see section 7.2.4) will allow you to save the estimated coefficients (`_b`) and standard errors (`_se`) from a moving-window regression. As posed by a user on Statalist, what if you want to compute a quantity that depends on the full estimated variance–covariance matrix of the regression? Those quantities cannot be saved by `rolling`. For instance, the regression

```
. regress y L(1/4).x
```

estimates the effects of the last four periods' values of x on y.[1] We might naturally be interested in the sum of the lag coefficients, because it provides the *steady-state* effect of x on y. This computation is readily performed with `lincom`. If this regression is run over a moving window, how might we access the information needed to perform this computation?

The solution. A solution is available in the form of a wrapper program that can be called by `rolling`. We write our own r-class program, `myregress`, which returns the quantities of interest: the estimated sum of lag coefficients and its standard error. The program takes as arguments the varlist of the regression and two required options: `lagvar()`, the name of the distributed lag variable, and `nlags()`, the highest-order lag to be included in the `lincom` computation.[2] We build up the appropriate expression for the `lincom` command and return its results to the calling program.

---

1. See section 2.4.3 for a discussion of Stata's time-series operators.
2. This logic assumes that the current value of the regressor is not included in the `lincom` computation. That can easily be modified by using `L(0/4)`.

```
. type myregress.ado
*! myregress v1.0.0  CFBaum 11aug2008
program myregress, rclass
        version 10.1
        syntax varlist(ts) [if] [in], LAGVar(string) NLAGs(integer)
        regress `varlist' `if' `in'
        local nl1 = `nlags' - 1
        forvalues i = 1/`nl1' {
                local lv "`lv' L`i'.`lagvar' + "
        }
        local lv "`lv'  L`nlags'.`lagvar'"
        lincom `lv'
        return scalar sum = `r(estimate)'
        return scalar se = `r(se)'
end
```

As with any program to be used under the control of a prefix command, it is a good idea to execute the program directly to test it to ensure that its results are those you could calculate directly with lincom.

```
. use wpi1, clear
. myregress wpi L(1/4).wpi t, lagvar(wpi) nlags(4)
```

Source	SS	df	MS		
Model	108199.912	5	21639.9823		
Residual	59.2997117	114	.520172909		
Total	108259.211	119	909.741272		

	Number of obs =	120
	F(5, 114) =	41601.52
	Prob > F =	0.0000
	R-squared =	0.9995
	Adj R-squared =	0.9994
	Root MSE =	.72123

wpi	Coef.	Std. Err.	t	P>|t|	[95% Conf. Interval]	
wpi						
L1.	1.43324	.0947574	15.13	0.000	1.245526	1.620954
L2.	-.3915563	.1648926	-2.37	0.019	-.7182073	-.0649053
L3.	.1669584	.1693717	0.99	0.326	-.1685657	.5024825
L4.	-.2276451	.0960385	-2.37	0.019	-.4178967	-.0373936
t	.0184368	.0071336	2.58	0.011	.0043052	.0325684
_cons	.2392324	.1641889	1.46	0.148	-.0860246	.5644894

(1) L.wpi + L2.wpi + L3.wpi + L4.wpi = 0

wpi	Coef.	Std. Err.	t	P>|t|	[95% Conf. Interval]	
(1)	.9809968	.0082232	119.30	0.000	.9647067	.9972869

```
. return list
scalars:
                r(se) =  .0082232176260432
               r(sum) =  .9809968042273991
```

```
. lincom L.wpi+L2.wpi+L3.wpi+L4.wpi

 ( 1)  L.wpi + L2.wpi + L3.wpi + L4.wpi = 0
```

wpi	Coef.	Std. Err.	t	P>|t|	[95% Conf. Interval]	
(1)	.9809968	.0082232	119.30	0.000	.9647067	.9972869

Having validated the wrapper program by comparing its results with those from lincom, we can now invoke it with rolling:

```
. rolling sum=r(sum) se=r(se), window(30): myregress wpi L(1/4).wpi t,
> lagvar(wpi) nlags(4)
(running myregress on estimation sample)

Rolling replications (95)
————+—— 1 ——+—— 2 ——+—— 3 ——+—— 4 ——+—— 5
..................................................   50
.............................................
```

We can graph the resulting series and its approximate 95% standard-error bands with twoway rarea ([G] **graph twoway rarea**) and tsline (see figure 12.1):

```
. tsset end, quarterly
        time variable:  end, 1967q2 to 1990q4
                delta:  1 quarter
. label var end Endpoint
. generate lo = sum - 1.96 * se
. generate hi = sum + 1.96 * se
. twoway rarea lo hi end, color(gs12)
> title("Sum of moving lag coefficients, approx. 95% CI") ||  tsline sum,
> legend(off) scheme(s2mono)
```

(Continued on next page)

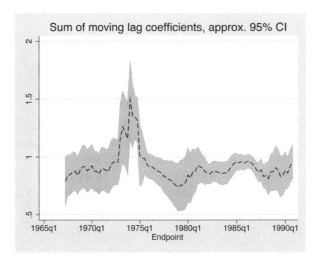

Figure 12.1. Rolling `lincom` estimates

12.2 Generalization of egen function pct9010() to support all pairs of quantiles

The problem. In section 11.10, we developed a useful `egen` function that will produce a percentile range for a specified variable. It can operate under the control of a by-list, juxtaposing the percentile range for each subset of the data specified in the by-list with those observations. It is a "one-trick pony" though, because the 90th and 10th percentiles are wired into the function.

The solution. It might be useful to take advantage of the `egen pctile()` function's ability to compute any percentiles of the specified variable. To achieve this, we modify `_gpct9010.ado` to the more general `_gpctrange.ado`. We add two options to the `egen` function: `lo()` and `hi()`. If not specified, those options take on 25 and 75, respectively, and compute the interquartile range. If specified, they must be checked for validity.

```
. type _gpctrange.ado
*! _gpctrange v1.0.0  CFBaum 11aug2008
program _gpctrange
        version 10.1
        syntax newvarname =/exp [if] [in] [, LO(integer 25) HI(integer 75)  *]

        if  'hi' > 99 | 'lo' < 1 {
                display as error ///
                        "Percentiles 'lo' 'hi' must be between 1 and 99."
                error 198
        }
        if 'hi' <= 'lo' {
                display as error ///
                        "Percentiles 'lo' 'hi' must be in ascending order."
                error 198
        }
        tempvar touse phi plo
        mark 'touse' 'if' 'in'
        quietly {
                egen double 'phi' = pctile('exp') if 'touse', 'options' p('hi')
                egen double 'plo' = pctile('exp') if 'touse', 'options' p('lo')
                generate 'typlist' 'varlist' = 'phi' - 'plo' if 'touse'
        }
end
```

The computations are then generalized merely by passing the contents of the `lo()` and `hi()` options to `egen pctile()`.[3]

```
. sysuse auto, clear
(1978 Automobile Data)
. bysort rep78: egen iqr = pctrange(price) if inrange(rep78, 3, 5)
(15 missing values generated)
. bysort rep78: egen p8020 = pctrange(price) if inrange(rep78, 3, 5),
> hi(80) lo(20)
(15 missing values generated)
. tabstat iqr if inrange(rep78, 3, 5), by(rep78)
Summary for variables: iqr
    by categories of: rep78 (Repair Record 1978)
```

rep78	mean
3	2108
4	2443
5	1915
Total	2174.22

---

3. Although I have suggested avoiding the use of `egen` within ado-files, its `pctile()` function is not easily replicated in ado-file code.

```
. tabstat p8020 if inrange(rep78, 3, 5), by(rep78)

Summary for variables: p8020
      by categories of: rep78 (Repair Record 1978)

 rep78 |      mean
-------+----------
     3 |    6231.5
     4 |      3328
     5 |      1915
-------+----------
 Total |  4540.915
```

This version of the **egen** function is much more generally useful and more likely to find a place in your toolbox.

12.3   Constructing a certification script

The problem. Assume you would like to establish that an ado-file you have written is providing the correct results and that any further modifications to the program will not harm its reliability. The best way to do that is to use Stata's *certification script* procedures (**help cscript**).

The solution. This solution draws heavily on the excellent discussion of statistical software certification in Gould (2001). We will develop a certification script for the last version of **pctrange**, the ado-file we developed in the previous chapter. That version of the program (as modified in section 11.8) is

```
. type pctrange.ado
*! pctrange v1.0.6  CFBaum 11aug2008
program pctrange, rclass byable(recall)
        version 10.1
        syntax varlist(min=1 numeric ts) [if] [in] [, noPRINT ///
        FORmat(passthru) MATrix(string)]
        marksample touse
        quietly count if 'touse'
        if 'r(N)' == 0 {
                error 2000
        }
        local nvar : word count 'varlist'
        if 'nvar' == 1 {
                local res range p7525 p9010 p9505 p9901
                tempname 'res'
                quietly summarize 'varlist' if 'touse', detail
                scalar 'range' = r(max) - r(min)
                scalar 'p7525' = r(p75) - r(p25)
                scalar 'p9010' = r(p90) - r(p10)
                scalar 'p9505' = r(p95) - r(p5)
                scalar 'p9901' = r(p99) - r(p1)
                if "'print'" != "noprint" {
                        display as result _n ///
        "Percentile ranges for 'varlist', N = 'r(N)'"
                        display as txt "75-25: " 'p7525'
```

```
                              display as txt "90-10: " 'p9010'
                              display as txt "95-05: " 'p9505'
                              display as txt "99-01: " 'p9901'
                              display as txt "Range: " 'range'
                     }
                     foreach r of local res {
                              return scalar 'r' = ''r''
                     }
                     return scalar N = r(N)
            }
            else {
                     tempname rmat
                     matrix 'rmat' = J('nvar',5,.)
                     local i 0
                     foreach v of varlist 'varlist' {
                              local ++i
                              quietly summarize 'v' if 'touse', detail
                              matrix 'rmat'['i',1] = r(max) - r(min)
                              matrix 'rmat'['i',2] = r(p75) - r(p25)
                              matrix 'rmat'['i',3] = r(p90) - r(p10)
                              matrix 'rmat'['i',4] = r(p95) - r(p5)
                              matrix 'rmat'['i',5] = r(p99) - r(p1)
                              local rown "'rown' 'v'"
                     }
                     matrix colnames 'rmat' = Range P75-P25 P90-P10 P95-P05 P99-P01
                     matrix rownames 'rmat' = 'rown'
                     if "'print'" != "noprint" {
                              local form ", noheader"
                              if "'format'" != "" {
                                       local form "'form' 'format'"
                              }
                              matrix list 'rmat' 'form'
                     }
                     if "'matrix'" != "" {
                              matrix 'matrix' = 'rmat'
                     }
                     return matrix rmat = 'rmat'
            }
            return local varname 'varlist'
     end
```

Following Gould (2001, 35), we create a directory in our current working directory called
/bench to hold certification scripts, and we create a subdirectory for this script called
/pctrange. In the bench directory, we create testall.do:

```
clear
discard
set more off
cd first
quietly log using pctrange, replace
do test1
quietly log close
```

We begin our certification process by copying auto.dta into the pctrange directory.
As Gould suggests (2001, 45), certification scripts should be self-contained, not referring
to files outside the directory in which they reside.

A certification script should run to normal completion if no errors are encountered. A script that aborts provides an indication that one or more tests in the script have failed. We first want to test the syntax of the program and verify that appropriate error messages are produced if the program is called incorrectly. We can trigger errors without aborting our do-file by using the `rcof` (`help rcof`) command. For instance, `pctrange` requires at least one numeric variable and cannot handle string variables. It makes no sense to call the program without some observations in the data. We can test these conditions with

```
. cd bench/pctrange
/Users/baum/doc/ITSP/dof.8824/bench/pctrange/
. do test1
. cscript "test script for ITSP:" pctrange
─────────────────────────────────────────BEGIN test script for ITSP: pctrange
. use auto
(1978 Automobile Data)
. rcof "pctrange" != 0
. rcof "pctrange price make" != 0
. drop _all
. rcof "pctrange price mpg" != 0
.
end of do-file
```

Let's now check that if *exp* and in *range* conditions are handled properly. To do this, we must compare the output of the program (as provided in the list of saved results) by using the `assert` command and the `mreldif()` matrix function. We test for absolute equality of the matrices produced, because there should be no possibility for rounding error in the two methods of computing the same quantities if our program is working properly. `pctrange` produces only the `r(rmat)` matrix when more than one variable appears in its varlist. We create a new script, `test2`, that incorporates the tests applied above.

```
. cd bench/pctrange
/Users/baum/doc/ITSP/dof.8824/bench/pctrange/
. do test2
. cscript "test script for ITSP:" pctrange
─────────────────────────────────────────BEGIN test script for ITSP: pctrange
. use auto
(1978 Automobile Data)
. rcof "pctrange" != 0
. rcof "pctrange price make" != 0
. drop _all
. rcof "pctrange" != 0
. use auto
(1978 Automobile Data)
. quietly pctrange price mpg if foreign
. matrix check1 = r(rmat)
```

```
. keep if foreign
(52 observations deleted)

. quietly pctrange price mpg

. mat check2 = r(rmat)

. scalar checka = mreldif(check1, check2)

. assert checka == 0

. use auto, clear
(1978 Automobile Data)

. quietly pctrange price mpg in 22/55

. matrix check3 = r(rmat)

. keep in 22/55
(40 observations deleted)

. quietly pctrange price mpg

. mat check4 = r(rmat)

. scalar checkb = mreldif(check3, check4)

. assert checkb == 0

. use auto, clear
(1978 Automobile Data)

. generate t = _n

. tsset t
        time variable:  t, 1 to 74
                delta:  1 unit

. generate dp = D.price
(1 missing value generated)

. generate lmpg = L.mpg
(1 missing value generated)

. quietly pctrange dp lmpg

. mat check5 = r(rmat)

. quietly pctrange D.price L.mpg

. mat check6 = r(rmat)

. scalar checkc = mreldif(check5, check6)

. assert checkc == 0

.

.
end of do-file
```

We have also included a check of the program's ability to handle time-series (`ts`) operators.

We now want to check some of the program's calculations against those produced by `summarize, detail`. We can do this with just one variable, in which case `pctrange` returns a set of scalars. We now use the `reldif()` scalar function, which allows us to check to see that computations from `summarize` results are appropriately close to those of `pctrange`. We do not check for equality because of the limitations of finite-precision arithmetic given that we are comparing the results of two different commands. The value of 10^{-13} is chosen to ensure that the two computed measures are sufficiently close. A larger value might have to be used on a different computing platform.

```
. cd bench/pctrange
/Users/baum/doc/ITSP/dof.8824/bench/pctrange/
. do test3
. cscript "test script for ITSP:" pctrange
────────────────────────────────────────────BEGIN test script for ITSP: pctrange
. // (earlier tests omitted)
. use auto, clear
(1978 Automobile Data)
. quietly pctrange price
. scalar enn = r(N)
. scalar p9901 = r(p9901)
. scalar p9505 = r(p9505)
. scalar p9010 = r(p9010)
. scalar p7525 = r(p7525)
. scalar range = r(range)
. assert "`r(varname)'" == "price"
. quietly summarize price, detail
. assert reldif(r(p99) - r(p1), p9901) < 1e-13
. assert reldif(r(p95) - r(p5), p9505) < 1e-13
. assert reldif(r(p90) - r(p10), p9010) < 1e-13
. assert reldif(r(p75) - r(p25), p7525) < 1e-13
. assert reldif(r(max) - r(min), range) < 1e-13
.
end of do-file
```

Finally, we should check whether **pctrange** produces the same results under the control of a **by** prefix. Like most Stata commands, **pctrange** only returns results from the last by-group. We add tests that verify that results from this by-group (where **foreign** = 1) are identical to those generated by limiting the sample to those observations.

```
. cd bench/pctrange
/Users/baum/doc/ITSP/dof.8824/bench/pctrange/
. do test4
. cscript "test script for ITSP:" pctrange
────────────────────────────────────────────BEGIN test script for ITSP: pctrange
. // (earlier tests omitted)
. use auto, clear
(1978 Automobile Data)
. quietly by foreign: pctrange price mpg
. matrix check7 = r(rmat)
. keep if foreign
(52 observations deleted)
. quietly pctrange price mpg
. mat check8 = r(rmat)
. scalar checkd = mreldif(check7, check8)
. assert checkd == 0
.
end of do-file
```

We did not use `tempnames` or `tempvars` in these scripts because of their simplicity. For a more complicated certification process, those tools might be useful to prevent the possibility of name collisions.

A certification script is a work in progress. As Gould (2001, 38) suggests, it need not be elegant, but it should be complete. As you think of additional conditions to be tested (or fix bugs in the ado-file), add them to the test script. The certification script should be run again every time a change is made to the code. If that change adds new features, tests should be added to the script putting those features through their paces.

The `savedresults` command (`help savedresults`) allows you to compare the results of two commands' return values for both r-class and e-class commands. This capability is useful when you have a validated version of a command and a new version with additional features, and you want to ensure that all the saved results produced by the new version match those of the old version.

If you are considering sharing your program with the Stata user community (for instance, by placing it in the Statistical Software Components archive or by making it available from your web site) you might want to include the certification script (and any datasets it references) as ancillary files in the package. Demonstrating that a user-written program passes several certification tests provides reassurance to other users that they can rely on the program's results.

12.4 Using the ml command to estimate means and variances

The problem. A Statalist user posed a question about the estimation of means and variances from subsamples of a normally distributed variable. He wanted to compute two nonlinear combinations of those estimates:

$$\beta = \frac{\sigma_1 - \sigma_2}{\sigma_1 + \sigma_2}$$

and

$$\alpha = 2\pi\sqrt{3}\left(\frac{\mu_1 - \mu_2}{\sigma_1 + \sigma_2}\right)$$

The user also wanted to estimate the quantity α given the assumption of a common variance, $\sigma = \sigma_1 = \sigma_2$.

The solution. This can readily be accomplished with `ml` as long as the user is willing to make a distributional assumption. We set up a variant of `mynormal_lf.ado`[4] that allows for separate means and variances, depending on the value of an indicator variable, which we access with global macro `subsample`:

---

4. See section 11.13.

```
. type meanvar.ado
*! meanvar v1.0.1  CFBaum 11aug2008
program meanvar
        version 10.1
        args lnf mu1 mu2 sigma1 sigma2
        qui replace 'lnf' = ln(normalden($ML_y1, 'mu1', 'sigma1')) ///
            if $subsample == 0
        qui replace 'lnf' = ln(normalden($ML_y1, 'mu2', 'sigma2')) ///
            if $subsample == 1
end
```

We now can set up the estimation problem. Because we do not have the user's data, we will use `auto.dta` and consider `foreign` as the binary indicator:

```
. sysuse auto, clear
(1978 Automobile Data)
. global subsample foreign
. ml model lf meanvar (mu1: price = ) (mu2: price = ) /sigma1 /sigma2
. ml maximize, nolog
initial:        log likelihood =      -<inf>  (could not be evaluated)
feasible:       log likelihood = -853.12316
rescale:        log likelihood = -710.29627
rescale eq:     log likelihood = -699.08529
```

	Number of obs	=	74
	Wald chi2(0)	=	.
Log likelihood = -695.14898	Prob > chi2	=	.

		Coef.	Std. Err.	z	P>|z|	[95% Conf. Interval]	
mu1							
	_cons	6072.423	425.3415	14.28	0.000	5238.769	6906.077
mu2							
	_cons	6384.682	546.1422	11.69	0.000	5314.263	7455.101
sigma1							
	_cons	3067.181	300.7618	10.20	0.000	2477.698	3656.663
sigma2							
	_cons	2561.634	386.1809	6.63	0.000	1804.733	3318.535

```
. estimates store unconstr
```

We use `estimates store` to save the results of the estimation under the name `unconstr`.

Estimates of the desired quantities can be readily computed, in point and interval form, with `nlcom`:

```
. nlcom ([sigma1]_b[_cons] - [sigma2]_b[_cons]) /
> ([sigma1]_b[_cons] + [sigma2]_b[_cons])

     _nl_1:  ([sigma1]_b[_cons] - [sigma2]_b[_cons]) / ([sigma1]_b[_cons] +
> [sigma2]_b[_cons])
```

	Coef.	Std. Err.	z	P>|z|	[95% Conf. Interval]	
_nl_1	.0898141	.089195	1.01	0.314	-.0850049	.264633

```
. nlcom 2*_pi*sqrt(3) * (([mu1]_b[_cons] - [mu2]_b[_cons]) /
> ([sigma1]_b[_cons] + [sigma2]_b[_cons]))

     _nl_1:  2*_pi*sqrt(3) * (([mu1]_b[_cons] - [mu2]_b[_cons]) / ([sigma1]_b
> [_cons] + [sigma2]_b[_cons]))
```

	Coef.	Std. Err.	z	P>|z|	[95% Conf. Interval]	
_nl_1	-.6037236	1.339398	-0.45	0.652	-3.228896	2.021449

We can verify that these maximum likelihood estimates of the subsample means and variances are correct by estimating the subsamples with ivreg2 (Baum, Schaffer, and Stillman 2007), available from the Statistical Software Components archive:

```
. ivreg2 price if !foreign
. ivreg2 price if foreign
```

12.4.1   Applying equality constraints in ml estimation

The second task in the problem statement is the computation of α subject to the constraint of equal variances in the two subsamples.[5] Define the constraint ([R] constraint), and include the constraints() option with ml model. As the output shows, the two sigma parameters are now forced to be equal. Given the nature of maximum likelihood estimation, this does not alter the point estimates of the mu parameters, but it alters their standard errors. We now can recompute the estimated quantity α.

```
. constraint 1 [sigma1]_cons = [sigma2]_cons
. ml model lf meanvar (mu1: price = ) (mu2: price = ) /sigma1 /sigma2, ///
> constraints(1)
```

---

5. In this form, the desired quantity could be calculated from a regression of price on the indicator variable. We use ml with constraints for pedagogical purposes.

```
. ml maximize, nolog
initial:       log likelihood =      -<inf>  (could not be evaluated)
feasible:      log likelihood = -1109.2082
rescale:       log likelihood = -711.67541
rescale eq:    log likelihood =  -709.0676
                                          Number of obs   =         74
                                          Wald chi2(0)    =          .
Log likelihood = -695.62494               Prob > chi2     =          .
 ( 1)  [sigma1]_cons - [sigma2]_cons = 0
```

	Coef.	Std. Err.	z	P>|z|	[95% Conf. Interval]	
mu1						
_cons	6072.423	405.766	14.97	0.000	5277.136	6867.71
mu2						
_cons	6384.682	623.8296	10.23	0.000	5161.998	7607.365
sigma1						
_cons	2926.02	240.5153	12.17	0.000	2454.619	3397.422
sigma2						
_cons	2926.02	240.5153	12.17	0.000	2454.619	3397.422

```
. estimates store constr
. nlcom 2*_pi*sqrt(3) * (([mu1]_b[_cons] - [mu2]_b[_cons]) /
> ([sigma1]_b[_cons] + [sigma2]_b[_cons]))

       _nl_1:  2*_pi*sqrt(3) * (([mu1]_b[_cons] - [mu2]_b[_cons]) / ([sigma1]_b
> [_cons] + [sigma2]_b[_cons]))
```

	Coef.	Std. Err.	z	P>|z|	[95% Conf. Interval]	
_nl_1	-.5806946	1.38475	-0.42	0.675	-3.294756	2.133366

We use `estimates store` to save the results of this constrained maximization under
the name `constr`. Because we have estimated an unconstrained and a constrained form
of the same model, we can use a likelihood-ratio test (`lrtest`; see [R] **lrtest**) to evaluate
whether the constraints are rejected by the data:

```
. lrtest unconstr constr
Likelihood-ratio test                          LR chi2(1)  =       0.95
(Assumption: constr nested in unconstr)        Prob > chi2 =     0.3292
```

The large p-value on the χ^2 statistic indicates that the data do not reject the constraint
of equal error variances for domestic and foreign automobiles.

12.5   Applying inequality constraints in ml estimation

The problem. We know that certain parameters in a maximum likelihood estimation
should obey inequality restrictions. For instance, the σ parameter should be strictly

positive because it is derived from the estimated variance of the error process. How can
we ensure that these constraints are satisfied?

The solution. Consider the `mynormal_lf` program presented in section 11.13. As Gould,
Pitblado, and Sribney (2006, 56) suggest, a more numerically stable form of the likelihood function for linear regression estimates $\log \sigma$ rather than σ itself, allowing the
parameter to take on values on the entire real line. The likelihood function evaluator
then becomes

```
. type mynormal_lf.ado
*! mynormal_lf v1.0.1  CFBaum 11aug2008
program mynormal_lf
  version 10.1
  args lnf mu lnsigma
  quietly replace 'lnf' = ln(normalden($ML_y1,  'mu', exp('lnsigma')))
end
```

Because we still want to take advantage of the three-argument form of the `normalden()`
function, we pass the parameter `exp('lnsigma')` to the function, which expects to
receive σ itself as its third argument. But because the optimization takes place with
respect to parameter `lnsigma`, difficulties with the zero boundary are avoided.

To apply the same logic to the ado-file version, `mynormal.ado`, we must consider
one issue: when the `Replay` routine is invoked, the parameter displayed will be `lnsigma`
rather than `sigma`. We can deal with this issue by using the `diparm()` option of `ml`
`model`. We modify `mynormal.ado` accordingly:

```
. type mynormal.ado
*! mynormal v1.0.1  CFBaum 11aug2008
program mynormal
        version 10.1
        if replay()  {
                if ("'e(cmd)'" != "mynormal") error 301
                Replay '0'
        }
        else Estimate '0'
end

program Replay
        syntax [, Level(cilevel) ]
        ml display, level('level')
end

program Estimate, eclass sortpreserve
        syntax varlist [if] [in]  [,  vce(passthru) Level(cilevel) * ]
        mlopts mlopts, 'options'
        gettoken lhs rhs: varlist
        marksample touse
        local diparm diparm(lnsigma, exp label("sigma"))
        ml model lf  mynormal_lf (mu: 'lhs' = 'rhs') /lnsigma ///
            if 'touse', 'vce' 'mlopts' maximize 'diparm'
        ereturn local cmd "mynormal"
        ereturn scalar k_aux = 1
        Replay, level('level')
end
```

Now, when the model is fitted, the ancillary parameter `lnsigma` is displayed and transformed into the original parameter space as `sigma`.

```
. sysuse auto, clear
(1978 Automobile Data)

. mynormal price mpg weight turn
initial:       log likelihood =     -<inf>  (could not be evaluated)
feasible:      log likelihood = -811.54531
rescale:       log likelihood = -811.54531
rescale eq:    log likelihood = -808.73926
Iteration 0:   log likelihood = -808.73926
Iteration 1:   log likelihood = -729.21876  (not concave)
Iteration 2:   log likelihood = -708.48511  (not concave)
Iteration 3:   log likelihood = -702.39976
Iteration 4:   log likelihood = -678.51799
Iteration 5:   log likelihood = -677.74671
Iteration 6:   log likelihood = -677.74638
Iteration 7:   log likelihood = -677.74638
```

		Number of obs	=	74
		Wald chi2(3)	=	46.26
Log likelihood = -677.74638		Prob > chi2	=	0.0000

price	Coef.	Std. Err.	z	P>|z|	[95% Conf. Interval]	
mpg	-72.86501	79.06769	-0.92	0.357	-227.8348	82.10481
weight	3.524339	.7947479	4.43	0.000	1.966661	5.082016
turn	-395.1902	119.2837	-3.31	0.001	-628.9819	-161.3985
_cons	12744.24	4629.664	2.75	0.006	3670.27	21818.22
/lnsigma	7.739796	.0821995	94.16	0.000	7.578688	7.900904
sigma	2298.004	188.8948			1956.062	2699.723

This same technique can be used to constrain a regression parameter to be strictly positive (or strictly negative).[6] Imagine that we believe the first slope parameter in a particular regression model must be negative. We can no longer use the `lf` (linear-form) method but must switch to one of the more complicated methods: `d0`, `d1`, or `d2`. In the linear-form method, your likelihood function evaluator calculates the likelihood function in terms of the index $\mathbf{x}'\beta$. When using the `d...` methods, you must calculate the total likelihood within your routine. In these methods, you must use the `mleval` command to evaluate the coefficient vector passed to the likelihood function evaluation and use the `mlsum` command to sum the likelihood over the available observations.[7] We start with the simple `mynormal_d0` routine from Gould, Pitblado, and Sribney (2006, 98):

---

6. For an explanation of how this can be done in an ordinary least-squares regression context, see the response to the Stata frequently asked question "How do I fit a regression with interval constraints in Stata?" (http://www.stata.com/support/faqs/stat/intconst.html), written by Isabel Cañette.

7. See Gould, Pitblado, and Sribney (2006) for full details.

```
. type mynormal_lf_d0.ado
*! mynormal_lf_d0 v1.0.0  CFBaum 11aug2008
program mynormal_lf_d0
        version 10.1
        args todo b lnf
        tempvar mu
        tempname lnsigma
        mleval 'mu' = 'b', eq(1)
        mleval 'lnsigma' = 'b', eq(2) scalar
        quietly {
                mlsum 'lnf' = ln(normalden($ML_y1,'mu', exp('lnsigma')))
        }
end
```

We invoke it with `ml model`:

```
. ml model d0 mynormal_lf_d0 (mu: price = mpg weight turn) /lnsigma,
> maximize nolog diparm(lnsigma, exp label("sigma"))
initial:       log likelihood =     -<inf>  (could not be evaluated)
feasible:      log likelihood = -811.54531
rescale:       log likelihood = -811.54531
rescale eq:    log likelihood = -808.73926

. ml display
```

					Number of obs	=	74
					Wald chi2(3)	=	46.26
Log likelihood = -677.74638					Prob > chi2	=	0.0000

price	Coef.	Std. Err.	z	P>|z|	[95% Conf.	Interval]
mu						
mpg	-72.86507	79.06768	-0.92	0.357	-227.8349	82.10474
weight	3.524333	.7947475	4.43	0.000	1.966656	5.082009
turn	-395.1891	119.2836	-3.31	0.001	-628.9807	-161.3974
_cons	12744.22	4629.664	2.75	0.006	3670.244	21818.19
lnsigma						
_cons	7.739795	.0821994	94.16	0.000	7.578688	7.900903
sigma	2298.002	188.8945			1956.06	2699.72

We now break out the first slope coefficient from the linear combination and give it its own "equation". To refer to the variable `mpg` within the likelihood function evaluator, we must use a global macro[8] to impose the constraint of negativity on the `mpg` coefficient. We only need specify that the mean equation (for `mu`) is adjusted by subtracting the exponential of the parameter `a`:

---

8. See section 3.7.

```
. type mynormal_lf_d0_c1.ado
*! mynormal_lf_d0_c1 v1.0.0  CFBaum 11aug2008
program mynormal_lf_d0_c1
        version 10.1
        args todo b lnf
        tempvar xb mu
        tempname a lnsigma
        mleval 'a' = 'b', eq(1) scalar
        mleval 'xb' = 'b', eq(2)
        mleval 'lnsigma' = 'b', eq(3) scalar
        quietly {
                generate double 'mu' = 'xb' - exp('a')* $x1
                mlsum 'lnf' = ln(normalden($ML_y1,'mu', exp('lnsigma')))
        }
end
```

We can now fit the model subject to the inequality constraint imposed by the exp() function. Because maximizing likelihood functions with inequality constraints can be numerically difficult, we supply starting values to ml model from the unconstrained regression. We type

```
. global x1 mpg
. qui regress price mpg weight turn
. matrix b0 = e(b), ln(e(rmse))
. matrix b0[1,1] = ln(-1*b0[1,1])
. ml model d0 mynormal_lf_d0_c1 (a:) (mu: price = weight turn) /lnsigma,
> maximize nolog diparm(lnsigma, exp label("sigma"))
initial:       log likelihood =    -<inf>  (could not be evaluated)
feasible:      log likelihood = -1876.4862
rescale:       log likelihood = -972.49514
rescale eq:    log likelihood = -972.25368
. ml display
```

| | | | | | Number of obs | = | 74 |
| | | | | | Wald chi2(0) | = | . |
Log likelihood = -677.74638 | | | | | Prob > chi2 | = | . |

price	Coef.	Std. Err.	z	P>|z|	[95% Conf. Interval]
a					
_cons	4.288541	1.085214	3.95	0.000	2.161561 6.415521
mu					
weight	3.524365	.7947519	4.43	0.000	1.96668 5.08205
turn	-395.1895	119.2839	-3.31	0.001	-628.9817 -161.3974
_cons	12744.03	4629.705	2.75	0.006	3669.979 21818.09
lnsigma					
_cons	7.739796	.0821995	94.16	0.000	7.578688 7.900905
sigma	2298.005	188.8948			1956.062 2699.723

We use the nlcom command to transform the estimated coefficient back to its original space:

```
. nlcom -exp([a]_cons)
      _nl_1:  -exp([a]_cons)
```

price	Coef.	Std. Err.	z	P>|z|	[95% Conf. Interval]
_nl_1	-72.86009	79.06879	-0.92	0.357	-227.8321 82.1119

We have estimated $-\ln(\beta_{\mathrm{mpg}})$; `nlcom` back-transforms the estimated coefficient to β_{mpg} in point and interval form.

Variations on this technique are used throughout Stata's maximum likelihood estimation commands to ensure that coefficients take on appropriate values. For instance, the bivariate probit command ([R] **biprobit**) estimates a coefficient, ρ, the correlation between two error processes. It must lie within $(-1, +1)$. Stata's `biprobit` routine estimates the hyperbolic arctangent (`atanh()`) of ρ, which constrains the parameter itself to lie within the appropriate interval when back-transformed. A similar transformation can be used to constrain a slope parameter to lie within a certain interval of the real line.

12.6 Generating a dataset containing the single longest spell for each unit in panel data

The problem. In panel (longitudinal) data, many datasets contain *unbalanced panels*, with differing numbers of observations for different units in the panel. Some estimators commonly used in a panel-data context can work with unbalanced panels but expect to find a single *spell* for each unit; that is, they expect to find a time series without gaps. Finding and retaining the single longest spell for each unit within the panel is straightforward for just one variable. However, for our purposes, we want to apply this logic listwise and delete shorter spells if any of the variables in a specified varlist are missing. The solution will entail the creation of a new, smaller dataset in which only panel units with single spells are present.

To motivate this form of data organization, the Arellano–Bond dynamic panel-data estimator and its descendants (see [XT] **xtabond** and the user-written `xtabond2`[9]), consider that each spell of observations represents a panel unit.

The solution. We present a solution to this problem with an ado-file, `onespell.ado`. The program builds upon Nicholas J. Cox's excellent `tsspell` command.[10] The command examines one variable, optionally given a logical condition that defines a spell,[11] and creates three new variables: `_spell`, indicating distinct spells (taking on successive integer values); `_seq`, giving the sequence of each observation in the spell (taking on successive integer values); and `_end`, indicating the end of spells. If applied to panel

---

9. See `findit xtabond2`.
10. See `findit tsspell`.
11. See sections 4.4 and 8.4.

data rather than one time series, the routine automatically performs these observations for each unit of a panel.

In this first part of the program, we define the syntax of the ado-file. The program accepts a varlist of any number of numeric variables, accepts if *exp* and in *range* options, and requires that the user provide a filename in the `saving()` option so that the resulting edited dataset will be stored. Optionally, the user can specify a `replace` option (which must be spelled out). The `noisily` option is provided for debugging purposes. The `preserve` command allows us to modify the data and return to the original dataset.

The `tsset` command allows us to retrieve the names of the panel variable and time variable. If the data are not `tsset`, the program will abort. The `tsfill` command fills any gaps in the time variable with missing observations. We then use `marksample touse` to apply any qualifiers on the set of observations and define several `tempvar`s.

For ease of exposition, I do not list the entire ado-file here. Rather, the first piece of the code is displayed is displayed (as a text file), and the remainder (also as a text file) as a separate listing below a discussion of its workings.

```
. type onespell_part1.txt
*! onespell 1.1.1  CFBaum  13jan2005
* locate units with internal gaps in varlist and zap all but longest spell
program onespell, rclass
        version 10.1
        syntax varlist(numeric) [if] [in], Saving(string) [ REPLACE NOIsily]
        preserve
        quietly tsset
        local pv "`r(panelvar)'"
        local tv "`r(timevar)'"
        summarize `pv', meanonly
        local n1 = r(N)
        tsfill
        marksample touse
        tempvar testgap spell seq end maxspell keepspell wantspell
        local sss = cond("`noisily'" != "", "noisily", "quietly")
```

The real work is performed in the second half of the program. The temporary variable `testgap` is generated with the `cond()` function to define each observation as either its value of the panel variable (pv) or missing. Cox's `tsspell` is then invoked on the `testgap` variable with the logical condition that the variable is nonmissing. We explicitly name the three variables created by `tsspell` as temporary variables `spell`, `seq`, and `end`.

In the first step of pruning the data, we note that any observation for which `spell` = 0 can be discarded, along with any observations not defined in the `touse` restrictions. Now for each panel unit, we consider how many spells exist. If `spell > 1`, there are gaps in the usable data. The longest spell for each panel unit is stored in the `maxspell` temporary variable, produced by `egen max()` from the `seq` counter. Now for each panel unit, we generate a `keepspell` temporary variable, identified by the longest observed spell (`maxspell`) for that unit. We then can calculate the `wantspell` temporary variable with `egen max()`, which places the `keepspell` value in each observation of the desired spell. What if there are two (or more) spells of identical length? By convention, the latest spell is chosen.

We can now apply `keep` to retain only those observations for each panel unit associated with that unit's longest spell, that is, those observations for which `wantspell` equals the `spell` number. The resulting data are then `saved` to the file specified in the `saving()` option, optionally using `replace`, and finally the original data are `restored`.

```
. type onespell_part2.txt
        'sss' {
* testgap is panelvar if obs is usable, 0 otherwise
            generate 'testgap' = cond('touse', 'pv', .)
            tsspell 'testgap' if !missing('testgap'), spell('spell') ///
                seq('seq') end('end')
            drop if 'spell' == 0 | 'touse' == 0
* if 'spell' > 1 for a unit, there are gaps in usable data
* calculate max length spell for each unit and identify
* that spell as the one to be retained
            egen 'maxspell' = max('seq'), by('pv')
            generate 'keepspell' = cond('seq'=='maxspell', 'spell', 0)
            egen 'wantspell' = max('keepspell'), by('pv')
* in case of ties, latest spell of max length is selected
            list 'pv' 'tv' 'spell' 'seq' 'maxspell' 'keepspell' ///
        'wantspell', sepby('pv')
            summarize 'spell' 'wantspell'
            keep if 'wantspell' == 'spell'
            summarize 'pv', meanonly
            local n2 = r(N)
            drop \__*
        }
        display _n "Observations removed: " 'n1'-'n2'
        save 'saving', 'replace'
        restore
end
```

To illustrate, we modify the **grunfeld** dataset (this dataset was also used in section 9.2). The original dataset is a balanced panel of 20 years' observations on 10 firms. We remove observations from different variables in firms 2, 3, and 5, creating two spells in firms 2 and 3 and three spells in firm 5. We then apply `onespell`:

```
. webuse grunfeld, clear
. quietly replace invest = . in 28
. quietly replace mvalue = . in 55
. quietly replace kstock = . in 87
```

```
. quietly replace kstock = . in 94
. onespell invest mvalue kstock, saving(grun1) replace
Observations removed: 28
file grun1.dta saved
```

A total of 28 observations are removed. The tabulation shows that firms 2, 3, and 5 now have longest spells of 12, 14, and 6 years, respectively.

```
. use grun1, clear
. tab company
    company │     Freq.      Percent        Cum.
────────────┼───────────────────────────────────
          1 │        20        11.63       11.63
          2 │        12         6.98       18.60
          3 │        14         8.14       26.74
          4 │        20        11.63       38.37
          5 │         6         3.49       41.86
          6 │        20        11.63       53.49
          7 │        20        11.63       65.12
          8 │        20        11.63       76.74
          9 │        20        11.63       88.37
         10 │        20        11.63      100.00
────────────┼───────────────────────────────────
      Total │       172       100.00
```

Although this routine meets a specialized need, the logic behind it can be useful in several circumstances for data management.

13 Mata functions for ado-file programming

13.1 Mata: First principles

Mata is a full-fledged programming language that operates in the Stata environment. You can start Stata, enter Mata by using the `mata` command, and work interactively in that environment. You can even leave Mata by using the `end` command, issue Stata commands, and return to Mata's workspace, which will be unchanged until you issue the `mata: mata clear` command or exit Stata. But Mata is useful in another context: that of noninteractive programming. You can write Mata functions, callable from Stata do-files or ado-files, that perform useful tasks. The functions make use of Mata's facility with matrices and vectors, or they can use other elements of Mata's environment. For instance, Mata can be used for file input and output (with functions much like those in the C programming language) and is an excellent environment for string processing. Unlike some matrix programming languages, Mata matrices can contain either numeric or string elements (but not both).

For matrix programming, Mata is a more useful environment than Stata's older `matrix` commands. All versions of Stata have limits on matrix size (for instance, 800 rows or columns in Stata/IC), and these limits can be constraining. Creating a sizable matrix from Stata variables will at least double the amount of memory needed to work with those two representations of the data. In contrast, Mata can work with views, or what might be considered *virtual matrices*, containing Stata variables' data but not requiring additional storage beyond a trivial amount of overhead. The suite of matrix functions available in Mata is also much broader than that available in Stata's matrix language. Additionally, Mata explicitly supports complex arithmetic and data types.

Most importantly, Mata code is automatically compiled into *bytecode*.[1] If you write an ado-file program that involves many explicit loops (such as those referencing the subscripts of matrices), the commands within the loop must be interpreted each time through the loop. The equivalent Mata code is compiled once, "on the fly", so that subsequent passes through the code are executing machine instructions rather than reinterpreting your code. If you write a Mata function, you can store it in object-module form or place it in a library of object modules. The net effect is a considerable speed advantage for many routines coded in Mata rather than in the ado-file language.

---

1. For more information, see http://en.wikipedia.org/wiki/Bytecode.

The integration between Mata's and Stata's elements (variables, local and global macros, scalars, and matrices) is complete. Within Mata, you can access Stata's objects and alter them. When using Mata's view matrices, alterations to the Mata matrix automatically modify the Stata observations and variables that comprise the view matrix.

In this chapter, I discuss how you can use Mata functions as adjuncts to your ado-file programs. The techniques developed in earlier chapters remain useful. For instance, you still use Stata's `syntax` and `marksample` statements to parse a user's command and prepare the appropriate sample for Mata. After calling a Mata function, you will use Stata's `return` and `ereturn` facilities to return computed quantities to the user. The best candidates for translation to Mata code are the heavily computational aspects of your routine, particularly if their logic can be readily expressed in matrix algebra.

13.1.1 What you should learn from this chapter

- An understanding of Mata's operators, subscripts, loop functions, and conditionals
- How the components of a Mata function are defined
- How to call a Mata function from an ado-file
- How to use Mata's `st_` functions to exchange data with Stata
- How to construct arrays of temporary objects in Mata
- How to use Mata structures
- How to create compiled Mata functions or a Mata function library
- How to use user-contributed Mata functions

13.2 Mata fundamentals

Before delving into writing Mata functions, we must understand Mata's elements of syntax and several of its operators. Mata's command prompt is the colon (:) rather than Stata's full stop (.).

13.2.1 Operators

The comma (,) is Mata's *column-join operator*.

```
: r1 = (1, 2, 3)
```

creates a three-element row vector. We could also construct this vector by using the *row-range operator* (..), as in

```
: r1 = (1..3)
```

The backslash (\) is Mata's *row-join operator*.

```
: c1 = (4 \ 5 \ 6)
```

creates a three-element column vector. We could also construct this vector by using the *column-range operator* (::), as in

```
: c1 = (4::6)
```

We can combine the column-join and row-join operators.

```
: m1 = (1, 2, 3 \ 4, 5, 6 \ 7, 8, 9 )
```

creates a 3×3 matrix. The matrix could also be constructed with the row-range operator:

```
: m1 = ((1..3) \(4..6) \(7..9))
```

The prime, or apostrophe ('), is Mata's transpose operator, so

```
: r2 = (1 \ 2 \ 3 )'
```

is a row vector.[2] Numeric matrix elements can be real or complex, so 2 - 3i refers to the complex number $2 - 3 \times \sqrt{-1}$. The comma and backslash operators can be used on vectors and matrices as well as on scalars, so

```
: r3 = r1, c1'
```

will produce a six-element row vector, and

```
: c2 = r1' \ c1
```

is a six-element column vector.

The standard algebraic operators plus (+), minus (-), and multiply (*) work on scalars or matrices:

```
: g = r1'  +  c1
: h = r1 * c1
: j = c1 * r1
```

In this example, h will be the 1×1 dot product,[3] or inner product, of row vector r1 and column vector c1, while j is their 3×3 outer product.[4]

Mata's algebraic operators, including the forward slash (/) for division, can also be used in element-by-element computations when preceded by a colon.

```
: k = r1' :* c1
```

---

2. The transpose operator applied to a complex matrix produces the *conjugate transpose* so that $(a + bi)' = (a - bi)$.
3. For more information, see http://en.wikipedia.org/wiki/Dot_product.
4. For more information, see http://en.wikipedia.org/wiki/Outer_product.

will produce the three-element column vector, with elements as the product of the respective elements.

Mata's *colon operator* is powerful in that it will work on nonconformable objects as long as the expression can be interpreted sensibly. For example,

```
: r4 = ( 1, 2, 3 )
: m2 = ( 1, 2, 3 \ 4, 5, 6 \ 7, 8, 9 )
: m3 = r4 :+ m2
: m4 = m1 :/ r1
```

adds the row vector r4 to each row of the 3×3 matrix m2 to form m3, then divides each row of matrix m1 by the corresponding elements of row vector r1 to form m4.

Mata's scalar functions will also operate on elements of matrices.

```
: m5 = sqrt(m4)
```

will take the element-by-element square root of the 3×3 matrix m4, returning missing values where appropriate.

The matrix operator A # B produces the Kronecker, or direct, product[5] of matrices A and B.

13.2.2   Relational and logical operators

As in Stata, the equality relational operators are a == b and a != b. They will work whether a and b are conformable or even of the same type: a could be a vector and b a matrix. They return 0 or 1. Unary not ! returns 1 if a scalar equals zero and 0 otherwise, and it can be applied in a vector or matrix context, returning a vector or matrix of 0 and 1 values.

The remaining relational comparison operators (>, >=, <, and <=) can be used only on objects that are conformable and of the same general type (numeric or string). They return 0 or 1. The logical operators and (&) and or (|) can be applied only to real scalars. If preceded by the colon operator (:), they can be applied to matrices.

13.2.3   Subscripts

Subscripts in Mata use square brackets and can appear on either the left or the right of an algebraic expression. There are two forms: list subscripts and range subscripts.

With *list subscripts*, you can reference a single element of an array, as in x[i, j]. The i or j references can also be vectors, such as x[i, jvec], where jvec = (4, 6, 8) references those three columns of x. Missingness (signaled by full stops [.]) references all rows or columns, so x[i, .] or x[i,] extracts row i, and x[., .] or x[,] references the whole matrix.

---

5. For more information, visit http://en.wikipedia.org/wiki/Kronecker_product.

You can also use the row-range and column-range operators described above to avoid listing each consecutive element: `x[(1..4), .]` and `x[(1::4), .]` both reference the first four rows of `x`. The double-dot range creates a row vector, while the double-colon range creates a column vector. Either can be used in a subscript expression. Ranges can also decrement, so `x[(3::1), .]` returns the first three rows of `x` in reverse order.

Range subscripts use the notation `[| |]`. They can reference single elements of matrices but are not useful for that. More useful is the ability to say `x[| i, j \m, n |]`, which creates a submatrix starting at `x[i, j]` and ending at `x[m, n]`. The arguments can be specified as missing, so `x[| 1, 2 \4, .|]` will specify the submatrix ending in the last column, and `x[| 4, 4 \., .|]` will discard the first three rows and columns of `x`. Range subscripts can also be used on the left-hand side of an expression or to extract a submatrix.

```
: v = invsym(xx)[| 2, 2 \ ., .|]
```

discards the first row and column of the inverse of `xx`.

You need not use range subscripts because even the specification of a submatrix can be handled with list subscripts and range operators, but they are more convenient for submatrix extraction (and faster in terms of execution time). Mata's interpretation of expressions will involve much less computation where range subscripts are used effectively. As Gould (2007b, 115) suggests, "Never code `x[5::1000]` or `x[5..1000]`. Code `x[|5 \1000|]`. ... Never code `X[5::100, 3..20]`. Code `X[|5, 3 \100, 20|]`." These admonitions arise because range subscripts allow Stata to identify a submatrix to be accessed, and the contents of that submatrix can be accessed more efficiently when it is known that they are contiguous.

13.2.4   Populating matrix elements

The Mata standard function `J()` provides the same capability as its Stata counterpart: `J(`*nrows*`, `*ncols*`, `*value*`)` creates a matrix of *nrows* × *ncols*, with each element set to *value*. However, in a February 2008 update, this function was enhanced. It can now generate a matrix with a pattern in its rows or columns.[6] For example,

---

6. See Cox (2008).

```
. version 10.1

. mata:
                                          ─────── mata (type end to exit) ───────
: m1 = J(7, 3, (0.25, 0.50, 0.75))

: m1
        1     2     3     4     5     6     7     8     9

  1   .25    .5   .75   .25    .5   .75   .25    .5   .75
  2   .25    .5   .75   .25    .5   .75   .25    .5   .75
  3   .25    .5   .75   .25    .5   .75   .25    .5   .75
  4   .25    .5   .75   .25    .5   .75   .25    .5   .75
  5   .25    .5   .75   .25    .5   .75   .25    .5   .75
  6   .25    .5   .75   .25    .5   .75   .25    .5   .75
  7   .25    .5   .75   .25    .5   .75   .25    .5   .75

: end
```

In this example, the second argument, 3, does not define the number of columns in the result matrix but the number of replicates or copies of the row vector to be produced.

Other Mata functions have similar functionality when a vector takes the place of a matrix. For instance, the `rnormal()` function takes four arguments: r, c, m, and s, where m and s can be scalars, vectors, or matrices. `rnormal(5, 5, 0, 1)` returns a 5×5 matrix of draws from the standard normal distribution, defined by $m = 0$ and $s = 1$. But those two arguments need not be scalars:

```
. mata:
                                          ─────── mata (type end to exit) ───────
: m2 = rnormal(3, 2, (1, 5, 10), 1)

: m2
             1              2              3              4              5

  1   1.127679588    6.993281523    9.201610313    1.44491133     6.018139269
  2  -.2659044268    4.640042947    9.496285076     .8234787212   9.089975386
  3   2.395392041    5.429020732   10.68722028     1.254825584    5.144066489

             6

  1   11.28016579
  2   11.25508169
  3   10.70177392

: end
```

In this example, the function call produces a 3×6 matrix of normal draws, with the mean vector of (1, 5, 10) repeated twice, with a scalar s of 1.

13.2.5 Mata loop commands

Several constructs support loops in Mata. As in any matrix language, explicit loops should not be used where matrix operations can be substituted. The most common loop construct resembles that of the C programming language:

```
: for ( starting'value; ending'value; incr) {
:    ...
: }
```

where the three elements define the starting value, an ending value or bound, and an increment or decrement. The increment (++) and decrement (--) operators[7] can be used to manage counters. These operators are available in two forms: *preincrement* and *postincrement*. The preincrement operator, ++i, performs the operation before the evaluation of the expression in which it appears, while the postincrement operator, i++, performs the operation after the evaluation of the expression in which it appears. The predecrement and postdecrement operators are defined similarly. Thus, given a vector x,

```
: i = 1
: x[ ++i ] = 42
: i = 1
: x[ i++ ] = 42
```

yield different results; the former places the constant, 42, in the second element of x, while the latter places the constant in the first element of x. After either command, the value of i is 2.

To apply the increment and decrement operators in the context of the for loop, type

```
: for (i = 1; i <= 10; i++) {
:          printf("%g squared is %g \n", i, i^2)
: }
```

We could also type

```
: for (i = 10; i >0 ; i = i - 2) {
:          printf("%g squared is %g \n", i, i^2)
: }
```

If only one statement will be executed by the for loop, it can be included on the same line as the for command without enclosing curly braces ({}).

As an alternative to the for statement, you can also use do to specify a loop. do follows the syntax

```
: do {
:    ...
: } while (exp)
```

which will execute the statements at least once. As another alternative, you can use while:

```
: while (exp) {
:    ...
: }
```

---

7. See [M-2] **op_increment**.

which could be used, for example, to loop until convergence with an *exp* of (sse > 1.0e-5).

13.2.6 Conditional statements

To execute certain Mata statements conditionally, use `if` and `else`:

```
: if (exp) statement
```

or

```
: if (exp) statement1
: else statement2
```

or

```
: if (exp1) {
:     statements1
: }
: else if (exp2) {
:     statements2
: }
: else {
:     statements3
: }
```

You can also use the conditional a ? b : c , where a is a real scalar. If a evaluates to true (nonzero), the result is set to b, otherwise c.[8,9] For instance,

```
: if (k == 0)  dof = n-1
: else         dof = n-k
```

can be written as

```
: dof = ( k==0 ? n-1 : n-k )
```

For compatibility with old-style Fortran, there is a `goto` statement that conditionally branches control to a statement elsewhere in the program:

```
:   label: statement
:     ...
:    if (exp) goto label
```

The `goto` statement is generally not necessary, because such a construct can be rewritten in terms of `do`:

```
: do {
: ...
: } while (exp)
```

---

8. This function plays the same role as Stata's `cond()` function; see section 3.3.2.

9. Although the conditional operator operates only on real scalars, Ben Jann's generalization of it, `mm_cond()`, can operate on matrices. It is available in his `moremata` package (`findit moremata`).

The `goto` statement is most useful when a program is being translated from Fortran code.

13.3   Function components

Mata code that is to be called from an ado-file must be defined as a *Mata function*, which is the equivalent of a Stata program. Unlike a Stata program, a Mata function has an explicit *return type* and set of *arguments*. A function can be of the return type `void` if it does not need a `return` statement. Otherwise, a function can be typed in terms of two characteristics: the nature of its elements, or the *element type*, and its *organization type*. For instance,

```
: real scalar calcsum(real vector x)
```

defines the `calcsum()` function as a `real scalar` function. Its element type is `real` and its organization is `scalar`, so the function will return a real number. There are six element types: `real`, `complex`, `numeric`, `string`, `pointer`, and `transmorphic`. `numeric` encompasses either `real` or `complex`. I discuss the `pointer` element type in section 13.8. `transmorphic` allows for any of the other types; that is, a `transmorphic` object can be filled with any of the other element types.

There are five organization types: `matrix`, `vector`, `rowvector`, `colvector`, and `scalar`. Strictly speaking, the latter four are special cases of the `matrix` type. In Stata's matrix language, all matrices have two subscripts, neither of which can be zero. In Mata, all but the `scalar` organization type can have zero rows, columns, or both (see [M-2] **void**).

If you do not declare a function's element type or organization type, its return type is, by default, the `transmorphic matrix`. That return type can be transmuted into any type of object. Although function definitions are not always necessary, it is good programming practice to make them explicit in case you mistakenly attempt to return a different kind of object than declared. Explicit declarations also serve to document the function and lead to more-efficient code. More importantly, they help to prevent errors, because they ensure that the returned object is of the proper return type.

13.3.1   Arguments

The Mata function declaration includes an *argument list*. In the example above, there is one argument: `real vector x`. To call this function, you must provide one (and only one) argument, and that argument must be typed as a `real vector`, or an error will occur. Inside the `calcsum()` function, the contents of the vector will be known as x regardless of their representation elsewhere in Mata. Items whose types are defined as arguments should not be included in declaration statements within the function.

The names of arguments are required and arguments are positional; that is, the order of arguments in the calling sequence must match that in the Mata function. It is good programming practice to specify the arguments' element type and organization explicitly, because then Mata will detect an error if the arguments' characteristics do not match. Explicit declarations also serve to document the function and lead to more-efficient code. The list of arguments can include a vertical bar (|), which indicates that all following arguments are optional. For example,

```
: function mynorm(real matrix A, | real scalar power)
```

can be called with one matrix or with a matrix followed by a scalar. The optional arguments will generally be given missing values in the function. It is possible for all the arguments of a function to be optional.

13.3.2 Variables

Within each Mata function, variables can be explicitly declared. Unlike some programming languages, Mata does not require you to declare variables before use unless `matastrict` is set as the default. However, it is good programming practice to explicitly declare variables, just as I suggested above that functions' arguments should be declared. Once variables are declared, they cannot be misused. For instance, variables declared as one of the numeric types cannot be used to store string contents, and vice versa.

Variables within a Mata function have *local scope*; that is, they cannot be accessed outside that function, so their names cannot conflict with other objects elsewhere in Mata.[10] As with functions, Mata variables can be declared with each of the six element types and five organization types. For example,

```
: real scalar sum
: complex matrix lambda
```

Several reserved words cannot be used as the names of Mata variables. They are listed in [M-2] **reswords**.

13.3.3 Saved results

Unlike a Stata ado-file program—which can return any number of scalars, macros, and matrices in the list of saved results—a Mata function can return only one object in its `return` statement. If the function is to return multiple objects, you should use Mata's `st_` functions to return scalars, macros, matrices, or Stata variables, as is demonstrated below.[11]

---

10. You can use the `external` declaration to indicate that a variable in a Mata function has *global scope* and is visible within any other Mata function. This is the same distinction that exists between local and global macros in Stata's do- and ado-files. One use of external variables in Mata is to allow a function to remember something from one call to the next (for instance, whether the function has been initialized). See [M-5] **findexternal( )** for more details.

11. Mata functions can also modify one or more of their arguments, but I do not discuss that method in this book.

13.4   Calling Mata functions

To illustrate now to use a Mata function within a Stata ado-file program, let's imagine that we did not have an easy way of computing the minimum and maximum of the elements of a Stata variable[12] and wanted to do so with Mata, returning the two scalar results to Stata. Let's define the Stata command `varextrema`:

```
. type varextrema.ado
*! varextrema v1.0.0  CFBaum 11aug2008
program varextrema, rclass
    version 10.1
    syntax varname(numeric) [if] [in]
    marksample touse
    mata: calcextrema("'varlist'", "'touse'")
    display as txt " min ('varlist') = " as res r(min)
    display as txt " max ('varlist') = " as res r(max)
    return scalar min = r(min)
    return scalar max = r(max)
end
```

The elements of this ado-file should be familiar to you with the exception of the `mata:` statement. This r-class program takes one numeric variable as an argument and allows the sample to be limited by `if` *exp* and `in` *range* qualifiers, as defined in section 11.6. The `marksample touse` statement, also described in section 11.6, creates a `tempvar` that marks the observations to be included in further computations.

We then invoke Mata. Instead of the simple command `mata`, which would place Mata into interactive mode, we tell Mata to execute a `calcextrema()` function with the syntax `mata:` *function*(). The `calcextrema()` function takes two arguments. From the code above, they are the name of the Stata variable passed to `varsum` and the name of the tempvar `'touse'`. Both arguments are of type string because we are passing only the names of Stata variables, not their values. The file `varextrema.mata` contains the Mata code. For presentation purposes, we have placed the Mata code in a separate file. In practice, we would include the lines below in `varextrema.ado` or place the line `include varextrema.mata` after the `end` statement in `varextrema.ado`.[13]

---

12. `summarize, meanonly` or `egen`'s `min()` and `max()` functions can be used for this purpose.
13. See [P] **include**.

```
. type calcextrema.mata
version 10.1
mata:
mata set matastrict on
// calcextrema 1.0.0  CFBaum 11aug2008
void calcextrema(string scalar varname, ///
                 string scalar touse)
{
  real colvector x, cmm
  st_view(x, ., varname, touse)
  cmm = colminmax(x)
  st_numscalar("r(min)", cmm[1])
  st_numscalar("r(max)", cmm[2])
}
end
```

We give the `version 10.1` statement to instruct Stata to invoke Mata under version control for version 10.1. After switching to Mata with `mata:`,[14] we use `mata set matastrict on` to invoke Mata's strict mode, which will catch various errors in our code because it requires that we explicitly declare all variables in the Mata function. If a variable's name is misspelled, it will be flagged in Mata's strict mode.

Because we will be returning two scalars to Stata, we must either return them in a vector (by using Mata's `return` statement) or use Stata's `st_` functions. We choose to do the latter and declare the function as return type `void`. Two arguments are defined: the `string scalar`s named `varname` and `touse`. These names are arbitrary because they are local to the function. They are defined as arguments and need not be defined again. We define two items: `real colvector x`, which will hold the view of the Stata variable,[15] and a vector, `cmm`. The `st_view()` function defines the column vector `x` as the contents of `varname`, with the optional fourth argument applying the `touse` indicator to reflect any `if` *exp* or `in` *range* qualifiers. Thus Mata's column vector `x` contains the subset of observations of `varname` for which `touse == 1`. The second argument of `st_view()`, set to missing (.) in this example, indicates which observations are to be retrieved. When set to missing, all observations are specified. We are using the `touse` argument to perform that selection.

The `st_numscalar()` function can be used to retrieve or set a Stata `scalar`. In its two-argument form, it sets the value of its first argument to the value of its second argument. Mata's `colminmax()` function computes the column minima and maxima of a vector or matrix; when applied to a vector, it returns a two-element vector containing the minimum and maximum. We create the Stata scalars named `r(min)` and `r(max)` by defining them as elements of the vector `cmm`.

---

14. Note the distinction between `mata` and `mata:` on Stata's command line. The former accepts Mata commands and stays in Mata mode regardless of errors. The latter accepts Mata commands but exits Mata at the first error (similar to the default behavior of a do-file). When programming Mata functions, you should always use `mata:` to ensure that an error will abort the ado-file.

15. I discuss view matrices in section 13.5.1.

We can now return to the last lines of `varextrema.ado`. Back in Stata, the scalars `r(min)` and `r(max)` are now defined and can be included in `display` statements. The `varextrema` command displays the variable name and the extrema of the selected elements, and returns those scalars to the user in the saved results:

```
. sysuse auto, clear
(1978 Automobile Data)

. varextrema price if foreign
 min ( price ) = 3748
 max ( price ) = 12990

. return list

scalars:
               r(max) =   12990
               r(min) =   3748
```

13.5   Mata's st_ interface functions

The example above made use of two of Mata's *st_ interface functions*: `st_view()` and `st_numscalar()`. Mata's `st_` interface functions are among the most important Mata components if you are using Mata in conjunction with ado-file programming. They provide full access to any object accessible in Stata and support a seamless interchange of computations between Stata and Mata. They are usefully summarized in [M-4] **stata**.

13.5.1   Data access

The first category of `st_` interface functions provides access to data. The `st_nobs()` and `st_nvar()` functions return the scalars corresponding to the number of observations and variables in the current Stata dataset, respectively; the same information is provided by `describe` as `r(N)` and `r(k)`. The `st_data()` and `st_view()` functions allow you to access any rectangular subset of Stata's numeric variables (including the entire dataset) within Mata. You can access string variables with `st_sdata()` and `st_sview()`.

Although both `st_data()` and `st_view()` provide access to Stata's numeric variables, they differ in important ways. The `st_data()` function copies the specified data into a Mata matrix, while `st_view()` creates a *view matrix*: a virtual construct that allows you to use and update the specified data. The view matrix does not involve a copy of the data and so has a trivial overhead in terms of memory requirements. In contrast, the Mata matrix created by `st_data()` can require from 1 to 8 times as much memory ([D] **memory**) as the underlying variables.[16] For a large dataset, that can make a tremendous difference if a sizable chunk of the dataset is selected. Views do involve a minor computational overhead, but it is likely to be offset by the memory savings. If you do not use views but use `st_data()` to access Stata variables within Mata, you can

---

16. This range exists because the underlying variables could all be stored as `double`s in Stata, requiring eight bytes per observation. All Mata numeric matrices require eight bytes per observation. On the other hand, the Stata data could all be `byte` variables, which would each require much more storage in Mata.

still return the contents of the Mata matrix to Stata by using the st_store() function, as described below.

A major advantage of views must be handled with care: altering the contents of a view matrix actually changes the values of the underlying Stata variables. This may be exactly what you want to do, or it may not.[17] As we will see, a useful technique to generate the contents of a set of Stata variables involves creating those variables (as missing) in Stata, then calling Mata, creating a view on those variables (call it matrix V), and defining the contents of that matrix:

    V[. , .] = *result of computation*

It is essential, in this instance, that you specify V as shown. If you merely place V on the left-hand side of the expression, it would create a new Mata matrix called V and break the link to the existing contents of the view matrix. For a thorough discussion of these issues, see Gould (2005).

The view matrix makes it possible to change the values of Stata variables within Mata but requires that you create those variables (as placeholders, at least) before invoking Mata. If you would like to generate results in Mata and return them to Stata but the number of result variables depends upon the computations, there is another way to approach the problem. Then you may want to use st_addvar() ([M-5] **st_addvar( )**) to create new variables in the Stata dataset from within Mata. The function call

    varnums = st_addvar(*type, name*)

allows you to specify the names of one or more new variables (*name* can be a quoted string or a vector of quoted string names), and for each new variable specify *type*, for instance, "float", "int", or "str5". The function returns the variable numbers in the row vector *varnums* and creates the new variables with missing contents. To fill them in, we could use st_store():

    newvarnum = st_addvar("int", "idnr")
    st_store(., newvarnum, *values*)

where *values* is the name of a column vector computed in Mata. If you are working with string data, you would use st_sstore() instead.

We can also combine the view matrix and st_addvar() techniques.

    st_view(z, ., st_addvar(("int", "float"), ("idnr", "bpmax")))
    z[., .] = ...

---

17. For instance, consider when you might use **preserve** and **restore** (see section 2.6) in ado-file programming to protect the current dataset's contents.

would create a view matrix, z, for two new variables, idnr and bpmax. We can then compute values for those variables within Mata and alter the contents of z, which will, in turn, fill in the new variables in Stata. For a thorough discussion of these techniques, see Gould (2006a).

There are situations where you may choose not to use views. For instance, if you are going to explicitly loop through observations in the dataset and every use you will make of the view matrix involves scalar calculations, then views provide no advantage. The variant _st_data(), which returns just one observation, can be used. You should also take care to not make copies of views in your Mata code, because they more than double the memory requirements of the underlying data.

Mata matrices must contain either all numeric or all string elements. Therefore, you cannot create a view that mixes these element types. If you loaded the auto dataset and created a view matrix of the entire dataset, it would have missing columns corresponding to string variables (here only the first column, make, is missing). The functions st_sview(), st_sdata(), and st_sstore() allow you to create view matrices of string variables, create regular matrices of string variables, or return the contents of a Mata string matrix to Stata, respectively. You must be careful to ensure that the rows of a numeric matrix or string matrix (created by st_sview() or st_sdata()) are properly aligned. In particular, you should use the same touse indicator to specify the observations that are to be transferred to Mata in each function call. An additional interface function is very useful in this regard: st_subview(). As [M-5] **st_subview( )** describes, this function can be used to ensure that missing values are handled properly across variables or within a panel context.[18]

13.5.2   Access to locals, globals, scalars, and matrices

The interface functions st_local(), st_global(), st_numscalar(), st_strscalar(), and st_matrix() allow you to access Stata's objects. In their single-argument form, they copy the object into Mata. For instance,

```
mylocal = st_local("macroname")
```

will return the contents of local macro macroname to the string scalar mylocal. Access to global macros, Stata's numeric scalars, string scalars, and matrices works in the same manner. You can also access Stata's characteristics (see section 9.7) and r(), e(), s(), and c() results. The functions st_rclear(), st_eclear(), and st_sclear() can be used to delete all entries in r(), e(), and s(). A useful summary of these interface routines is provided in [M-5] **st_global( )**.

In Stata, local macros and global macros are strings, even if they contain one number. Mata, however, distinguishes between numeric scalars and string scalars. If you define a macro in Stata,

---

18. There are also specialized Mata functions available for the handling of panel data; see [M-5] **panelsetup( )** and section 14.3.

```
scalar year 2008
global theanswer 42
```

and want to access those objects in Mata, you must use

```
theyear = st_numscalar("year")
theanswer = strtoreal(st_global("theanswer"))
```

Unless the global is cast to real with `strtoreal()`, it cannot be used in a computation within Mata.

These same interface functions can be used to set or replace the contents of Stata objects. In the two-argument form,

```
st_local("macroname", newvalue)
```

will create a local macro of that name in the scope of the calling program or replace the contents of an existing macro of that name with contents of *newvalue*. In the code example in section 13.4, we used `st_numscalar()` to create the Stata scalar `r(min)` and define its value.

13.5.3   Access to Stata variables' attributes

Additional interface functions provide the ability to check Stata variables' types; rename Stata variables; and obtain (or set) variables' formats, variable labels, or value labels. You can add (or drop) observations (or variables) to or from the Stata dataset from within Mata, and create temporary variables or filenames.

The interface functions `st_varindex()` and `st_varname()` allow you to map Stata variables' names to and from variable indices, respectively. Sometimes, you may need use variable indices in Mata. This is a concept generally alien to Stata users. A variable index is the number of the variable in the current dataset. We recognize that the order of variables in the dataset matters when using a hyphenated varlist (see section 3.2.1), but we do not usually consider that "`rep78` is the 4th variable in `auto.dta`". Unlike Stata matrices (see section 3.10), whose rows and columns have both numbers and names, Mata matrices' rows and columns are only numbered. Sometimes, it is useful to write Mata code by stepping through a number of variables by their indices, or positions within the Stata dataset.

13.6   Example: st_ interface function usage

Let's imagine that you would like to center several numeric variables on their means, creating a new set of transformed variables. Surprisingly, official Stata does not contain a command that performs this function. Ben Jann's `center` command (`findit center`) plays this role and works in version 7 or higher, because it does not use Mata. We can address this problem with an ado-file program:

```
. type centervars.ado
*! centervars 1.0.0  CFBaum 11aug2008
program centervars, rclass
    version 10.1
    syntax varlist(numeric) [if] [in], GENerate(string) [DOUBLE]
    marksample touse
    quietly count if `touse'
    if `r(N)' == 0  error 2000
    foreach v of local varlist {
        confirm new var `generate'`v'
    }
    foreach v of local varlist {
        qui generate `double' `generate'`v' = .
        local newvars "`newvars' `generate'`v'"
    }
    mata: centerv("`varlist'", "`newvars'", "`touse'")
end
```

The syntax of this wrapper program is simple. It is called with a numeric varlist, supports if *exp* and in *range* qualifiers, and has a mandatory generate(*string*) option specifying a stub for new variable names, along the same lines as tabulate rep78, gen(repair).

The program ensures that a positive number of observations is available after the marksample command. We must do some work with these new variables. First of all, we must ensure that all names created by the option are valid Stata new variable names that are not already in use. We do this with confirm new var ([P] **confirm**), which will return an error if the name is not valid or is already in use. In a second loop over the varlist, we construct each name, create the variable of the specified type with generate, and build up the local macro newvars as the list of the new variable names.[19] We then call the Mata function with three arguments: the list of variables to be centered, the list of result variables, and the touse indicator. The Mata function, stored in centerv.mata, reads as

```
. type centerv.mata
version 10.1
mata:
mata set matastrict on
// centerv 1.0.0  CFBaum 11aug2008
void centerv(string scalar varlist, ///
             string scalar newvarlist, ///
             string scalar touse)
{
  real matrix X, Z
  st_view(X=., ., tokens(varlist), touse)
  st_view(Z=., ., tokens(newvarlist), touse)
  Z[., .] = X:- mean(X)
}
end
```

---

19. We use two loops to ensure that no new variables are created if any of the new variable names are faulty.

We use the `st_view()` interface function to define matrix X as containing the specified variables and observations, while matrix Z refers to the corresponding `newvars`. Because the `varlist` and `newvarlist` arguments are lists of variable names, we must use the `tokens()` ([M-5] **tokens()**) function to split the lists into their components for use in `st_view()`.[20]

Recall that changes to Z will alter the underlying Stata variables' values. The Z statement above uses Mata's colon operator to subtract the row vector produced by `mean(X)` from each row of X.[21] This example of Mata's extended definition of conformability illustrates how the colon operator can be used to perform an operation that would be a bit tricky in terms of matrix algebra.

We invoke the `centervars` command, using the stubs `c_` and `cf_` to prefix the new variables created:

```
. sysuse auto, clear
(1978 Automobile Data)

. centervars price mpg, gen(c_)

. centervars price mpg if foreign, gen(cf_) double

. summarize c*
```

Variable	Obs	Mean	Std. Dev.	Min	Max
c_price	74	-.0000154	2949.496	-2874.257	9740.743
c_mpg	74	-4.03e-08	5.785503	-9.297297	19.7027
cf_price	22	1.65e-13	2621.915	-2636.682	6605.318
cf_mpg	22	-6.46e-16	6.611187	-10.77273	16.22727

13.7   Example: Matrix operations

In this section, we outline the construction of a Stata command that uses Mata to achieve a useful task with time-series data: constructing averages of p consecutive values of a variable as consecutive observations. For instance, you may want to combine quarterly national income data with the average inflation rate during the quarter, with inflation reported monthly. Likewise, you may want to convert monthly data into an annual format, quarterly data into an annual format, or business-daily data into weekly data.[22] Sometimes (for instance, for graphical or tabular presentation) we may want to retain the original (high-frequency) data and add the lower-frequency series to the dataset.

To address these tasks, we might consider using various `egen` functions. For instance, assuming that we have monthly data with variable `t` as the month counter,

---

20. Mata's `tokens()` function performs the same function as Stata's `tokenize` command, discussed in section 7.3.
21. Mata's `mean()` function generates column means from a matrix.
22. For the Stata date frequencies of monthly, quarterly, or half-yearly, this problem has already been solved by the author's `tscollap` routine (Baum 2000). However, that routine (like `collapse`) destroys the original higher-frequency data, requiring an additional `merge` step to emulate the Mata routine developed in this section.

```
generate qtr = qofd(dofm(t))
bysort qtr: egen avgx = mean(x)
```

would create the quarterly average of the monthly series x. However, that value will be repeated for each month of the quarter. We could easily remove all but one value per quarter with

```
replace avgx = . if mod(t, 3) != 0
```

but that would intersperse the lower-frequency data values with missing values. Such a listing would be awkward for tabular or graphical presentation.

As a more straightforward solution, we design a Stata command, avgper, that takes one variable and optional if *exp* or in *range* conditions, along with a mandatory per() option, the number of periods to be averaged into a lower-frequency series. We could handle multiple variables or alternative transformations (for example, sums over the periods rather than averages) with an expanded version of this routine.

The Stata ado-file defines the program and then validates the per() argument. We require that the number of high-frequency observations is a multiple the number specified in per().

```
. type avgper.ado
capture program drop avgper
*! avgper 1.0.0  CFBaum 08oct2007
program avgper, rclass
        version 10.1
        syntax varlist(max=1 numeric) [in], per(integer)
        marksample touse
        quietly count if 'touse'
        if 'r(N)' == 0 {
                error 2000
        }
* validate per versus selected sample
        if 'per' <= 0 | 'per' >= 'r(N)' {
                display as error "per must be > 0 and < N of observations."
                error 198
        }
        if mod('r(N)','per' != 0) {
                display as error "N of observations must be a multiple of per."
                error 198
        }
* validate the new varname
        local newvar = "'varlist'A'per'"
        quietly generate 'newvar' = .
* pass the varname and newvarname to Mata
        mata: avgper("'varlist'", "'newvar'",  'per', "'touse'")
end
```

We attempt to create a new variable named *vname*A*n*, where *vname* is the specified variable and *n* is the value of per(). If that variable name is already in use, the routine exits with error. The variable is created with missing values because it is only a placeholder. With successful validation, we pass the arguments to the Mata avgper() function.

After these checks have been passed, we turn to the heart of the computation. The solution to the problem is a type of reshape. The N-element column vector into which *vname* has been copied can be reshaped into a matrix, v3, with q rows and per columns. $q = N\ /$ per is the number of averaged observations that will result from the computation. If we postmultiply the transpose of matrix v3 by a per-element column vector of ones, ι, we would compute the sum over the per values for each new observation. The average would be $1/$per times that vector. Thus we define the column vector's elements as divisor $= 1/$per. The resulting column vector, v3, is our averaged series of length q.

To illustrate, let x be the N elements of the Stata variable. Each per consecutive element becomes a column of the reshaped matrix:

$$
\begin{pmatrix}
x_1 \\
x_2 \\
\vdots \\
x_{\text{per}} \\
x_{\text{per}+1} \\
x_{\text{per}+2} \\
\vdots \\
x_{2\text{per}} \\
x_{2\text{per}+1} \\
x_{2\text{per}+2} \\
\vdots \\
\vdots \\
x_N
\end{pmatrix}
\implies
\begin{pmatrix}
x_{1,1} & x_{1,2} & \cdots & x_{1,q} \\
x_{2,1} & x_{2,2} & \cdots & x_{2,q} \\
\vdots & & \ddots & \vdots \\
x_{\text{per},1} & x_{\text{per},2} & \cdots & x_{\text{per},q}
\end{pmatrix}
$$

We then transpose the reshaped matrix and postmultiply by a per-element column vector to construct the per-period average:

$$
\begin{pmatrix}
x_{1,1} & x_{2,1} & \cdots & x_{\text{per},1} \\
x_{1,2} & x_{2,2} & \cdots & x_{\text{per},2} \\
\vdots & & \ddots & \vdots \\
x_{1,q} & x_{2,q} & \cdots & x_{\text{per},q}
\end{pmatrix}
\begin{pmatrix}
\frac{1}{\text{per}} \\
\frac{1}{\text{per}} \\
\vdots \\
\frac{1}{\text{per}}
\end{pmatrix}
=
\begin{pmatrix}
x_1^* \\
x_2^* \\
\vdots \\
x_q^*
\end{pmatrix}
$$

The column vector x^*, labeled as v3 in the Mata function, contains the averages of each per elements of the original Stata variable. The Mata code to achieve this task is succinct:

```
. type avgper.mata
version 10.1
mata:
void avgper(string scalar vname,
    string scalar newvname,
    real scalar per,
        string scalar touse)
{
        real matrix v1, v2, v3
        st_view(v1=., ., vname, touse)
        st_view(v2=., ., newvname)
        v3 = colshape(v1', per) * J(per, 1, 1/per)
        v2[ (1::rows(v3)), ] = v3
}
end
```

We use view matrices to access the contents of **vname** (the existing variable name specified in the Stata **avgper** command) and to access **newvname**, that is, our newly-created
'**newvar**' in the Stata code, in Mata. The **colshape()** function creates a matrix that is
$q \times$**per**, where q is the number of low-frequency observations to be created. Postmulti-
plying that matrix by a **per**-element column vector of 1/**per** produces the desired result
of a q-element column vector. That result, **v3** in Mata, is then written to the first q
rows of view matrix **v2**, which corresponds to the Stata variable '**newvar**'.

To test the Stata **avgper** command, we use the **urates** dataset, which is used in the
Time-Series Reference Manual. The dataset contains monthly unemployment rates for
several U.S. states. We apply **avgper** to one variable, **tenn** (the Tennessee unemployment rate), using both **per(3)** and **per(12)** to calculate quarterly and annual averages,
respectively:

```
. use urates, clear
. tsset
        time variable:  t, 1978m1 to 2003m12
                delta:  1 month
. avgper tenn, per(3)   // calculate quarterly averages
. avgper tenn, per(12) // calculate annual averages
. summarize tenn*
```

Variable	Obs	Mean	Std. Dev.	Min	Max
tenn	312	6.339744	2.075308	3.7	12.8
tennA3	104	6.339744	2.078555	3.766667	12.56667
tennA12	26	6.339744	2.078075	3.908333	11.83333

The **summarize** command shows that the original series and two new series have identical
means, which they must. To display how the new variables appear in the Stata data
matrix, we construct two date variables with the **tsmktim** command (Baum and Wiggins
2000) and list the first 12 observations of Tennessee's data. As you can verify, the routine
is computing the correct quarterly and annual averages.

```
. tsmktim quarter, start(1978q1) // create quarterly calendar var
        time variable:  quarter, 1978q1 to 2055q4
                delta:  1 quarter
. tsmktim year, start(1978)        // create annual calendar var
        time variable:  year, 1978 to 2289
                delta:  1 year
. list t tenn quarter tennA3 year tennA12 in 1/12, sep(3)
```

	t	tenn	quarter	tennA3	year	tennA12
1.	1978m1	5.9	1978q1	5.966667	1978	5.8
2.	1978m2	5.9	1978q2	5.766667	1979	5.791667
3.	1978m3	6.1	1978q3	5.733333	1980	7.3
4.	1978m4	5.9	1978q4	5.733333	1981	9.083333
5.	1978m5	5.8	1979q1	5.733333	1982	11.83333
6.	1978m6	5.6	1979q2	5.7	1983	11.45833
7.	1978m7	5.7	1979q3	5.733333	1984	8.55
8.	1978m8	5.7	1979q4	6	1985	7.983334
9.	1978m9	5.8	1980q1	6.166667	1986	8.041667
10.	1978m10	5.9	1980q2	7.066667	1987	6.591667
11.	1978m11	5.7	1980q3	8	1988	5.775
12.	1978m12	5.6	1980q4	7.966667	1989	5.108333

How might we perform this transformation for a whole set of variables? Rather than generalizing avgper to handle multiple variables, we just use a foreach loop over the variables:

```
. use urates, clear
. foreach v of varlist tenn-arkansas {
  2.          avgper `v', per(3)
  3. }
. summarize tenn-arkansasA3, sep(6)
```

Variable	Obs	Mean	Std. Dev.	Min	Max
tenn	312	6.339744	2.075308	3.7	12.8
missouri	312	5.78109	1.591313	2.9	10.6
kentucky	312	6.867949	2.029192	3.8	12.6
indiana	312	6.106731	2.413917	2.8	12.7
illinois	312	6.865064	1.965563	4.1	12.9
arkansas	312	6.833974	1.676967	4.2	10.5
tennA3	104	6.339744	2.078555	3.766667	12.56667
missouriA3	104	5.78109	1.590077	3	10.46667
kentuckyA3	104	6.867949	2.029395	3.9	12.36667
indianaA3	104	6.106731	2.416999	2.9	12.4
illinoisA3	104	6.865064	1.964652	4.2	12.76667
arkansasA3	104	6.833974	1.679908	4.266667	10.46667

By using Mata and a simple matrix expression, we have considerably simplified the computation of the lower-frequency series, and we can apply the routine to any combination of data frequencies (for example, business-daily data into weekly data) without concern for Stata's support of a particular time-series frequency.

13.7.1   Extending the command

It is straightforward to generalize the `avgper` command to handle several algebraic operations. Rather than averaging the series over `per` observations, we might want to sum the series over those observations. Alternatively, we might want to construct a new low-frequency series composed of the first (or last) observations within each interval. This will require an option in the ado-file, which we now rename `aggreg.ado`, to permit the user to specify the desired operation. The Stata ado-file code is now

```
. type aggreg.ado
*! aggreg 1.0.0  CFBaum 11aug2008
program aggreg, rclass
        version 10.1
        syntax varlist(max=1 numeric) [if] [in], per(integer) [func(string)]
        marksample touse
        quietly count if 'touse'
        if 'r(N)' == 0 {
                error 2000
        }
* validate per versus selected sample
        if 'per' <= 0 | 'per' >= 'r(N)' {
                display as error "per must be > 0 and < N of observations."
                error 198
        }
        if mod('r(N)','per') != 0 {
                display as error "N of observations must be a multiple of per."
                error 198
        }
* validate func option; default is average (code A)
    local ops A S F L
    local opnames average sum first last
    if "'func'" == "" {
        local op "A"
    }
    else {
        local nop : list posof "'func'" in opnames
        if !'nop' {
                display as err "Error: func must be chosen from 'opnames'"
                error 198
        }
        local op : word 'nop' of 'ops'
    }
* validate the new varname
        local newvar = "'varlist''op''per'"
    quietly generate 'newvar' = .
* pass the varname and newvarname to Mata
        mata: aggreg("'varlist'", "'newvar'",  'per', "'op'", "'touse'")
        end
```

We add the `func()` option, which takes the default of `average`. We add code to validate the other choices: `sum`, `first`, and `last`. We use the macro list function `list posof` to determine whether the value of `func()` is in the list of supported operations; if not, an error is generated. The chosen function, `op`, is designated by a letter from local `ops` and passed to the Mata routine.

```
. type aggreg.mata
version 10.1
mata: mata set matastrict on
mata:
// aggreg 1.0.0  CFBaum 11aug2008
void aggreg(string scalar vname,
            string scalar newvname,
              real scalar per,
              string scalar op,
              string scalar touse)
    {
    real colvector mult, v1, v2
    real matrix v3
    if (op=="A") {
        mult = J(per, 1, 1/per)
    }
    else if (op=="S") {
        mult = J(per, 1, 1)
    }
    else if (op=="F") {
            mult = J(per, 1, 0)
            mult[1] = 1
    }
    else if (op=="L") {
        mult = J(per, 1, 0)
        mult[per] = 1
    }
    st_view(v1=., ., vname, touse)
    st_view(v2=., ., newvname)
    v3 = colshape(v1', per) * mult
    v2[(1::rows(v3)), ] = v3
    }
end
```

In the Mata routine, we need only modify the Mata `avgper()` code by altering the per-element column vector that multiplies the reshaped data. As shown in section 13.7, the average is computed by setting elements of that vector to 1/per. We create a `colvector` named `mult`. For the average operation, we define it as before. For the sum, it is merely defined as ι, a vector of ones. For the first (last) operation, it is defined as a null vector with 1 in the first (last) position. With that change, the Mata `aggreg()` function can perform any of these four functions on the time series.

We can readily verify that this extended program generates the proper series for each of its four operations:

```
. use urates, clear
. aggreg tenn, per(3)  // calculate quarterly averages
. aggreg tenn, per(3) func(sum) // calculate quarterly sum
```

```
. aggreg tenn, per(3) func(first) // extract first month's value

. aggreg tenn, per(3) func(last) // extract first month's value

. tsmktim quarter, start(1978q1) // create quarterly calendar var
        time variable:  quarter, 1978q1 to 2055q4
            delta:  1 quarter

. list t tenn quarter tennA3 tennS3 tennF3 tennL3 in 1/12, sep(3)
```

	t	tenn	quarter	tennA3	tennS3	tennF3	tennL3
1.	1978m1	5.9	1978q1	5.966667	17.9	5.9	6.1
2.	1978m2	5.9	1978q2	5.766667	17.3	5.9	5.6
3.	1978m3	6.1	1978q3	5.733333	17.2	5.7	5.8
4.	1978m4	5.9	1978q4	5.733333	17.2	5.9	5.6
5.	1978m5	5.8	1979q1	5.733333	17.2	5.8	5.7
6.	1978m6	5.6	1979q2	5.7	17.1	5.7	5.6
7.	1978m7	5.7	1979q3	5.733333	17.2	5.7	5.8
8.	1978m8	5.7	1979q4	6	18	5.9	6.1
9.	1978m9	5.8	1980q1	6.166667	18.5	6	6.3
10.	1978m10	5.9	1980q2	7.066667	21.2	6.7	7.5
11.	1978m11	5.7	1980q3	8	24	8	8
12.	1978m12	5.6	1980q4	7.966667	23.9	8	8

13.8   Creating arrays of temporary objects with pointers

In Stata's ado-file language, it is common to create a set of temporary objects: variables, local macros, or matrices. Sometimes it is necessary to create a different number of temporary items each time we run the program. That is a simple task in Stata's ado-file language, as we have seen in several examples. In Mata, however, the solution is not immediately clear: we need to declare variables, but the number of variables is not fixed. In this section, we consider that problem, using an example from the econometric literature.

Why might we need such a facility? Consider a Stata program that accepts a positive integer argument of lags(), which could take on any number of values, and then computes a matrix from several Stata variables for each lag. In some applications, such as the computation of a heteroskedasticity- and autocorrelation-consistent covariance matrix,[23] two temporary objects will suffice, because the logic of that computation accumulates the matrices computed for each lag. We create one matrix in which to compute each term in the series and a second to hold the accumulation. In other contexts, however, we might need to use the values contained in each of those matrices in some formula. In ado-file code, we might write

---

23. See [TS] **newey**.

```
forvalues i = 1/'lags' {
      tempname phi'i'
      matrix 'phi'i'' = exp
}
```

which would define and compute the contents of a sequence of temporary matrices `phi1`, `phi2`, ..., `phi'lags'`. What would we need to do to perform this same task in a Mata context?

The solution involves the use of the sixth element type in Mata: the *pointer*. A pointer, as defined in [M-2] **pointers**, is an object that contains the *address* of another object. Pointers are commonly used to put a collection of items under one name and to pass functions to functions. A pointer to matrix X contains the address, or location in the computer's memory, of X. We do not really care to know what that number might be. It suffices to know that if we dereference the pointer, we are referring to the contents of the underlying object to which it points.[24] The dereferencing operator in Mata is *, so if p is a pointer variable, *p refers to the underlying object.

How do we create a pointer variable? With the operator &, which instructs Mata to store the address (rather than the contents) of its argument in a pointer variable.

```
: X = (1, 2 \ 3, 4)
: p = &X
```

creates a pointer variable, p, that contains the address of matrix X. If we then refer to the variable,

```
: Z = *p
```

matrix Z will now be set equal to matrix X. Likewise,

```
: r = (*p)[ 2, .]
```

will cause r to be created as a row vector equal to the second row of X. We use parentheses around *p to clarify that we refer to the second row of the object defined by dereferencing the pointer. Also the contents of p itself are not very useful:[25]

```
: p
0x2823b10c
```

Why then are pointers an important programming tool? If pointer variables could only be real scalars, they would not be useful to most Mata programmers. But like other Mata objects, pointer variables can be arrayed in vectors or matrices. That now gives us the tools necessary to deal with the problem laid out above.

We would like to set up a set of matrices, `phi1`, ..., `phi'lags'`, within Mata, where the Mata function will be called with the argument of `lags`. Within Mata, we now

---

24. Note the similarity to Stata's macros: regardless of the name of the macro, when we dereference it, we access its contents.
25. You are unlikely to get this same value if you use the same sequence of commands, because it depends on the state of Stata's memory at a particular time.

know how many matrices are to be created, and for ease of computation, we want them
to be named sequentially as `phi1`, `phi2`, ... We need to set up an array of pointers
that will contain the addresses of each of these matrices. As with other Mata objects,
we can fully declare the type of object we create, and it is recommended that you do
so:[26]

```
: pointer(real matrix) rowvector ap
: ap = J(1, nlag, NULL)
```

We declare a vector of pointers, named `ap`, to real matrices. We then declare the vector
as having `nlag` elements, each `NULL`. The keyword `NULL` is the equivalent of a missing
value for a pointer variable; it indicates that the pointer variable currently points to
nothing. Having set up this array, we can now fill in each of its elements:

```
for(i = 1; i <= nlag; i++) {
        ap[i] = &(matrix expression)
}
```

Within the loop, we compute some matrix expression, enclose it in parentheses, and
store its address in the *i*th element of the `ap` array. This matrix is anonymous in that
it is not given a name; we merely record its address in the appropriate pointer variable.
When we want to use that matrix, we refer to it as `*ap[i]`.

   With the same technique, we could define a two-dimensional array of pointers, such
as

```
pointer(real vector) matrix p2
```

which might be useful for associating a vector of coordinates with each point on a two-
dimensional grid. Any other element type can be specified as the target of pointers,
including pointers themselves:

```
: mata:
: void pointer2(real scalar bar1, real scalar bar2)
: {
: pointer(pointer(real matrix) rowvector) rowvector pp
: pp = J(1, bar2, NULL)
```

This function declares an array of pointers, `pp`, of length `bar2`. Each element of `pp`
points to a row vector of pointers to real matrices. We could create those real matrices
as anonymous, `bar1` in number for each element of `pp`, with the code:

```
: for(j = 1; j <= bar2; j++) {
:         pp[j] = &(J(1, bar1, NULL))
:         for(i = 1; i <= bar1; i++) {
:                 (*pp[j])[i] = &(J(i , j, i*j + j^2) )
:         }
: }
```

Within the loop over `j`, we define the *j*th element of the `pp` pointer array as the address
of an anonymous row vector of length `bar1`. Within the loop over `i`, we define the

---

26. See [M-2] **pointers**.

contents of each element of that anonymous row vector as the address of a matrix. The
matrix takes on different dimensions based on the values of i and j, illustrating that an
array of pointers to objects need not refer to objects of the same dimension (only the
same type, here given our declaration). At this point, we have filled in the lowest-level
objects, the real matrices themselves, and have established a hierarchy of pointers that
address each of those objects.

We can now examine the contents of these arrays with the statements

```
: for(k = 1; k <= bar2; k++) {
:         for(l = 1; l <= bar1; l++) {
:                 printf("j=%5.2f, i=%5.2f\n", k, l)
:                 *(*pp[k])[l]
:         }
:     }
: }
: end
```

Because there are two levels of pointers in the data structure, we must dereference twice
with the $*$ operator: once to recover the value held in **pp[k]**, giving us the kth vector of
anonymous objects, and a second time (working outward) to retrieve the ℓth element of
that vector, which is the matrix stored at that location. If we execute the function with
bar1 = 2 and **bar2 = 3**, we are specifying that there should be three elements in the
pp array, each of which points to a two-element vector. Each of those elements points
to a real matrix:

```
. mata: pointer2(2, 3)
  j=  1, i=  1
   2
  j=  1, i=  2
          1

     1 │   3
     2 │   3

  j=  2, i=  1
          1    2

     1 │   6    6

  j=  2, i=  2
[symmetric]
          1    2

     1 │   8
     2 │   8    8

  j=  3, i=  1
          1    2    3

     1 │  12   12   12

  j=  3, i=  2
          1    2    3

     1 │  15   15   15
     2 │  15   15   15
```

Pointer variables also can be used to refer to functions, permitting a routine to specify that alternative functions can be invoked depending on user input. Additional examples of the usefulness of pointer variables are in the following cookbook chapter.

13.9 Structures

An additional useful tool available to the Mata programmer is the *structure*. For an extended discussion of Mata structures, see Gould (2007a). Structures cannot be used interactively. Structures allow you to organize several scalars, vectors, and matrices (potentially of different types) and pass them to other routines as a single structure. For instance, we might consider a linear regression routine that would return a structure consisting of the coefficient vector e(b), the covariance matrix e(V), and several scalars, such as R-squared, N, k, and Root MSE. Rather than referring to each of those quantities individually, we could group them into a structure and pass the structure. This would make it easy to write a number of computational routines—each of which would produce a structure—and one output routine that would display the contents of that structure, with minimal overhead.

As an example, consider a routine that generates graphical elements in the form of vectors in 2-space. Each vector has an origin at a (x coordinate, y coordinate) pair, a length, and an angle, where 0 and 360 refer to east, 90 to north, and so on. Let's also imagine that each vector has a color associated with it when it is to be graphed. Mata structures are defined outside of functions and must be defined before any reference to them appears:

```
. type myvecstr.mata
mata: mata clear
mata: mata set matastrict on

version 10.1
mata:
struct mypoint {
        real vector coords
}
struct myvecstr {
        struct mypoint scalar pt
        real scalar length, angle
        string scalar color
}
end
```

The `myvecstr` structure contains an instance of the `mypoint` structure, so we must define the latter first. The `mypoint` structure merely contains a real vector of undefined length. The `myvecstr` structure includes an instance of `mypoint` named `pt`, as well as two real scalars (`length` and `angle`) and a string scalar, `color`.

We can now define a function that uses the structure. It will take five arguments and reference an instance of the `myvecstr` structure named `v`. Because the `xcoord` and `ycoord` arguments are to be stored as a vector in `mypoint`, we place them in `v.pt.coords`. The `v` and `pt` references refer to instances of the structure, while `coords` refers to the definition of `mypoint`'s contents. We also store the function's `len`, `ang`, and `color` arguments in the `v` structure. We can then pass the entire `v` structure to a `myvecsub()` subroutine function.

```
. type makevec.mata
version 10.1
mata:
function makevec(real scalar xcoord,
                 real scalar ycoord,
                 real scalar len,
                 real scalar ang,
                 string scalar color)
{
    struct myvecstr scalar v
    v.pt.coords = (xcoord, ycoord)
    v.length = len
    v.angle = ang
    v.color = color
    myvecsub(v)
}
end
```

The `myvecsub()` function takes the structure as an argument named e, because it can access the elements of the structure. Likewise, a function defined as type `struct` can return a structure to the calling function. `myvecsub()` extracts the x and y coordinates from the elements of the `e.pt.coords` vector and uses them to compute the `dist_from_origin` scalar.

```
. type myvecsub.mata
version 10.1
mata:
function myvecsub(struct myvecstr scalar e)
{
        real scalar dist_from_origin, xx, yy, ll, ang
        string scalar clr
        xx = e.pt.coords[1]
        yy = e.pt.coords[2]
        dist_from_origin = sqrt(xx^2 + yy^2)
        printf("\n The %s vector begins %7.2f units from the ///
    origin at (%f, %f)", e.color, dist_from_origin, xx, yy)
        printf("\n It is %7.2f units long, at an angle of %5.2f
    degrees\n", e.length, e.angle)
}
end
```

We can now execute the function:

```
. mata:
─────────────────────────────── mata (type end to exit) ───────
: makevec(42, 30, 12, 45, "red")
 The red vector begins   51.61 units from the origin at (42, 30)
 It is   12.00 units long, at an angle of 45.00 degrees
: end
─────────────────────────────────────────────────────────────────
```

There are some additional things to note about structures. Structures are compiled into functions of the same name, so the examples above have created `mypoint()` and `myvecstr()` functions. If you are saving your Mata functions in compiled form, these functions should be saved as well. Also, although an instance of a structure is named (as is v in `makevec()` above), you cannot examine the structure's contents by using its name (as you can with Mata scalars, vectors, and matrices). A reference to the structure's name will print only its address in memory. If you want to examine the contents of a structure, use the `liststruct()` function.

Structures have many uses in Mata. You can program without them, but they will often make it much easier to write complicated code. For instance, you can test two structures for equality, which will be satisfied if and only if their members are equal. You can create vectors or matrices of structures, and use pointers to structures. For full details of the capabilities of Mata structures, see [M-2] **struct** and the excellent discussion in Gould (2007a).

13.10   Additional Mata features

13.10.1   Macros in Mata functions

Stata's local macros (see section 3.6) can be used to advantage in Mata functions but with a important caveat. Macro evaluation takes place in Mata as it does in Stata commands with one significant difference. When a Mata function is first defined, it is compiled into bytecode. In that compilation process, the values of any local (or global) macros are substituted for their names, that is, the macros are evaluated. However, once the function is compiled, those values are hardcoded into the Mata function, just as any other definition would be. For example,

```
. type gbpval.mata
version 10.1
mata:
real scalar function gbpval(real scalar dollar)
{
    real scalar usdpergbp
    usdpergbp = 2.08
    return(dollar / usdpergbp)
}
end
```

would define a function that converts U.S. dollars to British pounds sterling at a fixed exchange rate of U.S. $2.08 per pound sterling. Rather than placing that constant in the Mata function, we could type

```
. type gbpval2.mata
local usdpergbp 2.08
mata:
real scalar function gbpval2(real scalar dollar)
{
        return(dollar / 'usdpergbp')
}
end
```

We can invoke the function with

```
. mata: gbpval2(100)
  48.07692308
```

But what happens if we change the local macro?

```
. local usdpergbp 2.06
. mata: gbpval2(100)
  48.07692308
```

Why does Mata ignore the new value of the macro? Unlike Stata commands, which would use the current value of the local macro, the macro's value has been compiled into the `gbpval2` function and can only be changed by recompiling that function. This is an important distinction and explains why you cannot use local macros as counters within Mata functions as you can in ado-files or do-files.

Local macros in Mata functions serve to define objects, either numeric values or string values, that should be constants within the function. One useful task that they can serve is the replacement of a lengthy string—such as the URL of a filename on the Internet—with a symbol. The value of that symbol is the URL. If the filename changes, you need only update the local macro's definition rather than change it in several places in the Mata function. Local macros can also be used to good advantage to define abbreviations for commonly used words or phrases. For instance, we can use the abbreviation RS to avoid repeatedly typing `real scalar`:

```
. type gbpval3.mata
version 10.1
local RS  real scalar
mata:
'RS' function gbpval3('RS' dollar)
{
        'RS' usdpergbp
        usdpergbp = 2.08
        return(dollar / usdpergbp)
}
end
```

In summary, local macros can be used in Mata functions, but you must remember that they are expanded only when the function is first compiled by Mata.

13.10.2   Compiling Mata functions

We have illustrated Mata functions in conjunction with ado-files by placing the Mata code in-line, that is, within the ado-file. If placed in-line, the Mata code will be compiled "just in time" the first time that the ado-file is called in your Stata session. Subsequent calls to the ado-file will not require compilation of the Mata function (or functions). An exchange on Statalist[27] suggests that if the Mata functions amount to fewer than 2,000 lines of code, incorporating them in-line will be acceptable in terms of performance.

There is one disadvantage of the in-line strategy, though. What if you have multiple ado-files that all reference the same Mata function but can be called independently? Then you should consider saving a compiled version of the function that can then be accessed by all ado-files. Copying the Mata code into each ado-file works as a short-term solution but will create persistent problems if you update the Mata functions and fail to remember to update each ado-file. As a better solution, you should place the Mata functions into a separate `.mata` file and create a `.mo` object file. Say that you write the `crunch` Mata function and store the code defining that function in `crunch.mata`. You can then issue the Mata command with the following syntax:

---

27. See http://stata.com/statalist/archive/2005-08/msg00358.html.

```
mata mosave crunch() [ , dir(path) replace]
```

This will create the object file `crunch.mo`, by default in the current working directory. That has its disadvantages (as described in section 2.2.1). You may want to specify the option `dir(PERSONAL)` or `dir(PLUS)` on the `mata mosave` command, which will write the `.mo` file to those directories, respectively. The routine will then be found by a calling ado-file regardless of your current working directory. To implement creation of the object file for the function, you could include the `mata mosave` line in the `.mata` file. For example, `crunch.mata` might contain

```
version 10.1
mata:
real scalar crunch(...)
{
...
}
end
mata: mata mosave crunch(), dir(PERSONAL) replace
```

You could then define the function and create the object file with the Stata command

```
do crunch.mata
```

which would create `crunch.mo` in your `PERSONAL` directory.

13.10.3   Building and maintaining an object library

Creating a separate `.mo` file for each Mata function in your `PERSONAL` directory will make those functions available to any ado-file. But if you have developed many Mata functions or want to provide those functions to other users in a convenient way, you should know about Mata function libraries. A function library, or `.mlib`, can hold up to 500 compiled functions. You create a Mata function library with the following command syntax:

```
mata mlib create libname [ , dir(path) replace]
```

By convention, function libraries' names should start with the letter `l`. As with `mata mosave`, it is important to locate the library on the ado-path, which you can do with the `dir()` option.

To include a Mata function in the library, you must first compile it by loading it into Mata's memory. Following the strategy above, you could rewrite the last line of `crunch.mata` to be

```
mata: mata mlib add libname fcnname()
```

where *libname* refers to your existing library and *fcnname* is the Mata function name or names. Using wildcards, you can add a number of functions with one command. For example, by referring to `cr*()`, you specify all functions with names starting with `cr`.

The trailing parentheses are required. Then the Stata command `do crunch.mata` will compile the function and store it in the library.

One limitation exists, though. With `mata mlib add`, you can add new functions to a library. However, you cannot update functions stored in the library. If you must remove a function or replace it with a newer version, you must delete and re-create the library and load all its previous contents. Accordingly, it is a good idea to avoid altering a library with interactive commands but instead write do-files that create and load all functions in the library, as suggested in the last section of [M-3] **mata mlib**.

When you start Stata and first enter the Mata environment, Mata searches the standard ado-path for all available Mata libraries. If you make changes to a library during a Stata session or download a package that contains a Mata library, those libraries' contents will not be immediately available. You must first refresh Mata's list of available libraries with the command

    : mata mlib index

which will trigger the automatic search for Mata libraries on your machine.[28]

Mata compiled functions (`.mo` files) and object libraries (`.mlib` files) are not backward compatible. A function compiled in Stata 10 or a library constructed under Stata 10 cannot be used by Stata 9.x, even though Mata was available in that version of Stata. To generate an object library usable in both Stata 9 and Stata 10 (and later versions), you must compile the functions and produce the library in Stata 9. Because several functions and capabilities were added to Mata in Stata 10, you may not be able to execute the same Mata code in Stata 9.x as in Stata 10.

13.10.4   A useful collection of Mata routines

Presently, the largest and most useful collection of user-written Mata routines is Ben Jann's `moremata` package, available from the Statistical Software Components archive. The package contains a function library, `lmoremata`, as well as full documentation of all included routines (in the same style as Mata's online function descriptions). Very importantly, the package also contains the full source code for each Mata routine, accessible with `viewsource` (Mata authors need not distribute the Mata code for their routines). Unlike ado-file code, which is accessible by its nature, Mata functions can be distributed in object-code (or function-library) form.

Routines in `moremata` currently include kernel functions; statistical functions for quantiles, ranks, frequencies, means, variances, and correlations; functions for sampling; density and distribution functions; root finders; matrix utility and manipulation functions; string functions; and input–output functions. Many of these functions provide functionality that is currently missing from official Mata and ease the task of various programming chores. Once you have downloaded the package from the Statistical Soft-

---

28. Quitting and restarting Stata will have the same effect but is not necessary.

ware Components archive, be sure to issue the `mata mlib index` command to ensure that the package's components can immediately be located by Mata.

14 Cookbook: Mata function programming

This cookbook chapter presents for Stata do-file programmers several recipes using the programming features described in the previous chapter. Each recipe poses a problem and a worked solution. Although you may not encounter this precise problem, you should be able to recognize its similarities to a task that you would like to automate in an ado-file and Mata function.

14.1 Reversing the rows or columns of a Stata matrix[1]

The problem. Mata's built-in function list contains many useful matrix operations, but I recently came upon one I needed: the ability to flip a matrix along its row or column dimensions. Either of these operations can readily be done as a Mata statement, but I would rather not remember the syntax—nor remember what it is meant to do when I reread the code.

The solution. I wrote a simple Mata function to perform this task and wrapped it in an ado-file.[2] To define the Stata command, the `flipmat` program takes the name of an existing Stata matrix. By default, the command reverses the order of the matrix's rows. Optionally, you can specify `horiz` to reverse the order of the matrix's columns. We use the `syntax` command's `name` option, an abbreviation for `namelist`. The `name` option specifies that one name must be provided. The `confirm` command comes in handy to ensure that the name provided is indeed that of a matrix. The command operates on the matrix in place, updating its contents.

```
. type flipmat.ado
capture program drop flipmat
*! flipmat 1.0.0  CFBaum 08oct2007
program flipmat
        version 10.1
        syntax name, [HORIZ]
        confirm matrix 'namelist'
        mata: mf_flipmat("'namelist'", "'horiz'")
end

. set more off
```

---

1. This recipe is adapted from Stata tip 37 (Baum 2006b).
2. My thanks to Mata's chief architect, Bill Gould, for improvements he suggested that make the code more general.

We now consider the Mata code. The basic logic is one line: we replace matrix X with its own contents, with rows (and columns if horiz is specified) in reverse order. One complication is that unlike Mata matrices, Stata matrices have row and column *stripes* that can contain two components labeling each row or column. For estimation results, for instance, the stripe element may include both an equation name and a variable name.[3] We do not want to wipe out these contents in the process, so we use the st_matrixrowstripe() and st_matrixcolstripe() functions to preserve the contents of the stripes, and reorder them in line with the functioning of flipmat.

```
. type mf_flipmat.mata
version 10.1
mata:
void function mf_flipmat(string scalar name, string scalar horiz)
{
        real matrix X, rs, cs
        X = st_matrix(name)
        rs = st_matrixrowstripe(name)
        cs = st_matrixcolstripe(name)
        if (horiz == "") {
                X = (rows(X)>1 ? X[rows(X)..1, .]: X)
                rs = (rows(rs)>1 ? rs[rows(rs)..1, .]: rs)
        }
        else {
                X = (cols(X)>1 ? X[., cols(X)..1]: X)
                cs = (rows(cs)>1 ? cs[rows(cs)..1, .]: cs)
        }
        st_matrix(name, X)
        st_matrixcolstripe(name, cs)
        st_matrixrowstripe(name, rs)
}
end
mata: mata mosave mf_flipmat(), dir(PERSONAL) replace
```

We follow the advice of section 13.10.2 and store the compiled Mata function as a .mo file in our PERSONAL directory.

For example, imagine that you have a set of course grades for homework (hw), midterm (mt), final exam (final), and total points (total) stored in a matrix, grades, with the rows identifying the students' ID numbers:

---

3. See [M-5] **st_matrix( )**.

```
. use course, clear
. mat list grades, noheader
            hw      mt   final   total
    121      6       8      21      35
    958      8       9      23      40
    196      7      17      24      48
    190      8      13      38      59
    921     17      24      66     107
     33     17      31      76     124
    806     19      29      80     128
    514     21      51      71     143
    526     22      40      96     158
    340     26      59     105     190
    101     30      49     119     198
    661     30      62     124     216
    856     32      66     128     226
    581     34      57     135     226
    276     34      64     133     231
    287     33      64     137     234
    703     35      63     136     234
    210     37      78     134     249
     81     37      79     142     258
    366     40      82     160     282
    705     42      85     167     294
     66     43      85     173     301
    424     45      87     177     309
    964     47      86     190     323
    345     54     106     223     383
```

The matrix is currently stored in ascending order of total points scored. If you would like to reverse that, we would type

(Continued on next page)

```
. flipmat grades

. mat list grades, noheader
         hw     mt  final  total
345      54    106    223    383
964      47     86    190    323
424      45     87    177    309
 66      43     85    173    301
705      42     85    167    294
366      40     82    160    282
 81      37     79    142    258
210      37     78    134    249
703      35     63    136    234
287      33     64    137    234
276      34     64    133    231
581      34     57    135    226
856      32     66    128    226
661      30     62    124    216
101      30     49    119    198
340      26     59    105    190
526      22     40     96    158
514      21     51     71    143
806      19     29     80    128
 33      17     31     76    124
921      17     24     66    107
190       8     13     38     59
196       7     17     24     48
958       8      9     23     40
121       6      8     21     35
```

If you wanted to display the total points as the first column, we would then type

```
. flipmat grades, horiz

. mat list grades, noheader
       total  final     mt     hw
345     383    223    106     54
964     323    190     86     47
424     309    177     87     45
 66     301    173     85     43
705     294    167     85     42
366     282    160     82     40
 81     258    142     79     37
210     249    134     78     37
703     234    136     63     35
287     234    137     64     33
276     231    133     64     34
581     226    135     57     34
856     226    128     66     32
661     216    124     62     30
101     198    119     49     30
340     190    105     59     26
526     158     96     40     22
514     143     71     51     21
806     128     80     29     19
 33     124     76     31     17
921     107     66     24     17
190      59     38     13      8
196      48     24     17      7
958      40     23      9      8
121      35     21      8      6
```

Finally, note that this Mata routine works only with a Stata matrix, and a Stata matrix can contain only numeric (or missing) values. However, the technique used will operate on any Mata matrix, including complex and string matrices. An example of its application to a Mata string matrix is given in Baum (2006b).

14.2 Shuffling the elements of a string variable

The problem. Suppose you want to randomly shuffle the characters in each observation of a Stata string variable.

The solution. Mata's `jumble()` function fits the bill, using a simple ado-file program to call it:

```
. type shufstr.ado

*! shufstr 1.0.0  CFBaum 11aug2008
program shufstr
version 10.1
syntax varlist(string max=1)
mata: shufstr("'varlist'")
end
```

The program takes one string variable as an argument and passes its name to the Mata `shufstr()` function. The Mata code is equally simple:

```
. type shufstr.mata

version 10.1
mata: mata set matastrict on
mata: mata clear
// shufstr 1.0.0  CFBaum 11aug2008
mata:
void function shufstr(string vector vname)
{
        string matrix S
        real scalar i
        st_sview(S, ., vname)
        for(i = 1; i <= rows(S);  i++) {
            S[i, .] = char(jumble(ascii(S[i, .])')')
        }
}
end

mata: mata mosave shufstr(), dir(PERSONAL) replace
```

We create a view matrix containing the string variable and define each row in turn as the `jumble()` of its characters. Because `jumble()` is designed to shuffle the rows of a matrix,[4] we translate the string into a vector of ASCII codes[5] using the `ascii()` function. In Mata, this is the simplest way to split a string into its characters. The `char()` function translates the vector of shuffled ASCII codes back to one character string.

---

4. As [D] **sort** indicates, `jumble(1::52)` would shuffle a deck of cards.

5. The ASCII character codes express each character as an integer between 1 and 255. For instance, the ASCII code for b is 42.

To illustrate, we extract the second word of the `make` variable in `auto.dta` for the first 15 automobiles and shuffle that word:

```
. sysuse auto, clear
(1978 Automobile Data)

. generate model = ""
(74 missing values generated)

. quietly describe

. forvalues i = 1/'=r(N)' {
  2.          local temp = make['i']
  3.          local model: word 2 of 'temp'
  4.          qui replace model = "'model'" in 'i'
  5. }

. keep in 1/15
(59 observations deleted)

. keep make model price

. generate shufmod = model

. set seed 20080906

. shufstr(shufmod)

. list, sep(0) noobs
```

make	price	model	shufmod
AMC Concord	4,099	Concord	oCndocr
AMC Pacer	4,749	Pacer	cPera
AMC Spirit	3,799	Spirit	rStpii
Buick Century	4,816	Century	Cnetury
Buick Electra	7,827	Electra	eaElrtc
Buick LeSabre	5,788	LeSabre	ebaeSLr
Buick Opel	4,453	Opel	eOlp
Buick Regal	5,189	Regal	Raleg
Buick Riviera	10,372	Riviera	Reivira
Buick Skylark	4,082	Skylark	yakkrlS
Cad. Deville	11,385	Deville	Deielvl
Cad. Eldorado	14,500	Eldorado	oEldaord
Cad. Seville	15,906	Seville	leSeliv
Chev. Chevette	3,299	Chevette	eCethevt
Chev. Impala	5,705	Impala	amplaI

14.3   Firm-level correlations with multiple indices with Mata

The problem. Let's revisit the problem posed in section 10.1, in which a user with firm-level panel data would like to compute the correlations between firm returns and a set of index fund returns, and determine which of those correlations is the highest for each firm. In the earlier presentation, we solved this problem with do-file code and Stata matrices. Here we consider how it might be done more efficiently with Mata.

The solution. We first set up an ado-file to make the problem more general by allowing different variable names to be specified for the index fund returns, firm returns, and firm identifier. The program, `maxindcorr`, takes a varlist of the names of index fund returns and three required options: `ret()`, the name of the firm returns variable; `firmid()`, the firm identifier variable; and `gen()`, a name that is used as a stub for four new variable names. These four variables will contain, for each firm, the firm's ID, its mean return, the number of quotes available, and the number of the index fund with which the highest correlation appears. For example, if the `gen(foo)` option is given, the variables created will be foo*firmid*, `foomu`, `foon`, and `foomax`. After setting up a trading day calendar, we are ready to pass control to the Mata `indcorr()` function.[6]

```
. type maxindcorr.ado

*! maxindcorr 1.0.0  CFBaum 11aug2008
program maxindcorr
        version 10.1
        syntax varlist(numeric), RET(varname numeric) FIRMid(varname) GEN(string)
* validate new variable names
        confirm new variable 'gen''firmid'
        confirm new variable 'gen'max
        confirm new variable 'gen'mu
        confirm new variable 'gen'n

        tempvar trday
* establish trading day calendar using firmid variable
        bysort 'firmid': gen 'trday' = _n
        qui tsset 'firmid' 'trday'
        qui generate 'gen'max = .
        qui generate 'gen'mu = .
        qui generate 'gen'n = .
        qui generate 'gen''firmid' = .
        qui levelsof 'firmid'
        local firms 'r(levels)'
        local nf : word count 'r(levels)'
        forv i = 1/'nf' {
                local fid : word 'i' of 'firms'
                qui replace 'gen''firmid' = 'fid' in 'i'
        }
* create varlist of indices..ret
        local vl "'varlist' 'ret'"
* pass to Mata routine
        mata: indcorr("'firmid'", "'vl'","'gen'")
end
```

The Mata routine sets up views for the firm identifier (`ind`) and the data matrix (`pdata`). We then use the `panelsetup()` function to tell Mata that these are panel data indexed by the firm identifier, the first column of `pdata`. The resulting matrix, `info`, merely contains, for each panel, the beginning and ending observation numbers of its elements. We then create views for the three new variables to be created.

These panel data are in the long format,[7] with `nf` firms' time-series observations. The main component of the routine is a loop over the `nf` panels found in `eye`. The `panelsubmatrix()` function extracts the appropriate rows of `eye` for the *i*th panel, and Mata's `correlation()` function produces the correlation matrix. Because the

---

6. The `quietly` prefix could be used on a block of statements rather than on the individual statements, as shown here. However, that will suppress the listing of the statements in the block, which we avoid for pedagogical reasons.

7. See section 5.5.

firm return variable is the last column of **eye**, all but the last element of the last row of **eye** are the correlations of interest. In the do-file code of section 10.1, we stored those correlations in a matrix. Because we want to keep track only of the maximum correlation, we use **maxindex()** to locate that element and store it directly in the result variable, the view **highcorr**. Likewise, we invoke **panelsubmatrix()** to compute the firm's mean return (with **mean()**) and number of quotes, and we store those in the result variables.

```
. type indcorr.mata
mata: mata clear
version 10.1
mata: mata set matastrict on
mata:
// indcorr 1.0.0  CFBaum 11aug2008
void function indcorr(string scalar ind,
                      string scalar vn,
                      string scalar newvar)
{
        real matrix pdata, info, highcorr, sigma, z, enn, w
        real vector muret, ret
        real scalar nf, nv, nv1, i, imax
        string scalar mu, maxc, enname
        st_view(ind, ., ind)
        st_view(pdata, ., tokens(vn))
        info = panelsetup(ind, 1)
        nf = rows(info)
        nv = cols(pdata)
        nv1 = nv-1
        maxc = newvar + "max"
        st_view(highcorr, 1::nf, maxc)
        mu = newvar + "mu"
        st_view(muret, 1::nf, mu)
        enname = newvar + "n"
        st_view(enn, 1::nf, enname)
// compute correlations between index columns and last column (ret)
        for(i = 1; i <= nf; i++) {
                sigma = correlation(panelsubmatrix(pdata, i, info))
                ret = sigma[nv,  1::nv1]
                maxindex(ret, 1, imax, w)
                highcorr[i] = imax
// calculate mean return and number of quotes for this panel
                z = panelsubmatrix(pdata[.,nv], i, info)
                muret[i] = mean(z)
                enn[i] = rows(z)
        }
}
end
mata: mata mosave indcorr(), dir(PERSONAL) replace
```

We can now invoke the routine on the data used in section 10.1:

```
. use ch14.3, clear
. maxindcorr index1-index9, ret(ret) firm(permno) gen(set1)
. summarize set1*
```

Variable	Obs	Mean	Std. Dev.	Min	Max
set1max	291	4.917526	2.665243	1	9
set1mu	291	-.0005513	.0061444	-.0775419	.0080105
set1n	291	1071.261	369.9875	8	1259
set1permno	291	58499.78	26550.12	10051	92655

```
. gsort -set1mu
. label def ind 1 Kappa 2 Lambda 3 Nu 4 Xi 5 Tau 6 Upsilon 7 Phi 8 Chi 9 Psi
. label values set1max ind
. list set1permno set1mu set1n set1max in 1/50, noobs sep(0)
```

set1pe~o	set1mu	set1n	set1max
23317	.0080105	8	Nu
53196	.0037981	465	Tau
67345	.0033149	459	Upsilon
90879	.0028613	1001	Psi
80439	.0027118	1259	Chi
64629	.0026724	1259	Upsilon
87165	.0025065	1259	Chi
76177	.002376	531	Nu
85073	.0023348	945	Chi
24441	.0023073	1259	Lambda
81063	.0022981	489	Chi
37284	.0021567	1259	Nu
14593	.0019581	1259	Kappa
80778	.0019196	941	Chi
68347	.0019122	85	Kappa
84827	.0018903	1259	Chi
84107	.0017555	987	Chi
22250	.0017191	1259	Lambda
76139	.0017182	1259	Phi
23819	.0016474	1259	Lambda
66384	.0016361	1259	Upsilon
38156	.001558	1259	Nu
88845	.0015497	1259	Lambda
85631	.0015028	1259	Chi
84788	.0015013	1259	Chi
77077	.0014899	1006	Phi
48653	.0014851	1259	Xi
92284	.0014393	1259	Psi
81877	.0014366	454	Chi
81178	.0014339	1186	Chi
81282	.0014084	1259	Phi
80114	.0014042	1259	Chi
56937	.0014028	1259	Tau
75819	.0013873	1259	Xi
22293	.0013792	1259	Chi
10085	.0013747	1072	Kappa
76712	.0013615	1259	Nu
77713	.0013607	127	Phi
12265	.0012653	858	Kappa

76779	.0012513	648	Lambda
83422	.0012314	1259	Chi
76224	.0012159	1259	Phi
47888	.0012045	1259	Xi
86569	.0011911	1259	Chi
77803	.0011858	1173	Phi
82272	.0011574	1259	Kappa
83976	.0011543	1259	Kappa
75510	.0011498	1259	Phi
15720	.0011436	1259	Lambda
90352	.0011411	1259	Psi

This more general routine illustrates how an ado-file and Mata subroutine can be used to provide a more efficient solution that can immediately be applied to a similar problem. Here is another consideration: the do-file routine of section 10.1 can run afoul of Stata's matsize limits (`help limits`); there are no such limitations in Mata. Memory requirements can be reduced in `indcorr()` by using `panelsubview()` rather than `panelsubmatrix()`. As those functions' documentation indicates, `panelsubmatrix` is preferable if speed of execution is a greater concern than memory use.

14.4   Passing a function to a Mata function

The problem. You need a Mata routine to apply a mathematical function, but you would like to specify that function when you call the Mata routine rather than having every possible function coded in Mata. In short, you would like to pass a function to a function, as described in [M-2] **ftof**.

The solution. We illustrate this by modifying the `aggreg` routine presented in section 13.7.1 (because this is a major change, we rename it `aggreg2`). We add one more optional argument to its syntax: `trans(`*function*`)`, indicating that the original variable should be transformed before it is aggregated. The list of available transformations is defined in local macro `trops` as `abs exp log sqrt`. If no transformation is specified, the default identity transformation, `f(x) = x`, is applied.

Passing functions to functions makes use of pointers, as defined in section 13.8. Before we begin, one issue must be addressed. As [M-2] **ftof** states, you cannot pass a built-in function to a function, and the mathematical functions are built in. We get around this problem by defining our own trivial versions of each math function, for example,

```
: function mf_abs(x) return(abs(x))
```

We now are ready to set up the passing mechanism. We parse the `trans()` option and define the name of the function to be passed in local macro `trfn`. The call to Mata `aggreg2()` function includes an argument, `&'trfn'()`, that specifies the desired transformation in terms of a pointer to the function.

```
. type aggreg2.ado
*! aggreg2 1.0.0  CFBaum 11aug2008
program aggreg2, rclass
        version 10.1
        syntax varname(numeric) [if] [in], per(integer) ///
               [func(string) trans(string)]
        marksample touse
        quietly count if `touse'
        if `r(N)' == 0 {
                error 2000
        }
* validate per versus selected sample
        if `per' <= 0 | `per' >= `r(N)' {
                display as error "per must be > 0 and < N of observations."
                error 198
        }
        if mod(`r(N)',`per' != 0) {
                display as error "N of observations must be a multiple of per."
                error 198
        }
* validate func option; default is average (code A)
        local ops A S F L
        local opnames average sum first last
        if "`func'" == "" {
            local op  "A"
        }
        else {
        local nop : list posof "`func'" in opnames
        if !`nop' {
                display as err "Error: func must be chosen from `opnames'"
                error 198
        }
        local op : word `nop' of `ops'
        }
* validate trans option; default is none (identity)
        local trops abs exp log sqrt
        if "`trans'" == "" {
            local trfn  "mf_iden"
        }
        else {
            local ntr : list posof "`trans'" in trops
            if !`ntr' {
                    display as err "Error: trans must be chosen from `trops'"
                    error 198
            }
            local trfn "mf_`trans'"
        }
* validate the new varname
        local newvar = "`varlist'`op'`trans'`per'"
        quietly generate `newvar' = .
* pass the varname and newvarname to Mata
        mata: aggreg2("`varlist'", "`newvar'",  `per', "`op'", ///
              &`trfn'(), "`touse'")
end
```

In the Mata function, the argument defining the name of the function is specified as

```
: pointer(real scalar function) scalar f
```

and is invoked to transform the raw data in vector v1 into vector v1t:

```
: v1t = (*f)(v1)
```

The Mata code reads as follows:

```
. type aggreg2.mata
version 10.1
mata: mata clear
mata: mata set matastrict on
mata:
function mf_abs(x) return(abs(x))
function mf_exp(x) return(exp(x))
function mf_log(x) return(log(x))
function mf_sqrt(x) return(sqrt(x))
function mf_iden(x) return(x)

// aggreg2 1.0.0  CFBaum 11aug2008
void aggreg2(string scalar vname,
            string scalar newvname,
            real scalar per,
            string scalar op,
            pointer(real scalar function) scalar f,
            string scalar touse)
{
        real matrix v1, v1t, v2, v3
        real colvector mult
        if (op == "A") {
            mult = J(per, 1, 1/per)
        }
        else if (op == "S") {
            mult = J(per, 1, 1)
        }
        else if (op == "F") {
                mult = J(per, 1, 0)
                mult[1] = 1
        }
        else if (op == "L") {
            mult = J(per, 1, 0)
            mult[per] = 1
        }
        st_view(v1=., ., vname, touse)
        v1t = (*f)(v1)
        st_view(v2=., ., newvname)
        v3 = colshape(v1t', per) * mult
        v2[(1::rows(v3)),] = v3
}
end
mata: mata mosave aggreg2(), dir(PERSONAL) replace
```

We can now test our routine on the **urates** data used in section 13.7 and verify that it is applying the correct transformations:

```
. use urates, clear
. aggreg2 tenn, per(3) trans(log) // calculate quarterly averages of log(tenn)
. aggreg2 tenn, per(3) func(sum) trans(sqrt) // calculate quarterly sum of
> sqrt(tenn)
. tsmktim quarter, start(1978q1) // create quarterly calendar var
        time variable:  quarter, 1978q1 to 2055q4
                delta:  1 quarter
. list t tenn quarter tennAlog3 tennSsqrt3 in 1/12, sep(3)
```

	t	tenn	quarter	tennAl~3	tennSs~3
1.	1978m1	5.9	1978q1	1.786065	7.327801
2.	1978m2	5.9	1978q2	1.751859	7.203743
3.	1978m3	6.1	1978q3	1.746263	7.183253
4.	1978m4	5.9	1978q4	1.746062	7.182891
5.	1978m5	5.8	1979q1	1.746263	7.183253
6.	1978m6	5.6	1979q2	1.740364	7.162218
7.	1978m7	5.7	1979q3	1.746263	7.183253
8.	1978m8	5.7	1979q4	1.791667	7.348299
9.	1978m9	5.8	1980q1	1.818953	7.44945
10.	1978m10	5.9	1980q2	1.954307	7.9728
11.	1978m11	5.7	1980q3	2.079442	8.485281
12.	1978m12	5.6	1980q4	2.075249	8.467548

14.5   Using subviews in Mata

The problem. Suppose you use view matrices in Mata to avoid making copies of Stata's variables in Mata in your large dataset. You want to create subsets of those view matrices and minimize memory usage in doing so.

The solution. The solution involves the use of *subviews* in Mata. Like a view matrix, a subview is merely a reference to the specified components of another matrix. That matrix can be a view matrix or a regular matrix in Mata. In either case, you do not want to make a copy of the original matrix. You create subviews with the Mata st_subview() functions.

Let's say you want to form matrices within Mata for two sets of variables in your dataset. You could use st_view() functions, specifying lists of variable names to be included:

```
. sysuse auto, clear
(1978 Automobile Data)

. generate en = _n

. mata:
```
── mata (type end to exit) ──────
```
: st_view(Z1=., ., ("en", "price", "mpg", "turn"), 0)

: st_view(Z2=., ., ("en", "weight", "length", "rep78"), 0)

: Z1[ 6, .]
          1       2       3       4

  1       6    5788      18      43


: Z2[ 6, .]
          1       2       3       4

  1       8    3280     200       3


: end
```

In this example, view matrices Z1 and Z2 do not have the same number of rows because of the presence of five missing values in rep78. The fourth argument to st_view() ensures that missing values in any of the variables cause those observations to be excluded. But what if the observations are no longer properly aligned after removing those with missing values? Then the matrices Z1 and Z2 will be misaligned in terms of the original observation numbers. As you can see, the sixth row of Z1 contains the sixth observation of auto.dta, but the sixth row of Z2 contains the eighth observation of this Stata dataset.

You can avoid this difficulty by building one view matrix of all the needed variables,

```
: st_view(Z, ., ("en", "price", "mpg", "turn", "weight", "length", "rep78"), 0)
```

and creating the desired subsets as subviews:

```
: st_subview(Z1, Z, ., (1, 2, 3, 4))
: st_subview(Z2, Z, ., (1, 5, 6, 7))
```

In this syntax, the row vector of indices specifies that columns 1, 2, 3, and 4 of Z are to be included in view matrix Z1, and that columns 1, 5, 6, and 7 of Z are to be included in view matrix Z2. The st_view() matrix Z will exclude any observation with missing values on any of the variables, and the subview matrices Z1 and Z2 require no overhead (beyond a few bytes) because they are only references to the appropriate columns of Z. The subviews could even overlap, as above: the Stata variable en (the observation number) is included as the first column of each subview matrix.

A specialized version of subview() for use with panel data is panelsubview(); see section 14.3.

14.6   Storing and retrieving country-level data with Mata structures

The problem. You would like to store a collection of country-level data and be able to retrieve it by using Mata structures effectively.[8]

The solution. We set up a structure, `country`, that holds several string scalars, two numeric scalars, and a vector of geographic coordinates for its capital city. Each element of the structure must be declared by type.

```
. type country.mata
version 10.1
mata: mata set matastrict on
mata:
// country 1.0.0   CFBaum 11aug2008
struct country {
    string scalar isocode
    string scalar name
    real scalar population
    real scalar gdppc
    string scalar capital
    real vector latlong
}
end
```

We now can define a Mata function, `loadcty()`, that inserts data into the structure. It is defined as a `pointer` function so that it returns the address of that instance of the structure.

---

8. See section 13.9. For an extended discussion of Mata structures, see Gould (2007a).

```
. type loadcty.mata
version 10.1
mata: mata set matastrict on
mata:
// loadcty 1.0.0  CFBaum 11aug2008
pointer(real scalar) loadcty(
                          string scalar isocode,
                          string scalar name,
                          real scalar population,
                          real scalar gdppc,
                          string scalar capital,
                          real scalar latitudeD,
                          real scalar latitudeM,
                          real scalar longitudeD,
                          real scalar longitudeM)
{
        struct country scalar c
        c.isocode = isocode
        c.name = name
        c.population = population
        c.gdppc = gdppc
        c.capital = capital
        c.latlong = (latitudeD, latitudeM, longitudeD, longitudeM)
        return(&c)
}
end
```

To use the structure data, we define a Mata `compcty()` function that compares a pair of countries, taking as arguments their respective structures. We would like to compute the airline distance between capital cities. This measure is often used in so-called gravity models of international trade. The computation is available in Stata from Bill Rising's `sphdist` command, available in the Statistical Software Components archive. His command expects four variables, giving latitude and longitude values for two locations,[9] and generates a new variable containing the computed distances. We want to call this routine for one observation and receive one value in return. We use Mata's `stata()` function to give the Stata command that runs `sphdist`, and `st_view()` to move data back and forth between Mata and Stata.

---

9. The `sphdist` convention codes east longitude (east of Greenwich) as negative and west longitude as positive. Although it is an arbitrary convention, it is more common to code longitude in the same manner as time zones. For instance, for most of the year Boston is UCT−5, five hours behind London, and Berlin is UCT+1, one hour ahead of London. In the `compcty()` routine, we reverse the sign of longitude values.

```
. type compcty.mata
version 10.1
mata: mata set matastrict on
mata:
// compcty 1.0.0  CFBaum 11aug2008
void compcty(struct country scalar a,
             struct country scalar b)
{
        real scalar poprat, gdprat,dist
        real matrix latlong
        printf("\nComparing %15s and %-15s\n\n", a.name, b.name)
        poprat = a.population / b.population
        printf("Ratio of population:     %9.2f\n", poprat)
        gdprat = a.gdppc / b.gdppc
        printf("Ratio of per capita GDP: %9.2f\n", gdprat)
        printf("\nCapital of %15s: %-15s\n Lat. %5.2f deg. ///
    Long. %5.2f deg.\n", a.name, a.capital, a.latlong[1] + ///
    a.latlong[2] / 100, a.latlong[3] + a.latlong[4] / 100)
        printf("\nCapital of %15s: %-15s\n Lat. %5.2f deg. ///
    Long. %5.2f deg.\n", b.name, b.capital, b.latlong[1] + ///
    b.latlong[2] / 100, b.latlong[3] + b.latlong[4] / 100)
// store the latitude/longitude coordinates, reversing long. signs per sphdist
// convention
        st_view(latlong=., ., ("lat1","long1","lat2","long2"))
        latlong[1, 1] = a.latlong[1] + a.latlong[2]/60
        latlong[1, 2] = -1 * (a.latlong[3] + a.latlong[4]/60)
        latlong[1, 3] = b.latlong[1] + b.latlong[2]/60
        latlong[1, 4] = -1 * (b.latlong[3] + b.latlong[4]/60)
        stata("capture drop __dist")
// call Bill Rising's sphdist routine to compute the distance
        stata("sphdist, gen(__dist) lat1(lat1) lon1(long1) lat2(lat2) ///
            lon2(long2)")
        st_view(dist=., .,("__dist"))
        printf("\nDistance between capitals: %9.2f km.\n",dist[1])
}
end
```

We now can load the data structure with several countries' data.[10] Each invocation of the loadcty() function returns a pointer to that instance of the structure. We also set up the Stata variables lat1, long1, lat2, and long2, which will be accessed by sphdist.

---

10. Population, per capita gross domestic product (purchasing power parity) values, and geographic coordinates are taken from the World Factbook, published by the CIA and available at https://www.cia.gov/library/publications/the-world-factbook/.

```
. mata:
──────────────────────────────────────────── mata (type end to exit) ────
: addr1 = loadcty("US", "United States", 301.139947, 43800,"Washington DC", 38,
> 53, -77, -02)

: addr2 = loadcty("CA", "Canada", 33.390141, 35700, "Ottawa", 45, 24, -75, -43)

: addr3 = loadcty("TR", "Turkey", 71.158647, 9100, "Ankara", 39, 55, 32, 55)

: addr4 = loadcty("AR", "Argentina", 40.301927, 15200, "Buenos Aires", -34,
> -35, 58, 22)

: addr5 = loadcty("JP", "Japan", 127.433494, 33100, "Tokyo", 35, 40, 139, 45)

: end
──────────────────────────────────────────────────────────────────────────

. clear

. set obs 1
obs was 0, now 1

. generate lat1 = .
(1 missing value generated)

. generate long1 = .
(1 missing value generated)

. generate lat2 = .
(1 missing value generated)

. generate long2 = .
(1 missing value generated)
```

Now we can invoke the compcty() function for any pair of countries that we have
loaded into the data structure:

```
. mata:
──────────────────────────────────────────── mata (type end to exit) ────
: compcty(*addr1, *addr2)

Comparing   United States and Canada

Ratio of population:         9.02
Ratio of per capita GDP:     1.23

Capital of   United States: Washington DC
 Lat. 38.53 deg.  Long. -77.02 deg.

Capital of          Canada: Ottawa
 Lat. 45.24 deg.  Long. -75.43 deg.

Distance between capitals:    732.12 km.

: compcty(*addr1, *addr3)

Comparing   United States and Turkey

Ratio of population:         4.23
Ratio of per capita GDP:     4.81

Capital of   United States: Washington DC
 Lat. 38.53 deg.  Long. -77.02 deg.

Capital of          Turkey: Ankara
 Lat. 39.55 deg.  Long. 32.55 deg.

Distance between capitals:   8724.01 km.
```

```
: compcty(*addr3, *addr4)

Comparing          Turkey and Argentina

Ratio of population:          1.77
Ratio of per capita GDP:      0.60

Capital of          Turkey: Ankara
  Lat. 39.55 deg.  Long. 32.55 deg.

Capital of       Argentina: Buenos Aires
  Lat. -34.35 deg.  Long. 58.22 deg.

Distance between capitals:    8679.35 km.

: compcty(*addr1, *addr5)

Comparing   United States and Japan

Ratio of population:          2.36
Ratio of per capita GDP:      1.32

Capital of   United States: Washington DC
  Lat. 38.53 deg.  Long. -77.02 deg.

Capital of          Japan: Tokyo
  Lat. 35.40 deg.  Long. 139.45 deg.

Distance between capitals:  10897.34 km.

: end
```

Of course, the distance computation, a matter of spherical trigonometry implemented in sphdist, can be done directly in Mata. We construct a one-line function, deg2rad(), that converts a measurement in degrees into radians and implement the trigonometric formulas from sphdist in a Mata function, compcty2():

```
. type compcty2.mata
version 10.1
mata:
real scalar deg2rad(real scalar deg)
{
        return(deg * pi() / 180)
}
void compcty2(struct country scalar a, struct country scalar b)
{
        real scalar poprat, gdprat, lat1, lat2, lon1, lon2, costhet, res
        printf("\nComparing %15s and %-15s\n\n", a.name, b.name)
        poprat = a.population / b.population
        printf("Ratio of population:     %9.2f\n", poprat)
        gdprat = a.gdppc / b.gdppc
        printf("Ratio of per capita GDP: %9.2f\n", gdprat)
        printf("\nCapital of %15s: %-15s\n Lat. %5.2f deg. Long. ///
    %5.2f deg.\n", a.name, a.capital, a.latlong[1] + ///
    a.latlong[2] / 100, a.latlong[3] + a.latlong[4] /100)
        printf("\nCapital of %15s: %-15s\n Lat. %5.2f deg.  Long. ///
    %5.2f deg.\n", b.name, b.capital, b.latlong[1] + ///
    b.latlong[2] / 100, b.latlong[3] + b.latlong[4] / 100)
// convert the latitude/longitude coordinates to radians
```

```
        lat1 = deg2rad(a.latlong[1] + a.latlong[2]/60)
        lon1 = deg2rad(a.latlong[3] + a.latlong[4]/60)
        lat2 = deg2rad(b.latlong[1] + b.latlong[2]/60)
        lon2 = deg2rad(b.latlong[3] + b.latlong[4]/60)
        costhet = sin(lat1) * sin(lat2) + cos(lat1) * cos(lat2) * ///
              cos(lon2 - lon1)
        if (costhet == 1 | (lat1 == lat2 & lon1 == lon2)) {
              res = 0
        }
        else if (costhet == 1) {
              res = 20000
        }
        else {
              res = (pi() / 2 - atan(costhet / sqrt(1 - ///
    costhet^2))) * 20000 / pi()
        }
        printf("\nDistance between capitals: %9.2f km.\n",res)
    }
    end
```

Selected results from the pure Mata-based distance computation can be compared with those from the routine invoking `sphdist`:

```
.do compcty2.mata

. mata:
                                        ———— mata (type end to exit) ————
: compcty2(*addr1, *addr3)
Comparing   United States and Turkey
Ratio of population:         4.23
Ratio of per capita GDP:     4.81
Capital of    United States: Washington DC
 Lat. 38.53 deg.  Long. -77.02 deg.
Capital of            Turkey: Ankara
 Lat. 39.55 deg.  Long. 32.55 deg.
Distance between capitals:   8724.01 km.
: compcty2(*addr3, *addr4)
Comparing            Turkey and Argentina
Ratio of population:         1.77
Ratio of per capita GDP:     0.60
Capital of            Turkey: Ankara
 Lat. 39.55 deg.  Long. 32.55 deg.
Capital of         Argentina: Buenos Aires
 Lat. -34.35 deg.  Long. 58.22 deg.
Distance between capitals:   8679.35 km.
: end
```

14.7   Locating nearest neighbors with Mata

The problem. A Statalist user wanted to know how to find "nearest neighbors" in geographical terms; that is, which observations are spatially proximate to each observation in the dataset? This can be generalized to a broader problem: which observations are closest in terms of similarity of several variables? This might be recognized as a problem of calculating a propensity score (see Leuven and Sianesi's `psmatch2` in the Statistical Software Components archive), but we would like to approach it from first principles with a Mata routine.

The solution. In response to the original posting in which the user had latitude and longitude coordinates for each observation, David M. Drukker proposed a Mata-based solution using the `minindex()` function.[11] We modify his solution below to allow a match to be defined in terms of a set of variables on which a close match will be defined. The quality of the match can then be evaluated by calculating the correlation between the original variable's observations and its values of the identified "nearest neighbor". That is, if we consider two units (patients, cities, firms, households) with similar values of $x_1, \ldots, x_m$, how highly correlated are their values of y?

Although the original example is geographical, the underlying task is found in many disciplines where a control group of observations is to be identified, each of which is the closest match to one of the observations of interest. For instance, in finance, you can have a sample of firms that underwent a takeover. For each firm, you would like to find a "similar" firm (based on several characteristics) that did not undergo a takeover. Those pairs of firms are nearest neighbors. In our application, we will compute the Euclidian distance (L2-norm) between the standardized values of pairs of observations.

To implement the solution, we first construct a Stata ado-file defining the `nneighbor` program, which takes a varlist of one or more measures that are to be used in the match.[12] In our application, we can use any number of variables as the basis for defining the nearest neighbor. The user must specify `y`, a response variable; `matchobs`, a variable to hold the observation numbers of the nearest neighbor; and `matchval`, a variable to hold the values of `y` belonging to the nearest neighbor.

After validating any `if` *exp* or `in` *range* conditions with `marksample`, the program confirms that the two new variable names are valid, then generates those variables with missing values. The latter step is necessary because we construct view matrices in the Mata function related to those variables, which must already exist. We then call the Mata function, `mf_nneighbor()`, and compute one statistic from its results: the correlation between the `y()` variable and the `matchvals()` variable, measuring the similarity of these `y()` values between the observations and their nearest neighbors.

---

11. See http://stata.com/statalist/archive/2007-10/msg00365.html.
12. In this case, that could be two variables, latitude and longitude, as in the previous section, where we measured distance between capital cities.

```
. type nneighbor.ado
*! nneighbor 1.0.1  CFBaum 11aug2008
program nneighbor
        version 10.1
        syntax varlist(numeric) [if] [in], ///
            Y(varname numeric) MATCHOBS(string) MATCHVAL(string)

        marksample touse
        qui count if `touse'
        if r(N) == 0 {
                error 2000
        }
// validate new variable names
        confirm new variable `matchobs'
        confirm new variable `matchval'
        qui     generate long `matchobs' = .
        qui generate `matchval' = .
        mata: mf_nneighbor("`varlist'", "`matchobs'", "`y'", ///
                "`matchval'", "`touse'")
        summarize `y' if `touse', meanonly
        display _n "Nearest neighbors for `r(N)' observations of `y'"
        display    "Based on L2-norm of standardized vars: `varlist'"
        display    "Matched observation numbers: `matchobs'"
        display    "Matched values: `matchval'"
        qui correlate `y' `matchval' if `touse'
        display    "Correlation[ `y', `matchval' ] = " %5.4f `r(rho)'
end
```

We now construct the Mata function. The function uses a view on the `varlist`, constructing view matrix `X`. Because the scale of those variables affects the Euclidian distance (L2-norm) calculation, the variables are standardized in matrix `Z` by using Ben Jann's `mm_meancolvar()` function from the `moremata` package (see section 13.10.4).[13] Views are then established for the `matchobs` variable (`C`), the response variable (`y`), and the `matchvals` variable (`ystar`).

For each observation and variable in the normalized varlist, the L2-norm of distances between that observation and the entire vector is computed as `d`. The heart of the function is the call to `minindex()`.[14] As David Drukker stated in his Statalist posting, this function is a fast, efficient calculator of the minimum values of a variable. Its fourth argument can deal with ties; for simplicity, I do not discuss ties here. We request the closest two values, in terms of the distance `d`, to each observation, recognizing that each observation is its own nearest neighbor. The observation numbers of the two nearest neighbors are stored in vector `ind`. Therefore, the observation number desired is the second element of the vector, and `y[ind[2]]` is the value of the nearest neighbor's response variable. Those elements are stored in `C[i]` and `ystar[i]`, respectively.

---

13. The `mm_meancolvar()` function avoids the overhead of computing the full covariance matrix of `X`.

14. There is also a `maxindex()` function; see section 14.3.

```
. type mf_nneighbor.mata
mata: mata clear
mata: mata set matastrict on
version 10.1
mata:
// mf_nneighbor 1.0.0   CFBaum 11aug2008
void function mf_nneighbor(string scalar matchvars,
                           string scalar closest,
                           string scalar response,
                           string scalar match,
                           string scalar touse)
{
        real matrix X, Z, mc, C, y, ystar
        real colvector ind
        real colvector w
        real colvector d
        real scalar n, k, i, j
        string rowvector vars, v
        st_view(X, ., tokens(matchvars), touse)
// standardize matchvars with mm_meancolvar from moremata
        mc = mm_meancolvar(X)
        Z = (X:- mc[1, .]):/ sqrt(mc[2, .])
        n = rows(X)
        k = cols(X)
        st_view(C, ., closest, touse)
        st_view(y, ., response, touse)
        st_view(ystar, ., match, touse)
// loop over observations
        for(i = 1; i <= n; i++) {
// loop over matchvars
            d = J(n, 1, 0)
            for(j = 1; j <= k; j++) {
                d = d + (Z[., j]:- Z[i, j]):^2
            }
            minindex(d, 2, ind, w)
            C[i] = ind[2]
            ystar[i] = y[ind[2]]
            }
}
end
mata: mata mosave mf_nneighbor(), dir(PERSONAL) replace
```

We now can try out the routine. We use the `airquality` dataset used in earlier examples. It contains statistics for 41 U.S. cities' air quality (so2, or sulphur dioxide concentration), as well as several demographic factors. To test our routine, we first apply it to one variable: population (`pop`). Examining the result, we can see that it is properly selecting the city with the closest population value as the nearest neighbor:

```
. use airquality, clear
. sort pop
```

```
. nneighbor pop, y(so2) matchobs(mo1) matchval(mv1)
```

Nearest neighbors for 41 observations of so2
Based on L2-norm of standardized vars: pop
Matched observation numbers: mo1
Matched values: mv1
Correlation[so2, mv1] = 0.0700

```
. list pop mo1 so2 mv1, sep(0)
```

	pop	mo1	so2	mv1
1.	71	2	31	36
2.	80	1	36	31
3.	116	4	46	13
4.	132	3	13	46
5.	158	6	56	28
6.	176	7	28	94
7.	179	6	94	28
8.	201	7	17	94
9.	244	10	11	8
10.	277	11	8	26
11.	299	12	26	31
12.	308	11	31	26
13.	335	14	10	14
14.	347	13	14	10
15.	361	14	9	14
16.	448	17	18	23
17.	453	16	23	18
18.	463	17	11	23
19.	497	20	24	14
20.	507	21	14	17
21.	515	22	17	61
22.	520	21	61	17
23.	529	24	14	29
24.	531	23	29	14
25.	540	24	26	29
26.	582	27	10	30
27.	593	26	30	10
28.	622	29	56	10
29.	624	28	10	56
30.	716	31	12	16
31.	717	30	16	12
32.	744	33	29	28
33.	746	32	28	29
34.	751	33	65	28
35.	757	34	29	65
36.	844	37	9	47
37.	905	36	47	9
38.	1233	39	10	35
39.	1513	38	35	10
40.	1950	39	69	35
41.	3369	40	110	69

However, the response variable's values are very weakly correlated with those of the
`matchvar`. Matching cities on the basis of one attribute does not seem to imply that
they will have similar values of air pollution. We thus exercise the routine on two

broader sets of attributes: one adding `temp` and `wind`, and the second adding `precip` and `days`, where `days` measures the mean number of days with poor air quality.

```
. nneighbor pop temp wind, y(so2) matchobs(mo3) matchval(mv3)
Nearest neighbors for 41 observations of so2
Based on L2-norm of standardized vars: pop temp wind
Matched observation numbers: mo3
Matched values: mv3
Correlation[ so2, mv3 ] = 0.1769
. nneighbor pop temp wind precip days, y(so2) matchobs(mo5) matchval(mv5)
Nearest neighbors for 41 observations of so2
Based on L2-norm of standardized vars: pop temp wind precip days
Matched observation numbers: mo5
Matched values: mv5
Correlation[ so2, mv5 ] = 0.5286
```

We see that with the broader set of five attributes on which matching is based, there is a much higher correlation between the `so2` values for each city and those for its nearest neighbor.

14.8   Computing the seemingly unrelated regression estimator for an unbalanced panel

The problem. Stata's seemingly unrelated regression estimator (`sureg`) estimates a set of equations, using the matrix of residual correlations for each equation to produce a refined estimate of its parameter vector. The `sureg` estimator can be considered as a panel-data estimator that operates in the wide form. It is common for panel data to be unbalanced,[15] and Stata's `xt` commands handle unbalanced panels without difficulty. However, `sureg` discards any observation that is missing in any of its equations.[16] We would like to use `sureg` without losing these observations. If we use `correlate` *varlist*, observations missing for any of the variables in the varlist will be dropped from the calculation. In contrast, the pairwise correlation command `pwcorr` *varlist* will compute correlations from all available observations for each pair of variables in turn. The logic of `sureg` is that of `correlate`; but the correlations of residuals used by `sureg` can be, as in `pwcorr`, calculated on a pairwise basis, allowing the estimator to be applied to a set of equations that can cover different time periods (as long as there is meaningful overlap).

---

15. See section 5.3.
16. An alternative solution to this problem was provided (using Stata's `xtgee` command) by McDowell (2004).

The solution. A poster on Statalist raised this issue and questioned whether `sureg` could be used to work with an unbalanced panel. In this section, I illustrate how Mata can be used to handle this sophisticated estimation problem. I worked with the code of official Stata's `reg3` ([R] **reg3**) command to extract the parsing commands that set up the problem,[17] and then I wrote an ado-file to perform the computations; it calls a Mata function that produces the estimates.

For brevity, I do not reproduce the parsing code taken from `reg3.ado`. The syntax of my program, `suregub`, is identical to that of `sureg` but does not support all of its options. The basic syntax is

suregub (*depvar1 varlist1*) (*depvar2 varlist2*) ... (*depvarN varlistN*)

When the parsing code has executed, it has populated a local macro, `eqlist`, of equations to be estimated and a set of local macros, `ind1` ...`indN`, that contains the right-hand-side variables for each of the N equations.

We first generate the ordinary least-squares residuals for each equation by running `regress` and `predict, residual`. We also find the maximum and minimum observation indices for each set of residuals with the `max()` and `min()` functions, respectively.

Temporary matrix `sigma` will contain the pairwise correlations of the equations' residuals. They are produced with calls to `correlate`. We then invoke the Mata `mm_suregub()` function, passing it the number of equations (`neq`), the list of equations (`eqlist`), and the computed `sigma` matrix. The remainder of `suregub.ado` is

---

17. Official Stata's `sureg.ado` is essentially a wrapper for `reg3.ado`.

```
                // generate residual series
                local minn = .
                local maxn = 0
                forvalues i = 1/'neq' {
                        local dv : word 'i' of 'eqlist'
                        local eq'i' = "'dv' 'ind'i''"
                        qui {
                        regress 'dv' 'ind'i''
                        tempvar touse'i' es eps'i'
                        predict double 'eps'i'' if e(sample), resid
                        generate byte 'touse'i'' = cond(e(sample), 1, .)
                        summarize 'eps'i'', meanonly
                        local maxn = max('maxn', r(N))
                        local minn = min('minn', r(N))
                        }
                }
                tempname sigma
                matrix 'sigma' = J('neq', 'neq', 0)
                // generate pairwise correlation matrix of residuals;
                // for comparison with sureg, use divisor N
                local neq1 = 'neq' - 1
                forvalues i = 1/'neq1' {
                        forvalues j = 2/'neq'  {
                                qui correlate 'eps'i'' 'eps'j'', cov
                                mat 'sigma'['i', 'i'] = r(Var_1) * (r(N) - 1) / (r(N))
                                mat 'sigma'['j', 'j'] = r(Var_2) * (r(N) - 1) / (r(N))
                                mat 'sigma'['i', 'j'] = r(cov_12) * (r(N) - 1) / (r(N))
                                mat 'sigma'['j', 'i'] = 'sigma'['i', 'j']
                        }
                }
                mata: mm_suregub('neq', "'eqlist'", "'sigma'")
                display _newline "Seemingly unrelated regression for an unbalanced panel"
                display _newline "Minimum observations per unit = 'minn'"
                display     "Maximum observations per unit = 'maxn'"
                mat b = r(b)
                mat V = r(V)
                ereturn clear
                ereturn post b V
                ereturn local cmd "suregub"
                ereturn local minobs 'minn'
                ereturn local maxobs 'maxn'
                ereturn display
                end
```

Following the Mata function's execution, we retrieve the estimated coefficient vector
(returned as `r(b)`) and covariance matrix (returned as `r(V)`) to Stata matrices b and
V, respectively. Those matrices are then posted to the **ereturn** values with **ereturn
post**. The **ereturn display** command triggers the standard estimation output routine,
which now presents the results of estimation from mm_suregub().

The Mata function is more complicated than those we have seen in earlier examples. As introduced in section 13.8, I use pointers, here row vectors of pointers to real matrices. We also must work with the coefficient names and matrix stripes[18] attached to Stata matrices so that the display of estimation results will work properly. The first section of the function loops over equations, setting up the appropriate contents of the dependent variable (`yy[i]`) and the right-hand-side variables (`xx[i]`) for each equation in turn.[19]

In the second loop over equations, the elements of the full $\mathbf{X'X}$ matrix are computed as scalar multiples of an element of the inverse of `sigma` times the cross-product of the ith and jth equations' regressor matrices. The full $\mathbf{y'y}$ vector is built up from scalar multiples of an element of the inverse of `sigma` times the cross-product of the ith equation's regressors and the jth equation's values of `yy`. When these two matrices are assembled, the least-squares solution is obtained with `invsym()`, and the appropriate matrix row and column stripes are defined for the result matrices `r(b)` and `r(V)`.

```
. type mm_suregub.mata
version 10.1
mata: mata clear
mata: mata set matastrict on
mata:
// mm_suregub 1.0.0  CFBaum 11aug2008
void mm_suregub(real scalar neq,
                string scalar eqlist,
                string scalar ssigma)
{
        real matrix isigma, tt, eqq, iota, XX, YY, xi, xj, yj, vee
        real vector beta
        real scalar nrow, ncol, i, ii, i2, jj, j, j2
        string scalar lt, touse, le, eqname, eqv
        string vector v, vars, stripe
        pointer (real matrix) rowvector eq
        pointer (real matrix) rowvector xx
        pointer (real matrix) rowvector yy

        eq = xx = yy = J(1, neq, NULL)
        isigma = invsym(st_matrix(ssigma))
        nrow = 0
        ncol = 0
        string rowvector coefname, eqn
        string matrix mstripe
// equation loop 1
        for(i = 1; i <= neq; i++) {
                lt = "touse" + strofreal(i)
                touse = st_local(lt)
                st_view(tt, ., touse)
                le = "eq" + strofreal(i)
                eqv = st_local(le)
                vars = tokens(eqv)
                v = vars[|1, .|]
// pull in full matrix, including missing values
                st_view(eqq, ., v)
                eq[i] = &(tt:* eqq)
// matrix eq[i] is [y|X] for ith eqn
```

---

18. See section 14.1.
19. A constant term is assumed to be present in each equation.

```
                    eqname = v[1]
                    stripe = v[2::cols(v)], "_cons"
                    coefname = coefname, stripe
                    eqn = eqn, J(1, cols(v), eqname)
// form X, assuming constant term
                    nrow = nrow + rows(*eq[i])
                    iota = J(rows(*eq[i]), 1, 1)

                    xx[i] = &((*eq[i])[| 1,2 \ .,. |], iota)
                    ncol = ncol + cols(*xx[i])
// form y
                    yy[i] = &(*eq[i])[.,1]
            }
        XX = J(ncol, ncol, 0)
        YY = J(ncol, 1, 0)
        ii = 0
// equation loop 2
        for(i=1; i<=neq; i++) {
                    i2 = cols(*xx[i])
                    xi = *xx[i]
                    jj = 0
                    for(j=1; j<=neq; j++) {
                            xj = *xx[j]
                            j2 = cols(*xx[j])
                            yj = *yy[j]
                            XX[| ii+1, jj+1 \ ii+i2, jj+j2 |] = isigma[i, j]:*  ///
                                    cross(xi, xj)
                            YY[| ii+1, 1 \ ii+i2, 1 |] =                        ///
                                    YY[| ii+1, 1 \ ii+i2, 1 |] +               ///
                                    isigma[i, j]:* cross(xi, yj)
                            jj = jj + j2
                    }
                    ii = ii + i2
            }
// compute SUR beta (X' [Sigma^-1 # I] X)^-1 (X' [Sigma^-1 # I] y)
        vee = invsym(XX)
        beta = vee * YY
        st_matrix("r(b)", beta')
        mstripe=eqn', coefname'
        st_matrixcolstripe("r(b)", mstripe)
        st_matrix("r(V)", vee)
        st_matrixrowstripe("r(V)", mstripe)
        st_matrixcolstripe("r(V)", mstripe)
    }
    end
    mata: mata mosave mm_suregub(), dir(PERSONAL) replace
```

To validate the routine, we first apply it to a balanced panel, for which it should replicate standard **sureg** results if it has been programmed properly. Because we have already verified that **suregub** passes that test using the **grunfeld.dta** dataset, we modify that dataset to create an unbalanced panel:

```
. use grunfeld, clear
. drop in 75/80
(6 observations deleted)
. drop in 41/43
(3 observations deleted)
. drop in 18/20
(3 observations deleted)
. keep if company <= 4
(120 observations deleted)
. drop time
. reshape wide invest mvalue kstock, i(year) j(company)
(note: j = 1 2 3 4)
Data                                 long    ->    wide
─────────────────────────────────────────────────────────────────
Number of obs.                         68    ->      20
Number of variables                     5    ->      13
j variable (4 values)             company    ->    (dropped)
xij variables:
                                   invest    ->    invest1 invest2 ... invest4
                                   mvalue    ->    mvalue1 mvalue2 ... mvalue4
                                   kstock    ->    kstock1 kstock2 ... kstock4
─────────────────────────────────────────────────────────────────

. list year invest*, sep(0)
```

	year	invest1	invest2	invest3	invest4
1.	1935	317.6	209.9	.	40.29
2.	1936	391.8	355.3	.	72.76
3.	1937	410.6	469.9	.	66.26
4.	1938	257.7	262.3	44.6	51.6
5.	1939	330.8	230.4	48.1	52.41
6.	1940	461.2	361.6	74.4	69.41
7.	1941	512	472.8	113	68.35
8.	1942	448	445.6	91.9	46.8
9.	1943	499.6	361.6	61.3	47.4
10.	1944	547.5	288.2	56.8	59.57
11.	1945	561.2	258.7	93.6	88.78
12.	1946	688.1	420.3	159.9	74.12
13.	1947	568.9	420.5	147.2	62.68
14.	1948	529.2	494.5	146.3	89.36
15.	1949	555.1	405.1	98.3	.
16.	1950	642.9	418.8	93.5	.
17.	1951	755.9	588.2	135.2	.
18.	1952	.	645.5	157.3	.
19.	1953	.	641	179.5	.
20.	1954	.	459.3	189.6	.

We can then invoke **suregub** to generate point and interval estimates:

```
. suregub (invest1 mvalue1 kstock1) (invest2 mvalue2 kstock2)
> (invest3 mvalue3 kstock3) (invest4 mvalue4 kstock4)
```

Seemingly unrelated regressions for an unbalanced panel

Min obs per unit = 14
Max obs per unit = 20

	Coef.	Std. Err.	z	P>|z|	[95% Conf. Interval]	
invest1						
mvalue1	.0787979	.0220396	3.58	0.000	.035601	.1219948
kstock1	.245538	.044413	5.53	0.000	.1584901	.3325858
_cons	64.69007	96.85787	0.67	0.504	-125.1479	254.528
invest2						
mvalue2	.1729584	.0640754	2.70	0.007	.047373	.2985439
kstock2	.4150819	.1279443	3.24	0.001	.1643156	.6658482
_cons	-53.8353	128.2932	-0.42	0.675	-305.2853	197.6147
invest3						
mvalue3	.0522683	.0191105	2.74	0.006	.0148124	.0897243
kstock3	.1071995	.0287962	3.72	0.000	.0507599	.163639
_cons	-39.32897	37.37061	-1.05	0.293	-112.574	33.91607
invest4						
mvalue4	.0632339	.0142336	4.44	0.000	.0353364	.0911313
kstock4	.1487322	.0825525	1.80	0.072	-.0130678	.3105321
_cons	12.50393	11.02009	1.13	0.257	-9.09504	34.1029

Several additional features could be added to `suregub` to more closely match the behavior of `sureg`. That effort is left to the ambitious reader. The full code for `suregub.ado` is available in the electronic supplement to this book.

14.9   A GMM-CUE estimator using Mata's optimize() functions (with Mark E. Schaffer)

The problem. We would like to implement the continuously updated generalized method-of-moments estimator (GMM-CUE) of Hansen, Heaton, and Yaron (1996) in Mata.[20] This is an estimator of a linear instrumental-variables (IV) model that requires numerical optimization for its solution. We have implemented this estimator for `ivreg2` in Stata's ado-file language using the maximum-likelihood commands (`ml`). Although that is a workable solution, it can be slow for large datasets with many regressors and instruments. In Stata version 10, a full-featured suite of optimization commands is available in Mata as `optimize()` ([M-5] **optimize( )**). We implement a simple IV-GMM estimator in Mata and use that as a model for implementing GMM-CUE.

The solution. The two-step generalized method-of-moments (GMM) estimator for a linear IV regression model reduces to standard IV if we assume an independent and identically distributed (i.i.d.) error process or if the equation is exactly identified with the

---

20. The rationale for this estimator is presented in Baum, Schaffer, and Stillman (2007, 477–479).

number of instruments equal to the number of regressors.[21] The following ado-file, `mygmm2s.ado`, accepts a dependent variable and three additional optional variable lists: for endogenous variables, included instruments, and excluded instruments. A constant is automatically included in the regression and in the instrument matrix. There is only one option, `robust`, which specifies whether we are assuming i.i.d. errors or allowing for arbitrary heteroskedasticity. The routine calls the Mata `m_mygmm2s()` function and receives results back in the saved results. Estimation results are assembled and `posted` to the official locations so that we can make use of Stata's `ereturn display` command and enable the use of postestimation commands, such as `test` and `lincom`.

```
. type mygmm2s.ado
*! mygmm2s 1.0.2 MES/CFB 11aug2008
program mygmm2s, eclass
        version 10.1
/*
  Our standard syntax:
  mygmm2s y, endog(varlist1) inexog(varlist2) exexog(varlist3)  [robust]
  where varlist1 contains endogenous regressors
        varlist2 contains exogenous regressors (included instruments)
        varlist3 contains excluded instruments
  Without robust, efficient GMM is IV. With robust, efficient GMM is 2-step
  efficient GMM, robust to arbitrary heteroskedasticity.
  To accommodate time-series operators in the options, add the "ts"
*/
        syntax varname(ts) [if] [in] [, endog(varlist ts) inexog(varlist ts) ///
                exexog(varlist ts) robust ]

        local depvar 'varlist'
/*
  marksample handles the variables in 'varlist' automatically, but not the
  variables listed in the options 'endog', 'inexog' and so on. -markout- sets
  'touse' to 0 for any observations where the variables listed are missing.
*/
        marksample touse
        markout 'touse' 'endog' 'inexog' 'exexog'
// These are the local macros that our Stata program will use
        tempname b V omega

// Call the Mata routine. All results will be waiting for us in "r()" macros
// afterwards.
        mata: m_mygmm2s("'depvar'", "'endog'", "'inexog'", ///
                        "'exexog'", "'touse'", "'robust'")
// Move the basic results from r() macros into Stata matrices.
        mat 'b' = r(beta)
        mat 'V' = r(V)
        mat 'omega' = r(omega)
// Prepare row/col names.
// Our convention is that regressors are [endog   included exog]
// and instruments are                  [excluded exog  included exog]
// Constant is added by default and is the last column.
```

---

21. See Baum, Schaffer, and Stillman (2007, 467–469).

```
        local vnames 'endog' 'inexog' _cons
        matrix rownames 'V' = 'vnames'
        matrix colnames 'V' = 'vnames'
        matrix colnames 'b' = 'vnames'
        local vnames2 'exexog' 'inexog' _cons
        matrix rownames 'omega' = 'vnames2'
        matrix colnames 'omega' = 'vnames2'
// We need the number of observations before we post our results.
        local N = r(N)
        ereturn post 'b' 'V', depname('depvar') obs('N') esample('touse')
// Store remaining estimation results as e() macros accessible to the user.
        ereturn matrix omega 'omega'
        ereturn local depvar = "'depvar'"
        ereturn scalar N = r(N)
        ereturn scalar j = r(j)
        ereturn scalar L = r(L)
        ereturn scalar K = r(K)
        if "'robust'" != "" {
            ereturn local vcetype "Robust"
        }
        display _newline "Two-step GMM results" _col(60) "Number of obs = " e(N)
        ereturn display
        display "Sargan-Hansen J statistic: " %7.3f e(j)
        display "Chi-sq(" %3.0f e(L)-e(K) ")        P-val = " ///
                %5.4f chiprob(e(L)-e(K), e(j)) _newline
end
```

The Mata function receives the names of variables to be included in the regression and creates view matrices for Y (the dependent variable), X1 (the endogenous variables), X2 (the exogenous regressors or included instruments), and Z1 (the excluded instruments). The st_tsrevar() ([M-5] **st_tsrevar( )**) function is used to deal with Stata's time-series operators in any of the variable lists.

```
. type m_mygmm2s.mata
mata:mata clear
version 10.1
mata: mata set matastrict on
mata:
// m_mygmm2s 1.0.0 MES/CFB 11aug2008
void m_mygmm2s(string scalar yname,
            string scalar endognames,
            string scalar inexognames,
            string scalar exexognames,
            string scalar touse,
            string scalar robust)
{
        real matrix Y, X1, X2, Z1, X, Z, QZZ, QZX, W, omega, V
        real vector cons, beta_iv, beta_gmm, e, gbar
        real scalar K, L, N, j

// Use st_tsrevar in case any variables use Stata's time-series operators.
        st_view(Y, ., st_tsrevar(tokens(yname)), touse)
        st_view(X1, ., st_tsrevar(tokens(endognames)), touse)
        st_view(X2, ., st_tsrevar(tokens(inexognames)), touse)
        st_view(Z1, ., st_tsrevar(tokens(exexognames)), touse)
```

```
// Our convention is that regressors are [endog    included exog]
// and instruments are                   [excluded exog   included exog]
// Constant is added by default and is the last column.
        cons = J(rows(X2), 1, 1)
        X2 = X2, cons
        X = X1, X2
        Z = Z1, X2

        K = cols(X)
        L = cols(Z)
        N = rows(Y)

        QZZ = 1/N * quadcross(Z, Z)
        QZX = 1/N * quadcross(Z, X)

// First step of 2-step feasible efficient GMM: IV (2SLS).  Weighting matrix
// is inv of Z'Z (or QZZ).
        W = invsym(QZZ)
        beta_iv = (invsym(X'Z * W * Z'X) * X'Z * W * Z'Y)
// By convention, Stata parameter vectors are row vectors
        beta_iv = beta_iv'
// Use first-step residuals to calculate optimal weighting matrix for 2-step
> FEGMM
        omega = m_myomega(beta_iv, Y, X, Z, robust)
// Second step of 2-step feasible efficient GMM: IV (2SLS).  Weighting matrix
// is inv of Z'Z (or QZZ).
        W = invsym(omega)
        beta_gmm = (invsym(X'Z * W * Z'X) * X'Z * W * Z'Y)
// By convention, Stata parameter vectors are row vectors
        beta_gmm = beta_gmm'

// Sargan-Hansen J statistic: first we calculate the second-step residuals
        e = Y - X * beta_gmm'
// Calculate gbar = 1/N * Z'*e
        gbar = 1/N * quadcross(Z, e)
        j = N * gbar' * W * gbar

// Sandwich var-cov matrix (no finite-sample correction)
// Reduces to classical var-cov matrix if Omega is not robust form.
// But the GMM estimator is "root-N consistent", and technically we do
// inference on sqrt(N)*beta.  By convention we work with beta, so we adjust
// the var-cov matrix instead:
        V = 1/N * invsym(QZX' * W * QZX)

// Easiest way of returning results to Stata: as r-class macros.
        st_matrix("r(beta)", beta_gmm)
        st_matrix("r(V)", V)
        st_matrix("r(omega)", omega)
        st_numscalar("r(j)", j)
        st_numscalar("r(N)", N)
        st_numscalar("r(L)", L)
        st_numscalar("r(K)", K)
}
end

mata: mata mosave m_mygmm2s(), dir(PERSONAL) replace
```

This function in turn calls an additional Mata function, m_myomega(), to compute the appropriate covariance matrix. This is a **real matrix** function, because it will return its result, the real matrix omega, to the calling function. Because we will reuse the m_myomega() function in our GMM-CUE program, we place it in a separate file, m_myomega.mata, with instructions to compile it into a .mo file (see section 13.10.3).

```
. type m_myomega.mata
mata: mata clear
version 10.1
mata: mata set matastrict on
mata:
//  m_myomega 1.0.0 MES/CFB 11aug2008
real matrix m_myomega(real rowvector beta,
                      real colvector Y,
                      real matrix X,
                      real matrix Z,
                      string scalar robust)
{
        real matrix QZZ, omega
        real vector e, e2
        real scalar N, sigma2

// Calculate residuals from the coefficient estimates
                N = rows(Z)
                e = Y - X * beta'

                if (robust=="") {
// Compute classical, non-robust covariance matrix
                    QZZ = 1/N * quadcross(Z, Z)
                    sigma2 = 1/N * quadcross(e, e)
                    omega = sigma2 * QZZ
                }
                else {
// Compute heteroskedasticity-consistent covariance matrix
                    e2 = e:^2
                    omega = 1/N * quadcross(Z, e2, Z)
                }
                _makesymmetric(omega)
                return (omega)
}
end
mata: mata mosave m_myomega(), dir(PERSONAL) replace
```

This gives us a working Mata implementation of an IV estimator and an IV-GMM estimator (accounting for arbitrary heteroskedasticity), and we can verify that its results match those of ivregress ([R] **ivregress**) or our own ivreg2.

To implement the GMM-CUE estimator,[22] we clone mygmm2s.ado to mygmmcue.ado. The ado-file code is similar:

---

22. See Baum, Schaffer, and Stillman (2007, 477–480).

```
      . type mygmmcue.ado
      *! mygmmcue 1.0.2 MES/CFB 11aug2008
      program mygmmcue, eclass
              version 10.1
              syntax varname(ts) [if] [in] [ , endog(varlist ts) ///
                      inexog(varlist ts) exexog(varlist ts) robust ]
              local depvar 'varlist'

              marksample touse
              markout 'touse' 'endog' 'inexog' 'exexog'
              tempname b V omega

              mata: m_mygmmcue("'depvar'", "'endog'", "'inexog'", ///
                              "'exexog'", "'touse'", "'robust'")

              mat 'b' = r(beta)
              mat 'V' = r(V)
              mat 'omega'=r(omega)
      // Prepare row/col names
      // Our convention is that regressors are [endog    included exog]
      // and instruments are                  [excluded exog  included exog]
              local vnames 'endog' 'inexog' _cons
              matrix rownames 'V' = 'vnames'
              matrix colnames 'V' = 'vnames'
              matrix colnames 'b' = 'vnames'
              local vnames2 'exexog' 'inexog' _cons
              matrix rownames 'omega' = 'vnames2'
              matrix colnames 'omega' = 'vnames2'

              local N = r(N)
              ereturn post 'b' 'V', depname('depvar') obs('N') esample('touse')

              ereturn matrix omega 'omega'
              ereturn local depvar = "'depvar'"
              ereturn scalar N = r(N)
              ereturn scalar j = r(j)
              ereturn scalar L = r(L)
              ereturn scalar K = r(K)

              if "'robust'" != "" ereturn local vcetype "Robust"

              display _newline "GMM-CUE estimates" _col(60) "Number of obs = " e(N)
              ereturn display
              display "Sargan-Hansen J statistic: " %7.3f e(j)
              display "Chi-sq(" %3.0f e(L)-e(K) ")        P-val = " ///
                      %5.4f chiprob(e(L)-e(K), e(j)) _newline

      end
```

We now consider how the Mata function must be modified to incorporate the numerical
optimization routines. We must first use Mata's **external** ([M-2] **declarations**) declaration to specify that the elements needed within our objective function evaluator are
visible to that routine. We could also pass those arguments to the evaluation routine,
but treating them as **external** requires less housekeeping.[23] As in the standard two-
step GMM routine, we derive first-step estimates of the regression parameters from a
standard GMM estimation.

We then use Mata's **optimize()** functions to set up the optimization problem.
The **optimize_init()** call, as described in Gould (2007a), sets up a Mata struc-

---

23. As discussed in [M-5] **optimize( )**, you can always pass up to nine extra arguments to evaluators.
    However, you must then keep track of those arguments and their order.

ture, in our case named S, containing all elements of the problem.[24]   In a call to
`optimize_init_evaluator()`, we specify that the evaluation routine is a Mata
`m_mycuecrit()` function.[25] We call `optimize_init_which()` to indicate that we are
minimizing (rather than maximizing) the objective function. After that call, we use
`optimize_init_evaluatortype()` to indicate that our evaluation routine is a type d0
evaluator. Finally, we call `optimize_init_params()` to provide starting values for the
parameters from the IV coefficient vector `beta_iv`. The `optimize()` function invokes the
optimization routine, returning its results in the parameter row vector `beta_cue`. The
optimal value of the objective function is retrieved with `optimize_result_value()`.

```
. type m_mygmmcue.mata
mata: mata clear
version 10.1
mata: mata set matastrict on
mata:
//  m_mygmmcue 1.0.0 MES/CFB 11aug2008
void m_mygmmcue(string scalar yname,
                string scalar endognames,
                string scalar inexognames,
                string scalar exexognames,
                string scalar touse,
                string scalar robust)
{
        real matrix X1, X2, Z1, QZZ, QZX, W, V
        real vector cons, beta_iv, beta_cue
        real scalar K, L, N, S, j

// In order for the optimization objective function to find various variables
// and data they have to be set as externals.  This means subroutines can
// find them without having to have them passed to the subroutines as arguments.
// robustflag is the robust argument recreated as an external Mata scalar.
        external Y, X, Z, e, omega, robustflag
        robustflag = robust

        st_view(Y, ., st_tsrevar(tokens(yname)), touse)
        st_view(X1, ., st_tsrevar(tokens(endognames)), touse)
        st_view(X2, ., st_tsrevar(tokens(inexognames)), touse)
        st_view(Z1, ., st_tsrevar(tokens(exexognames)), touse)
// Our convention is that regressors are [endog          included exog]
// and instruments are              [excluded exog  included exog]
// The constant is added by default and is the last column.
        cons = J(rows(X2), 1, 1)
        X2 = X2, cons
        X = X1, X2
        Z = Z1, X2

        K = cols(X)
        L = cols(Z)
        N = rows(Y)

        QZZ = 1/N * quadcross(Z, Z)
        QZX = 1/N * quadcross(Z, X)
```

─────────────────────────

24. See section 13.9.
25. We do this by providing a pointer to the function; see section 13.8.

```
// First step of CUE GMM: IV (2SLS).  Use beta_iv as the initial values for
// the numerical optimization.
        W = invsym(QZZ)
        beta_iv = invsym(X'Z * W *Z'X) * X'Z * W * Z'Y
// Stata convention is that parameter vectors are row vectors, and optimizers
// require this, so must conform to this in what follows.
        beta_iv = beta_iv'
// What follows is how to set out an optimization in Stata.  First, initialize
// the optimization structure in the variable S.  Then tell Mata where the
// objective function is, that it's a minimization, that it's a "d0" type of
// objective function (no analytical derivatives or Hessians), and that the
// initial values for the parameter vector are in beta_iv.  Finally, optimize.
        S = optimize_init()
        optimize_init_evaluator(S, &m_mycuecrit())
        optimize_init_which(S, "min")
        optimize_init_evaluatortype(S, "d0")
        optimize_init_params(S, beta_iv)
        beta_cue = optimize(S)
// The last omega is the CUE omega, and the last evaluation of the GMM
// objective function is J.
        W = invsym(omega)
        j = optimize_result_value(S)

        V = 1/N * invsym(QZX' * W * QZX)

        st_matrix("r(beta)", beta_cue)
        st_matrix("r(V)", V)
        st_matrix("r(omega)", omega)
        st_numscalar("r(j)", j)
        st_numscalar("r(N)", N)
        st_numscalar("r(L)", L)
        st_numscalar("r(K)", K)
}
end
mata: mata mosave m_mygmmcue(), dir(PERSONAL) replace
```

Let's now examine the evaluation routine. Given values for the **beta** parameter vector, it computes a new value of the **omega** matrix (Ω, the covariance matrix of orthogonality conditions) and a new set of residuals **e**, which are also a function of **beta**. The j statistic, which is the minimized value of the objective function, is then computed, depending on the updated residuals **e** and the weighting matrix $W = \Omega^{-1}$, a function of the updated estimates of **beta**.

```
. type m_mycuecrit.mata
mata: mata clear
version 10.1
mata: mata set matastrict on
mata:

// GMM-CUE evaluator function.
// Handles only d0-type optimization; todo, g and H are just ignored.
// beta is the parameter set over which we optimize, and
// j is the objective function to minimize.

// m_mycuecrit 1.0.0 MES/CFB 11aug2008
void m_mycuecrit(todo, beta, j, g, H)
```

```
{
        external Y, X, Z, e, omega, robustflag
        real matrix W
        real vector gbar
        real scalar N

        omega = m_myomega(beta, Y, X, Z, robustflag)
        W = invsym(omega)
        N = rows(Z)
        e = Y - X * beta'
// Calculate gbar=Z'*e/N
        gbar = 1/N * quadcross(Z,e)
        j = N * gbar' * W * gbar
}
end
mata: mata mosave m_mycuecrit(), dir(PERSONAL) replace
```

Our Mata-based GMM-CUE routine is now complete. To validate both the two-step GMM routine and its GMM-CUE counterpart, we write a simple certification script[26] for each routine. First, let's check to see that our two-step routine works for both i.i.d. and heteroskedastic errors:

```
. cscript mygmm2s adofile mygmm2s
─────────────────────────────────────────────────────────BEGIN mygmm2s

-> which mygmm2s
./mygmm2s.ado
*! mygmm2s 1.0.2 MES/CFB 11aug2008

. set more off

. use griliches76, clear
(Wages of Very Young Men, Zvi Griliches, J.Pol.Ec. 1976)

. quietly ivreg2 lw s expr tenure rns smsa  (iq=med kww age mrt), gmm2s

. savedresults save ivhomo e()

. mygmm2s lw, endog(iq) inexog(s expr tenure rns smsa) exexog(med kww age mrt)
Two-step GMM results                           Number of obs = 758
```

lw	Coef.	Std. Err.	z	P>|z|	[95% Conf. Interval]	
iq	-.0115468	.0052169	-2.21	0.027	-.0217717	-.001322
s	.1373477	.0169631	8.10	0.000	.1041007	.1705947
expr	.0338041	.007268	4.65	0.000	.019559	.0480492
tenure	.040564	.0088553	4.58	0.000	.023208	.0579201
rns	-.1176984	.0353037	-3.33	0.001	-.1868924	-.0485043
smsa	.149983	.0314254	4.77	0.000	.0883903	.2115757
_cons	4.837875	.3448209	14.03	0.000	4.162038	5.513711

```
Sargan-Hansen J statistic:  61.137
Chi-sq(  3  )        P-val = 0.0000

. savedresults compare ivhomo e(),
> include(macros: depvar scalar: N j matrix: b V) tol(1e-7) verbose
comparing macro  e(depvar)
comparing scalar e(N)
comparing scalar e(j)
```

---

26. See section 11.12.

```
comparing matrix e(b)
comparing matrix e(V)

. quietly ivreg2 lw s expr tenure rns smsa  (iq=med kww age mrt), gmm2s robust

. savedresults save ivrobust e()

. quietly mygmm2s lw, endog(iq) inexog(s expr tenure rns smsa)
> exexog(med kww age mrt) robust

. savedresults compare ivrobust e(),
> include(macros: depvar scalar: N j matrix: b V) tol(1e-7) verbose
comparing macro  e(depvar)
comparing scalar e(N)
comparing scalar e(j)
comparing matrix e(b)
comparing matrix e(V)
```

Our `mygmm2s` routine returns the same results as `ivreg2` (Baum, Schaffer, and Stillman 2007) for the several objects included in the `savedresults compare` validation command.[27] Now we construct and run a similar script to validate the GMM-CUE routine:

```
. cscript mygmmcue adofile mygmmcue
─────────────────────────────────────────────────────────────BEGIN mygmmcue
-> which mygmmcue
./mygmmcue.ado
*! mygmmcue 1.0.2 MES/CFB 11aug2008

. set more off

. set rmsg on
r; t=0.00 9:39:34

. use griliches76, clear
(Wages of Very Young Men, Zvi Griliches, J.Pol.Ec. 1976)
r; t=0.00 9:39:34

. quietly ivreg2 lw s expr tenure rns smsa  (iq=med kww age mrt), cue
r; t=7.18 9:39:41

. savedresults save ivreg2cue e()
r; t=0.00 9:39:41

. mygmmcue lw, endog(iq) inexog(s expr tenure rns smsa) exexog(med kww age mrt)
Iteration 0:   f(p) =  61.136598
Iteration 1:   f(p) =  32.923655
Iteration 2:   f(p) =   32.83694
Iteration 3:   f(p) =  32.832195
Iteration 4:   f(p) =  32.831616
Iteration 5:   f(p) =  32.831615

GMM-CUE estimates                                   Number of obs = 758
```

lw	Coef.	Std. Err.	z	P>|z|	[95% Conf. Interval]	
iq	-.0755427	.0132447	-5.70	0.000	-.1015018	-.0495837
s	.3296909	.0430661	7.66	0.000	.245283	.4140989
expr	.0098901	.0184522	0.54	0.592	-.0262755	.0460558
tenure	.0679955	.0224819	3.02	0.002	.0239317	.1120594
rns	-.3040223	.0896296	-3.39	0.001	-.4796931	-.1283515
smsa	.2071594	.0797833	2.60	0.009	.050787	.3635318
_cons	8.907018	.8754361	10.17	0.000	7.191194	10.62284

```
Sargan-Hansen J statistic:  32.832
```

---

27. These functions of `ivreg2` have been certified against official Stata's `ivregress`.

```
Chi-sq( 3  )           P-val = 0.0000
r; t=0.57 9:39:42

. savedresults compare ivreg2cue e(),
> include(macros: depvar scalar: N j matrix: b V) tol(1e-4) verbose
comparing macro  e(depvar)
comparing scalar e(N)
comparing scalar e(j)
comparing matrix e(b)
comparing matrix e(V)
r; t=0.01 9:39:42

. quietly ivreg2 lw s expr tenure rns smsa  (iq=med kww age mrt), cue robust
r; t=13.46 9:39:55

. savedresults save ivreg2cue e()
r; t=0.00 9:39:55

. mygmmcue lw, endog(iq) inexog(s expr tenure rns smsa)
> exexog(med kww age mrt) robust
Iteration 0:  f(p) =   52.768916
Iteration 1:  f(p) =   28.946591  (not concave)
Iteration 2:  f(p) =   27.417939  (not concave)
Iteration 3:  f(p) =   27.041838
Iteration 4:  f(p) =   26.508996
Iteration 5:  f(p) =   26.420853
Iteration 6:  f(p) =   26.420648
Iteration 7:  f(p) =   26.420637
Iteration 8:  f(p) =   26.420637
GMM-CUE estimates                                Number of obs = 758
```

lw	Coef.	Robust Std. Err.	z	P>|z|	[95% Conf. Interval]	
iq	-.0770701	.0147825	-5.21	0.000	-.1060433	-.048097
s	.3348492	.0469881	7.13	0.000	.2427542	.4269441
expr	.0197632	.0199592	0.99	0.322	-.019356	.0588825
tenure	.0857848	.0242331	3.54	0.000	.0382888	.1332807
rns	-.3209864	.091536	-3.51	0.000	-.5003937	-.1415791
smsa	.255257	.0837255	3.05	0.002	.091158	.419356
_cons	8.943698	.9742228	9.18	0.000	7.034257	10.85314

```
Sargan-Hansen J statistic:  26.421
Chi-sq( 3  )           P-val = 0.0000
r; t=0.68 9:39:56

. savedresults compare ivreg2cue e(),
> include(macros: depvar scalar: N j matrix: b V) tol(1e-4) verbose
comparing macro  e(depvar)
comparing scalar e(N)
comparing scalar e(j)
comparing matrix e(b)
comparing matrix e(V)
r; t=0.01 9:39:56
```

We find that the estimates produced by `mygmmcue` are reasonably close to those produced by the different optimization routine used by `ivreg2`. Because we sought to speed up the calculation of GMM-CUE estimates, we are pleased to see the timings displayed by `set rmsg on`. For the nonrobust estimates, `ivreg2` took 6.07 seconds, while `mygmmcue` took 0.51 seconds, a twelvefold speedup. For the robust CUE estimates, `ivreg2` required

12.69 seconds, compared with 0.64 seconds for mygmmcue, an amazing 20 times faster. Calculation of the robust covariance matrix using Mata's matrix operations is apparently much more efficient from a computational standpoint.

Although this Mata-based GMM-CUE routine for linear IV models is merely a first stab at taking advantage of Mata's efficiency, it is evident that the approach has great potential for the development of more readable and efficient code.

References

Baum, C. F. 2000. sts17: Compacting time series data. *Stata Technical Bulletin* 57: 44–45. In *Stata Technical Bulletin Reprints*, vol. 10, 369–370. College Station, TX: Stata Press.

———. 2006a. *An Introduction to Modern Econometrics Using Stata.* College Station, TX: Stata Press.

———. 2006b. Stata tip 37: And the last shall be first. *Stata Journal* 6: 588–589.

———. 2006c. Stata tip 38: Testing for groupwise heteroskedasticity. *Stata Journal* 6: 590–592.

———. 2007. Stata tip 40: Taking care of business. *Stata Journal* 7: 137–139.

Baum, C. F., and N. J. Cox. 2007. Stata tip 45: Getting those data into shape. *Stata Journal* 7: 268–271.

Baum, C. F., M. E. Schaffer, and S. Stillman. 2003. Instrumental variables and GMM: Estimation and testing. *Stata Journal* 3: 1–31.

———. 2007. Enhanced routines for instrumental variables/generalized method of moments estimation and testing. *Stata Journal* 7: 465–506.

Baum, C. F., and V. Wiggins. 2000. dm81: Utility for time series data. *Stata Technical Bulletin* 57: 2–4. In *Stata Technical Bulletin Reprints*, vol. 10, 29–30. College Station, TX: Stata Press.

Cook, R. D., and S. Weisberg. 1994. *An Introduction to Regression Graphics.* New York: Wiley.

Cox, N. J. 1999. dm70. Extensions to generate, extended. *Stata Technical Bulletin* 50: 9–17. In *Stata Technical Bulletin Reprints*, vol. 9, 34–45. College Station, TX: Stata Press.

———. 2000. dm70.1: Extensions to generate, extended: Corrections. *Stata Technical Bulletin* 57: 2. In *Stata Technical Bulletin Reprints*, vol. 10, 9. College Station, TX: Stata Press.

———. 2002a. Speaking Stata: How to face lists with fortitude. *Stata Journal* 2: 202–222.

————. 2002b. Speaking Stata: How to move step by: step. *Stata Journal* 2: 86–102.

————. 2002c. Speaking Stata: On numbers and strings. *Stata Journal* 2: 314–329.

————. 2003a. Speaking Stata: Problems with lists. *Stata Journal* 3: 185–202.

————. 2003b. Speaking Stata: Problems with tables, Part II. *Stata Journal* 3: 420–439.

————. 2003c. Stata tip 2: Building with floors and ceilings. *Stata Journal* 3: 446–447.

————. 2004a. Speaking Stata: Graphing agreement and disagreement. *Stata Journal* 4: 329–349.

————. 2004b. Speaking Stata: Graphing categorical and compositional data. *Stata Journal* 4: 190–215.

————. 2004c. Speaking Stata: Graphing distributions. *Stata Journal* 4: 66–88.

————. 2004d. Speaking Stata: Graphing model diagnostics. *Stata Journal* 4: 449–475.

————. 2004e. Stata tip 9: Following special sequences. *Stata Journal* 4: 223.

————. 2005a. Speaking Stata: Density probability plots. *Stata Journal* 5: 259–273.

————. 2005b. Speaking Stata: Smoothing in various directions. *Stata Journal* 5: 574–593.

————. 2005c. Speaking Stata: The protean quantile plot. *Stata Journal* 5: 442–460.

————. 2005d. Stata tip 17: Filling in the gaps. *Stata Journal* 5: 135–136.

————. 2005e. Stata tip 27: Classifying data points on scatter plots. *Stata Journal* 5: 604–606.

————. 2005f. Suggestions on Stata programming style. *Stata Journal* 5: 560–566.

————. 2006a. Speaking Stata: Graphs for all seasons. *Stata Journal* 6: 397–419.

————. 2006b. Stata tip 33: Sweet sixteen: Hexadecimal formats and precision problems. *Stata Journal* 6: 282–283.

————. 2006c. Stata tip 39: In a list or out? In a range or out? *Stata Journal* 6: 593–595.

————. 2007a. NJC_STUFF: Stata module documenting Stata programs and help files by Nicholas J. Cox. Statistical Software Components, Boston College Department of Economics. Downloadable from http://ideas.repec.org/c/boc/bocode/s456852.html.

————. 2007b. Speaking Stata: Identifying spells. *Stata Journal* 7: 249–265.

————. 2007c. Speaking Stata: Making it count. *Stata Journal* 7: 117–130.

————. 2007d. Stata tip 43: Remainders, selections, sequences, extractions: Uses of the modulus. *Stata Journal* 7: 143–145.

————. 2007e. Stata tip 50: Efficient use of summarize. *Stata Journal* 7: 438–439.

————. 2008. Stata tip 67: J() now has greater replicating powers. *Stata Journal* 8: 450–451.

Cox, N. J., and J. Weesie. 2001. dm88: Renaming variables, multiply and systematically. *Stata Technical Bulletin* 60: 4–6. In *Stata Technical Bulletin Reprints*, vol. 10, 41–44. College Station, TX: Stata Press.

————. 2005. dm88_1: Update: Renaming variables, multiply and systematically. *Stata Journal* 5: 607.

Delwiche, L. D., and S. J. Slaughter. 1998. *The Little SAS Book*. 2nd ed. Cary, NC: SAS Institute Inc.

Drukker, D. M. 2006. Importing Federal Reserve economic data. *Stata Journal* 6: 384–386.

Franklin, C. H. 2006. Stata tip 29: For all times and all places. *Stata Journal* 6: 147–148.

Gini, R., and J. Pasquini. 2006. Automatic generation of documents. *Stata Journal* 6: 22–39.

Gould, W. 2001. Statistical software certification. *Stata Journal* 1: 29–50.

————. 2005. Mata Matters: Using views onto the data. *Stata Journal* 5: 567–573.

————. 2006a. Mata Matters: Creating new variables—sounds boring, isn't. *Stata Journal* 6: 112–123.

————. 2006b. Mata Matters: Precision. *Stata Journal* 6: 550–560.

————. 2007a. Mata Matters: Structures. *Stata Journal* 7: 556–570.

————. 2007b. Mata Matters: Subscripting. *Stata Journal* 7: 106–116.

Gould, W., J. Pitblado, and W. Sribney. 2006. *Maximum Likelihood Estimation with Stata*. 3rd ed. College Station, TX: Stata Press.

Greene, W. H. 2008. *Econometric Analysis*. 6th ed. Upper Saddle River, NJ: Prentice–Hall.

Hand, D. J., F. Daly, A. D. Lunn, K. J. McConway, and E. Ostrowski. 1994. *A Handbook of Small Data Sets*. London: Chapman & Hall.

Hansen, L. P., J. Heaton, and A. Yaron. 1996. Finite-sample properties of some alternative GMM estimators. *Journal of Business and Economic Statistics* 14: 262–280.

Jann, B. 2005. Making regression tables from stored estimates. *Stata Journal* 5: 288–308.

———. 2007. Making regression tables simplified. *Stata Journal* 7: 227–244.

Kantor, D., and N. J. Cox. 2005. Depending on conditions: A tutorial on the cond() function. *Stata Journal* 5: 413–420.

Kernighan, B. W., and P. J. Plauger. 1978. *The elements of programming style.* New York: McGraw-Hill.

Kolev, G. I. 2006. Stata tip 31: Scalar or variable? The problem of ambiguous names. *Stata Journal* 6: 279–280.

McDowell, A. 2004. From the help desk: Seemingly unrelated regression with unbalanced equations. *Stata Journal* 4: 442–448.

Merton, R. C. 1980. On estimating the expected return on the market: An exploratory investigation. *Journal of Financial Economics* 8: 323–61.

Mitchell, M. N. 2008. *A Visual Guide to Stata Graphics.* 2nd ed. College Station, TX: Stata Press.

Newson, R. 2004. Stata tip 13: generate and replace use the current sort order. *Stata Journal* 4: 484–485.

Rabe-Hesketh, S., and B. S. Everitt. 2006. *A Handbook of Statistical Analyses Using Stata.* 4th ed. Boca Raton, FL: Chapman & Hall/CRC.

Rising, W. 2007. Creating self-validating datasets. 18th United Kingdom Stata Users Group meeting. Downloadable from http://ideas.repec.org/p/boc/usug07/18.html.

Wooldridge, J. M. 2002. *Econometric Analysis of Cross Section and Panel Data.* Cambridge, MA: MIT Press.

———. 2009. *Introductory Econometrics: A Modern Approach.* 4th ed. New York: Thomson Learning.

Author index

A

Azevedo, J. P. 62, 173, 226

B

Baum, C. F. 62, 119, 123, 132, 136, 162, 165, 173, 175, 226, 230, 236, 237, 242, 261, 307
Blasnik, M. 62, 173

C

Cañette, I. 264
Cook, R. D. 185
Cox, N. J. 10, 29, 39, 41, 42, 44, 45, 54, 57, 58, 62, 63, 66, 67, 71, 105, 112, 114, 118, 119, 153, 158, 162, 165, 169, 175, 185, 188, 198, 230, 235, 267, 275
Crow, K. 161

D

Daly, F. 59
Delwiche, L. D. 25
Drukker, D. M. 123, 327

E

Everitt, B. S. 59

F

Franklin, C. H. 112

G

Gini, R. 184, 188
Gould, W. . . 10, 234–236, 254, 263, 275, 284, 299, 307, 321
Greene, W. H. 240

H

Hand, D. J. 59

H

Hansen, L. P. 337
Heaton, J. 337

J

Jann, B. 52, 98, 128, 181, 278, 286, 305, 328

K

Kantor, D. 44
Kernighan, B. W. 27
Kolev, G. I. 59

L

Leuven, E. 327
Longton, G. 170
Lunn, A. D. 59

M

McConway, K. J. 59
McDowell, A. 331
Merton, R. C. 166
Mitchell, M. N. 185, 196

N

Newson, R. 66

O

Ostrowski, E. 59

P

Pasquini, J. 184, 188
Pitblado, J. 236, 263
Plauger, P. J. 27

R

Rabe-Hesketh, S. 59
Rising, W. 190, 230, 322

S
Schaffer, M. E. 236, 261, 337
Sianesi, B. 327
Slaughter, S. J. 25
Sribney, W. 236, 263
Stillman, S. 236, 261

W
Wada, R. 99
Weesie, J. 105
Weisberg, S. 185
Wiggins, V. 132
Wooldridge, J. M. 182

Y
Yaron, A. 337

Subject index

A

added-variable plot 185
ado-path 234, 304
adoupdate command 4, 30
Akaike information criterion 94
anova command 145
append command 80, 105, 107
args command 237, 240
ASCII text file 16, 178
assert command 19, 234
atanh() function 267
audit trail 79
autocorrelation 165
automated output 181
avplot command 185

B

_b saved results 249
backslash 74
Bayesian information criterion 94
binary file handling 179
biprobit command 267
Boolean condition 43
bootstrap prefix command . . 151, 242,
          244
built-in commands 86
built-in function 316
business-daily data 123, 213
by prefix 140
byable 47, 226
bysort prefix 140
byte data type 92
bytecode 271

C

calendar variable 88
capture command 14

casewise deletion 36, 92, 175
ceil() function 45
center command 52, 286
Center for Research in Security Prices
          191
certification script 234, 254, 345
char command 189
characteristics 188, 285
ckvar command 190
class programming 1
codebook command 20, 81
codebooks 20
collapse command 14, 53, 65, 100,
          135, 175, 203, 288
colon operator 274
column-join operator 174, 198, 272
column-range operator 273
comma-delimited file 18
comma-separated–values file 178
compiled function 303
complex arithmetic 271
compound double quotes 39
compress command 25
cond() function 44, 268
conditional statements 278
confirm command 287, 307
constraint command 261
continuous price series 209
contract command 14, 65, 175
correlate command . . 37, 89, 152, 192,
          331
correlation() Mata function 313
creturn command 57, 76
cross command 112
cross() Mata function 334
cscript command 254

CSV file...*see* comma-separated–values file

D

D. time-series operator..............13
daily data........................123
data dictionary....................22
data validation....................80
decode command....................39
decrement operator................277
delimiters.........................17
delta method.....................241
dereferencing......................54
describe command............35, 87
destring command 38
dialog programming.................1
dictionary file....................178
difference operator..............*see* D.
diparm() option...................263
directory separator 74
display command............31, 39
Do-file Editor.......................8
dofm() function 132
dot product 273
duplicates command 85, 111
dyex option 181
dynamic panel data...............267

E

e(b) matrix 92
e(sample) function 90, 92
e-class............................86
eform() option...................228
egen anycount() command 63
egen bom() command..............49
egen bomd() command.............49
egen command....37, 47, 63, 231, 252, 288
egen corr() command.............49
egen count() command...........48
egen eom() command.............49
egen eomd() command.............49
egen ewma() command.............49
egen filter() command..........49
egen gmean() command............49

egen group() command...........141
egen hmean() command...........49
egen iqr() command.............48
egen kurt() command............48
egen max() command.............48
egen mdev() command............48
egen mean() command........48, 211
egen median() command..........48
egen min() command.............48
egen mode() command............48
egen nvals() command..........169
egen pc() command..............48
egen pctile() command.....48, 232, 252
egen rank() command............48
egen record() command.........49
egen rndint() command.........49
egen rowfirst() command........47
egen rowlast() command.........47
egen rowmax() command..........47
egen rowmean() command.........47
egen rowmin() command..........47
egen rowmiss() command.....47, 63
egen rownonmiss() command..47, 63
egen rowsd() command...........47
egen rowtotal() command........47
egen sd() command..............48
egen semean() command..........49
egen skew() command............48
egen std() command.............48
egen tag() command............169
egen total() command.......48, 65
egen var() command.............49
egenmore package.........49, 63, 169
elasticity....................181, 194
elementwise operations 273
embedded spaces 17
.eps format.......................187
ereturn display command ..333, 338
ereturn list command........90, 93
ereturn post command..........333
estadd command 98, 129
estat ovtest command...........93
estat vce command 93
estimates command 93

estimates for command...........96
estimates notes command........97
estimates replay command.......94
estimates save command......94, 97
estimates stats command........91
estimates store command...93, 128, 260
estimates table command ... 94, 128
estimates use command...........94
estimation commands...............86
estout command 98, 128, 181
eststo command..................181
esttab command.............129, 181
Euclidian distance................327
extended macro functions..56, 159, 172
external declaration..............342
eydx option 182
eyex option..................181, 196

F
F. time-series operator.............13
FAQs..............................29
fdasave command.................178
fdause command..................16
file command...................178
file handle.......................179
file open command..............179
file read command..............180
file write command.............179
fillin command.................112
findexternal() Mata function....280
findit command...........4, 29, 217
finite-precision arithmetic.........257
fixed format....................17, 20
floor() function 45
foreach command........55, 154, 172
format() option....................39
forvalues command..............154
freduse command.................123
free format 17
function arguments................279
function library..................304

G
generalized method of moments....337

generate command 40
global command......56, 73, 259, 265
global macro..................73, 155
GMM-CUE 337
.gph file suffix................187, 204
graph combine command 186
graph display command 187
Graph Editor 188, 205
graph export command..........187
graph save command.............205
graph twoway connected command..
.............196
graph twoway rarea command ... 251
graph twoway rline command ... 196
graph twoway scatter command ....
.............196
graphics...........................185
groupwise heteroskedasticity 242
gsort command 66, 192

H
help files..........................228
HTML.........................98, 229

I
if qualifier.....................35, 36
implied decimal...................24
in qualifier 35
include command.................281
increment operator 277
indicator variable 42, 186
inequality constraints..............262
infile command......17, 23, 115, 118
infix command 22, 202
inlist() function 41, 84
inner product.....................273
inrange() function 41, 64, 83
insheet command 18, 38, 203
instrumental variables 337
int() function..................41, 46
integer division 41
interaction terms 143
interval estimates..................195
invsym() Mata function..........334
irecode() function................46

`ivreg2` command ... 236, 261, 337, 346
`ivregress` command 341

J
`J()` Mata function 172, 275
`jackknife` prefix 242
`joinby` command 115

K
Kronecker product 274

L
L. time-series operator.............. 13
L2-norm 327, 328
lag operator .... *see* L.
latent variable 131
LaTeX output 98, 173, 174, 181, 197
lead operator .... *see* F.
`levelsof` command 75, 172, 191
likelihood function 91
limits........................... 316
`lincom` command 94, 182, 232, 249
linear constraints 261
linear filter .... *see* `filter()`
list subscripts.................... 274
`liststruct()` Mata function 301
listwise deletion 175
`local` command 53
local macro 155
log price relative.................. 209
`log2html` command 230
`logit` command 130
long form........................ 100
longitudinal data 101
loop constructs 154, 276

M
Macintosh spreadsheet dates........ 20
macro.......................... 53–56
macro evaluation 54
macro list functions 57, 75, 159
`mad()` command 48
`makematrix` command 198
Mann–Whitney test 199
many-to-many merge 111

marginal effects................... 194
`marksample` command 223, 268
`mat2txt` command 62, 173
Mata 60, 271
`.mata` file suffix................. 303
`mata mlib add` command 305
`mata mlib create` command 304
`mata mosave` command 304
match-merge 109
`matrix accum` command........... 61
`matrix colnames` command 62, 92,
        172, 180, 308
`matrix` command 60, 172
matrix language 60
`matrix list` command 60, 92, 173
matrix programming............... 271
`matrix rownames` command 62, 92,
        172, 180, 308
matrix stripes.............. 308, 334
`matsize` command................ 316
`max()` function 66
`maxbyte()` function............... 42
maximum likelihood 261, 262, 337
`maxindex()` Mata function 314
`maxint()` function 42
`maxlong()` function 42
`mean()` Mata function 314
`merge` command 80, 106, 109, 203,
        207
`_merge` variable................... 109
method d0........................ 264
`mfx` command 128, 181, 195, 232
`min()` function 66
`minindex()` Mata function 328
`missing()` function................ 36
missing string value............... 37
missing values 43
`mkmat` command 61
`ml` command 337
`.mlib` file suffix.................. 304
`mlopts` parsing tool 238
`mm_cond()` Mata function 278
`mm_meancolvar()` Mata function ... 328
`.mo` file suffix.................... 303
`mod()` function................ 42, 132

`month()` function 132
`moremata` package 278, 305, 328
moving correlation 165
moving-window statistics 162
`mreldif()` function 256
`mvcorr` command 165
`mvdecode` command 37
`mvencode` command 37
`mvsumm` command 162

N

naturally coded indicator 142
nearest neighbor 327
nested loop 155, 157
nested macros 54
`nestreg` command 153
`net install` command 7
`njcstuff` package 188
`nlcom` command 241, 260, 266
`normalden()` function 263
null model 91
null pointer 297
`nullmat()` function 174, 199
numlist 35

O

object file 303
object-oriented programming 1
ODBC 26
`odbc` command 26
one-to-many merge 109, 118
one-to-one merge 110
`optimize_init_evaluator()` Mata
           function 343
`optimize_init_evaluatortype()`
           Mata function 343
`optimize_init()` Mata function . . . 342
`optimize_init_params()` Mata
           function 343
`optimize_init_which()` Mata
           function 343
`optimize()` Mata function 337, 343
`optimize_result_value()` Mata func-
           tion 343
optional arguments 280

options 221
`order` command 35, 156, 178
`.out` file suffix 178
outer join 115
outer product 273
`outfile` command 178
`outreg2` command 94, 99
`outtable` command 62, 173, 226
overlapping subsamples 146

P

p-value 199
panel data 70, 101, 267
panel variable 88
`panelsetup()` Mata function 313
`panelsubmatrix()` Mata function . . 313
`panelsubview()` Mata function 316
passing functions to functions 316
`pctrange` command 254
`.pdf` format 187
percentiles 195
`permute` command 151
`.png` format 187
point estimates 195
pointer function 321
pointer variables 296
pointers 316, 334
pointers to pointers 298
`post` command 177
`postfile` command 176
`predict` command 10, 232, 332
prefix commands 139
`preserve` command 14, 175, 268
price deflator 162
`probit` command 130
`program` command 218
program properties 228
propensity score 327
`psmatch2` command 327
`pwcorr` command 37, 331

Q

`qofd()` function 134
`qreg` command 147

R

r-class 86, 89
range subscripts 274
`ranksum` command.................199
`.raw` file suffix.....................178
`rc0` option.........................142
`rcof` command....................256
`real()` function 38
`recode` command 45, 156
`recode()` function..................45
record 66
recursive estimation 146
`reg3` command 332
`regress` command.................332
relational operators..................274
relative difference....................154
`reldif()` function 154, 257
`rename` command.............105, 107
`renvars` command.................105
`replace` command..................40
replay feature.......................238
reserved words.....................280
`reshape` command............100, 290
`restore` command........14, 175, 269
`return list` command.............86
return types 279
`reverse()` function................203
reverse recursive estimation........146
rich text format.................98, 99
`rnormal()` (Mata).................276
`robvar` command..................242
`rolling` prefix...........146, 162, 249
rolling-window estimation..........146
root finders........................305
`round()` function 46
row-join operator 174, 198, 273
row-range operator 272
rowwise functions...................47

S

S. time-series operator..............13
SAS XPORT Transport file..........178
`savedresults` command......259, 346
scalar..............................257
`scalar` command...................58

scheme programming 1
Schwarz criterion 94
`_se` saved results....................249
`search` command....................29
seasonal difference operator ..... *see* S.
seemingly unrelated regression 331
semielasticity 181
`separate` command...........114, 186
`serset` command..................205
`set rmsg` command 347
`set seed` command..........149, 243
`set trace` command 31
similarity...........................327
`simulate` command...........148, 242
SMCL 98, 228
`sort` command 50
sort order............................87
`sortpreserve` option...............238
space-delimited files.................17
spells..........................67, 267
`sphdist` command.................322
spreadsheet data....................20
SQL databases 26
SSC archive 29, 202, 259
`ssc` command 7, 47, 202
`ssc hot` command.................204
`st_addvar()` Mata function........284
`st_data()` Mata function..........283
`st_global()` Mata function.......285
`st_local()` Mata function........285
`st_matrix()` Mata function...285, 308
`st_matrixrowstripe()` Mata function
.............308
`st_matrixcolstripe()` Mata function
.............308
`st_nobs()` Mata function..........283
`st_numscalar()` Mata function ... 282, 285
`st_nvar()` Mata function..........283
`st_sdata()` Mata function....283, 285
`st_sstore()` Mata function...284, 285
`st_store()` Mata function........284
`st_strscalar()` Mata function....285
`st_subview()` Mata function..285, 319
`st_sview()` Mata function....283, 285

st_tsrevar() Mata function 339
st_varindex() Mata function 286
st_varname() Mata function 286
st_view() Mata function 282, 322
stack command 112
standardized values ... *see* egen std() command
stars, for significance 94
Stat/Transfer 25, 37, 178
stata() Mata function 322
Stata Journal 7, 29
Stata Technical Bulletin 7, 29
Statalist ... 30, 119, 126, 163, 167, 169, 191, 249
Statistical Software Components archive .... *see* SSC archive
statsby command 145
statsmat command 62, 175
stepwise command 153
stock returns 191
storage optimization 25
string() function 39
string missing values 37
strpos() function 203
strtoreal() Mata function 286
structure 299
stset command 189
substr() function 203
subview matrix 319
sum() function 42, 169
summarize command 87, 192
sureg command 331
svmat command 192
svy command 153
svyset command 153, 189
syntax command 218, 231, 307
sysdir command 6

T
tab-delimited file 18, 178
tabstat command 44, 62, 65, 175, 198
tabulate command 84, 142
tabulating results 94
tempname command 176, 200

temporary objects 295
temporary variable 223
tempvar command 195, 223, 268
test command 94, 232
text editors 16
text files 16
texteditors command 16
.tif format 187
time-series calendar 132
time-series operators 12, 257
tokenize command 157
tokens() Mata function 288
tostring command 39
transmorphic matrix 279
transpose 115
transpose operator 273
tscollap command 136, 288
tsfill command 112, 162, 268
tsline command 124, 163, 251
tsmktim command 132, 291
tsset command .. 13, 49, 88, 123, 132, 189, 268
tsspell command 67, 267
ttest command 89, 152, 199, 242
type d0 evaluator 343

U
unbalanced panel 112, 267, 331
update command 6, 30
update option 208

V
variable indices 286
varlist 35
vce(bootstrap) option 152
vce(jackknife) option 153
version command 218
veryshortlabel option 114
view matrices 271, 285, 319
viewsource command 86, 305
void matrices 279

W
which command 31, 86, 217
while command 154

wide form 99, 331
wildcards 35, 47, 103
word processors 16
wrapper program 249

X
`xi` command 142
`xpose` command 115
`xt` command 100, 331
`xtabond` command 267
`xtabond2` command 267
`xtile` command 46
`xtset` command . . 13, 88, 102, 181, 188